T0192573

Methods of Experimental Physics

Graduate Student Series in Physics

Series Editor:
Professor Derek Raine
Senior Lecturer, Department of Physics and Astronomy, University of Leicester

Methods of Experimental Physics
Mikhail I. Pergament

The Dynamics of Discrete Populations and Series of Events
Keith Iain Hopcraft, Eric Jakeman, and Kevin D. Ridley

Introduction to Surface and Superlattice Excitations, Second Edition
Michael G. Cottam and D.R. Tilley

Cosmology in Gauge Field Theory and String Theory
D. Bailin and Alexander Love

Introduction to Nuclear Reactions
C.A. Bertulani and P. Danielewicz

Gauge Theories in Particle Physics, Volume II: QCD and the Electroweak Theory, Third Edition
I.J.R. Aitchison and A.J.G. Hey

Gauge Theories in Particle Physics: Volume I: From Relativistic Quantum Mechanics to QED, Third Edition
I.J.R. Aitchison and A.J.G. Hey

Problems on Statistical Mechanics
D.A.R Dalvit, J. Frastai, and Ian Lawrie

Symmetries in Quantum Mechanics: From Angular Momentum to Supersymmetry (PBK)
M. Chaichian and R. Hagedorn

Supersymmetric Gauge Field Theory and String Theory
D. Bailin and Alexander Love

Introduction to Gauge Field Theory Revised Edition
D. Bailin and Alexander Love

Superfluidity and Superconductivity
D.R. Tilley and J. Tilley

Methods of Experimental Physics

M. I. Pergament

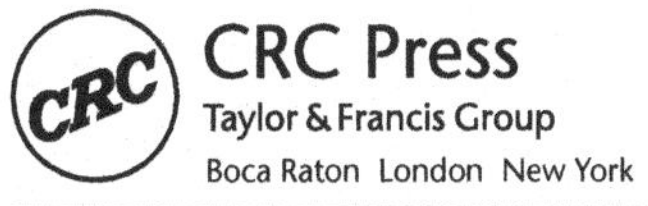

CRC Press is an imprint of the
Taylor & Francis Group, an **informa** business

CRC Press
Taylor & Francis Group
6000 Broken Sound Parkway NW, Suite 300
Boca Raton, FL 33487-2742

First issued in paperback 2019

ISBN-13: 978-0-7503-0608-9 (hbk)
ISBN-13: 978-0-367-88642-6 (pbk)

Library of Congress Cataloging-in-Publication Data

Pergament, M. I., author.
Methods of experimental physics / M.I. Pergament.
pages cm -- (Graduate student series in physics)
Includes bibliographical references and index.
ISBN 978-0-7503-0608-9 (hardcover : alk. paper) 1.
Physics--Experiments--Methodology. I. Title. II. Series: Graduate student series in physics (CRC Press)

QC33.P426 2015
530.072'4--dc23 2014027308

Visit the Taylor & Francis Web site at
http://www.taylorandfrancis.com

and the CRC Press Web site at
http://www.crcpress.com

To the memory of my late wife, Dr. Anna Khalilovna Pergament, a person of exceptional intellect, honor, talent, dignity, and kindness.

She was very patient with me during all the long hours of my work on this book and I benefitted greatly from her high professional expertise in mathematics. In particular, she contributed the most sophisticated of the computations that were beyond my own skills.

Contents

Preface

THE BOOK THAT YOU are holding in your hand is the result of many years of experience of experimental research and is based on a lecture course, "Methods of Experimental Physics," given by the author at the Problems of Physics and Energetics Department of the Moscow Institute of Physics and Technology (State University) (usually known as Phystech). The contents of the book can be divided into four parts: the general theory of systems for measuring and recording data; the equipment and methods for studying fast processes with the help of electronics and optics; the basic methods of experimental physics (Fourier optics, spectroscopy, interferometry and holography, electromagnetic waves, and also X-ray and corpuscular methods); and finally the methods of interpretation and data processing. The selection of the material has been based on the author's own experience and deep understanding of the knowledge that is required in physical laboratories when problems are posed during an experiment, when the experiment is conducted, and when the data are subsequently analyzed. These considerations are also important at the initial stages of design, testing, and certifying measurement equipment. The basic methods of experimental physics are explained in sufficient detail to assist the reader in further study of specialized scientific literature.

The book concentrates on the principles of the methods rather than excessive technical detail because the principles are more important for planning and conducting experiments as well as for designing the equipment used for measurements. This approach also allows the book to remain reasonably concise and to avoid becoming an engineering encyclopedia running into many volumes. The book gives special emphasis to widely used methods of information theory that provide powerful tools to adequately describe and analyze the qualities of measuring systems, to diagnose the quality of output signals, and to establish the correct criteria for solving inverse problems. The author has endeavored to make this book accessible to a wide spectrum of readers by following the principles of his mentor, Academician Lev Andreevich Artsimovich, which state, "always when it is necessary to choose between precision and simplicity of an explanation, choose simplicity."

This book is written for graduate and postgraduate students in physics and technical studies, for scientific researchers who plan and conduct experiments, and also for engineers who design, test, and certify measuring equipment and systems. If the readers of this book find it useful, the author will be pleased that his efforts have not been in vain.

Mikhail I. Pergament

Acknowledgments

I AM VERY GRATEFUL TO two of my former students who are now my colleagues, Professor Alexander Yu. Goltsov and Dr. Michael K. Rudnev, who translated this book into English. The final polishing of the translation was done by Dr. Peter E. Stott, a very close friend of mine. He was also the first reader and the first critic of the book. His deep knowledge and many years of experience in experimental physics allowed him to verify the terminology and to check the explanations. I cannot thank him enough for his help and advice.

I am also deeply in debt to my partners in scientific discussions about the methods of explanations of the more complicated physical phenomena: my colleagues Professors Nick G. Kovalsky and Alexander Yu. Goltsov as well as Drs. Ruslan V. Smirnov and Ildar K. Fasakhov. I would like to thank a representative of the young generation, my son Dr. Mikhail M. Pergament, who, for obvious reasons, unlike students and young colleagues, was not shy to criticize my book when he thought that the explanations were not very clear or precise. I am very grateful to engineers Valeriy V. Kryzhko and Alexander A. Volferts, specialists in radio electronics and computer technology who read Chapter 3 very carefully and made some very useful comments.

Finally, I express my thanks to Olga L. Dedova and Ekaterina Yu. Krasovskaya who prepared most of the figures for this book. Special thanks are due to my colleagues at Taylor & Francis for their patience and encouragement.

The author thanks the following individuals and organizations for granting permission to use the figures: Professor David T. Attwood (Figures 10.1, 10.20, and 10.22), Professor Igor S. Gruzman (Figure 5.8), Professor Rick Trebino (Figure 3.19), the Society of Motion Picture and Television Engineers (Figure 4.4), the European Fusion Development Agreement-Joint European Torus (EFDA-JET; Figure 9.12), and Bifo

Industries Ltd. (BIFO) company (Figures 4.10, 4.11, 4.13, and 4.14). Especially, the author thanks BIFO General Director Dr. Grigory G. Feldman and BIFO Scientific Director Dr. Vitaly B. Lebedev. The copyright for all these materials remains with the copyright holders. Every effort has been made to contact the copyright holders.

CHAPTER 1

Introduction

THE METHODS OF EXPERIMENTAL physics provide the way for the experimental study of nature, its phenomena, and its laws. It is clear that any experimental study can be carried out only on a solid knowledge base that has already been acquired in previous research—that is, on existing physical laws. In other words, in experimental physics, we usually take advantage of well-established physical laws and approaches for investigating unknown physical processes and phenomena. Investigation of any object implies gathering information on its parameters and their time evolution in the case when the object under investigation is nonstationary. In experimental physics, this information is obtained by measurements. The final goal of such measurements is to establish a set of functions describing the behavior in space and time of the object under investigation. To do this, one has to receive, transmit, record, and process the information about these functions. In most cases, this information is of an indirect nature, since as a rule it is impossible to directly measure the parameters of interest. For instance, suppose that we are interested in the velocity distribution of electrons within a plasma. A direct measurement of this distribution would present a rather difficult and sophisticated problem. However, we can probe the plasma volume with a powerful laser beam and measure the spectrum of laser light scattered by the plasma. As the spectrum of the scattered light (the intensity of scattered light vs. wavelength) is determined by the velocity distribution of the plasma electrons, we can immediately obtain the information of interest. In this case, the temperature of plasma is measured by Thomson scattering, although it should be emphasized that we actually do not measure the temperature but the spectrum of laser light scattered by the plasma! The temperature can then be calculated, taking advantage of the fact that the relationship between the electron temperature and the shape of the scattered spectrum is well known and unambiguous, and has been studied in very much detail. We have emphasized here that the temperature was measured using the Thomson scattering method because there are other methods for solving this particular problem; for instance, the temperature can also be determined using the plasma X-ray spectrum (measurements of the so-called *bremsstrahlung*). It is clear that the relationship between the shape of the plasma X-ray spectrum and the temperature of electrons is completely different in this case for the obvious reason that the physical mechanisms responsible for laser scattering and X-ray emission in plasma are quite different.

The above considerations give rise to the question: What does the expression "*method of measurement*" really mean? Usually, behind this concept, we imply the combination of physical phenomena that allow the determination and formalization of the mutual relationships between the parameters of an object and the physical quantities that can be measured directly. It may happen that the values or variables that can be used for direct measurements could be measured using different approaches. The *recording method* implies the combination of technical solutions that make it possible to determine, with the help of proper equipment, some quantities or functions that implicitly characterize the object under investigation. Usually, these experimentally obtained quantities or functions are referred to as the *primary experimental data*.

In most cases, these primary data are not of direct interest in themselves and are just used as the starting point for calculating the parameters of the object under investigation. This is why the process of any experimental study can be (rather conventionally) subdivided into two separate stages—the measurement itself and the *processing* or *interpretation* of the experimental data. By the term "*data processing*," we imply the combination of algorithms that allow us to calculate the parameters of the object under investigation taking advantage of the primary experimental data. The character of these experimental data always affects the selection of the appropriate processing method and the requirements for the primary data being determined by the accuracy which we need to adequately describe the functions that characterize the object under investigation—the final goal of our measurements. In turn, the possibility of ensuring proper accuracy is strongly affected by the character of the information received from the object, the specific features of the measuring equipment, and the method selected for processing the acquired experimental data. All these issues will be of primary concern on the long voyage through the methods of experimental physics. Three reefs lie in wait for us during this voyage: (1) the indirect character of experimental data, (2) the inverse problems, and (3) the stochastic nature of the physical world.

1.1 INDIRECT DATA AND INVERSE PROBLEMS

The terms "direct and inverse problems" have their origin in the direction in which we move relative to the flow of events. Direct problems correspond to movement along this direction of flow. In other words, we know the cause and look for the consequences. On the contrary, inverse problems are related to movement against this flow. In this case, we know the consequence and our goal is to take advantage of some indirect data to find the cause for this given consequence. Consider, for instance, the following situation: You have invited your girlfriend to a theater, but she has caught a cold, her temperature is high, she has a severe headache, and so on. The consequence is straightforward—you are unlikely to watch a play this evening. But consider another situation: You are waiting at the theater with a bouquet of flowers for your girlfriend as arranged, but she does not come. What is the reason? She may be sick and could not let you know, she may have been delayed by bad traffic, or she may have gone out with someone else—you know that John has persistently invited her to a restaurant for the past week. We come back to our subject from this sorrowful situation with the immediate conclusion—there are no reasons to expect a unique solution when dealing with inverse problems. It turns out that in similar situations we have to consider the whole aggregate of possible solutions. It is not a simple problem to select the correct one, especially in the absence of additional information. Unfortunately, in physics—and in our daily life as well—we have to solve inverse problems most of the time.

From the formal point of view, the relationship between the parameters of the object under investigation and the measured quantities (these are functions, as a rule) can be expressed in the following manner:

$$AZ = I \tag{1.1}$$

where:

Z is a function describing the object parameters [in the above example of Thomson scattering, Z is the electron temperature $T_e(\mathbf{r}, t)$, and $\mathbf{r}$ is the spacial vector, t is time]

I is a function whose values can be measured directly [in the above example, *I* is the spectrum of scattered laser light: $I = f(\lambda, z, t)$, where λ is the wavelength, z is the spatial coordinate along the probing direction, and t is the time]

A is an operator establishing a formal relationship between *Z* and *I*

The operator ***A*** characterizes the method of measurement that is being used. It is clear that different methods of measurements have to be described with different operators. It is also evident that the intensity of scattered radiation spectrum $I = f(\lambda, z, t)$ does not represent the primary experimental data. First, the spectrum has to be recorded. If *U* represents the primary experimental data and ***G*** is an operator describing the recording method, similar to Equation 1.1, one can write the following equation:

$$\boldsymbol{G}I = U \tag{1.2}$$

Dealing with the experimental data *U*, we usually start from Equation 1.2 and find *I*. After this we can obtain *Z* using Equation 1.1. There are two methods that are widely used for solving the system of Equations 1.1 and 1.2.

In the first method, an inverse operator, for instance, $\boldsymbol{G}^{-1}$, is constructed, and in principle, one can find *I* using the equation: $I = \boldsymbol{G}^{-1}U$. However, there are a number of problems in this approach due to the properties of the inverse operator $\boldsymbol{G}^{-1}$. The first problem arises in the correct selection of this operator for facilitating further calculations. Unfortunately, the inverse operator usually happens to be of the integro-differential type, which results in the unstable character of solutions obtained with its help. The function to be found is located inside the integral sign, while the right-hand side of the equation—that is the experimental data—is known only with a finite accuracy. The equations of this type are known as the *ill-posed* problems and were first investigated by Jacques Hadamard in his studies of Fredholm equations of the first type. Hadamard also showed for the first time that in such equations, "any arbitrary infinitely small variation of the equation's right-hand side can result in an infinitely large variation of the function comprised inside the integrand operator." *Apropos*, this strong and generally speaking correct statement significantly delayed the progress in development of the mathematical foundations of experimental data processing. Clearly, there is no point in attempting to solve an equation if it is known that the right-hand side of the equation is inaccurate and the solutions could be completely different compared to the perfect solution. The situation in this field changed dramatically after the work of A.N. Tikhonov, who developed the technique, now known as *Tikhonov regularization*, for solving ill-posed problems, which will be discussed frequently in this book. At this point, we outline only the principles of Tikhonov's approach and will later elaborate on the technical details of the method. When solving ill-posed problems in cases where one tries to use approximate data to find the exact solution of a problem, "any arbitrary infinitely small variation of the equation's right-hand side can result in an infinitely large variation of the function comprised inside the integrand operator," but fortunately, the problem can be resolved if approximate solutions are also admitted for consideration. Then a stable approximate solution does exist, gradually converging to the exact solution, provided the error on the right-hand side tends to 0.

The second method for solving the system of Equations 1.1 and 1.2 is as follows: Let us consider the model I^* that is some approximation for the function I. For instance, this approximation function may be selected as a serial sum of functions (polynomials, Fourier series, etc.). The coefficients in this expansion will be found when minimizing the following functional:

$$\|\boldsymbol{G}I^* - U\|$$

In fact, in this case, we compare $\boldsymbol{G}I^*$ and $\boldsymbol{G}I$ (since $U = \boldsymbol{G}I$). Indeed, the minimum of this functional will be obtained, provided I and I^* are fairly close to each other. Unfortunately, this method also leads to unstable solutions, and the only way to overcome this situation is found in reducing the number of terms in the expansion. In its turn, this technique inevitably deteriorates the quality of the solutions obtained. However, one should not have expected any other result, since the unstable character of ill-posed problems has its origin not in the mathematical procedures, but is determined by the fundamental physical problems, as will be discussed later in this book. From the point of view of communication theory, the variable I is the signal at the input of a measuring–recording system, and the variable $U = \boldsymbol{G}I$ is the output signal. The procedure of solving Equation 1.2 corresponds to the process of reconstruction of the input signal from the output signal (more exactly from its values that are measured at the output). Due to inevitable losses of information within the measuring–recording channel, the output signal carries less information than the input one. Moreover, the reconstructed signal certainly cannot contain more information than the output signal. Consequently, the reconstruction procedure in principle cannot provide more details on the input signal than on the measured output signal. The solution becomes unstable as soon as this limitation is violated by trying to "improve" the reconstruction quality.

1.2 EXPERIMENT AND STOCHASTICITY OF THE PHYSICAL WORLD

We all live in a probabilistic world. Preparing for a walk in the morning and thinking about an umbrella, we take a look out of the window, observe a clear sky, and conclude that a shower is unlikely. We try to estimate the probability of the stock market to rise or fall. Nevertheless, in everyday life, we consider the surrounding world as a well-determined one, but in truth this is very far from reality. Moreover, the majority of the world's population has become used to understanding the word "probability" as an intuitive feeling of whether some event is possible or not. Most people do not even consider probability to be a strict mathematical concept. In our discussion, we cannot afford to take this attitude.

Kolmogorov gave one of the most strict and comprehensive definitions of the concept of probability. In Kolmogorov's axiomatic approach, probability is introduced as a nonnegative measure in elementary event space. Although this concept is quite fruitful, it is rather difficult when applied to a phenomenological description of physical processes. For this reason, in this work we use another definition of probability.

Suppose we observe a set of events that happen N times, $N \gg 1$. Assume that in this set the event with feature A happens n times. If all the events in this set are independent, the probability of the event A is determined as follows:

$$P(A) = \lim_{N \to \infty} \frac{n}{N} \tag{1.3}$$

As one can easily see from the above equation, the inequalities $0 \leq P(A) \leq 1$ take place. The above equation is the estimate of the exact value of probability. This estimate converges to the exact value in terms of the probabilistic measure introduced by Kolmogorov. If the probability of some event to happen (for instance, that it will rain) is $P(A)$, then it is quite clear that the probability of dry weather is $1 - P(A)$.

Random events or variables are characterized in terms of their stochastic and spectral properties. Stochastic properties of random variables are described using the concept of *distribution function*. For any given x, the *probability distribution function* $F(x)$ is introduced as the probability of the random variable ξ to be smaller than x:

$$P(\xi < x) = F(x) \tag{1.4}$$

As one can easily see from the above equation, the following equation also holds: $P(\xi \geq x) = 1 - F(x)$, that is, the probability of ξ to lie in the interval $[x, \infty]$ is $1 - F(x)$.

Let us consider the concept of distribution function in more detail. Suppose that the manufacturer of flashlamps used in your camera guarantees that with 80% probability, the longevity of the lamp is not less than 350,000 (3.5×10^5) flashes. (Note that the probability must be written in the manual, otherwise it is not actually a guarantee. The term "longevity" without claiming the probability of this longevity to be realized means nothing.) Indeed the flashlamp can be just in two states—either it is broken or not. Consequently, the statement "with 80% probability the longevity of the lamp is not less than 350,000 (3.5×10^5) flashes" is completely equivalent to the following statement: "with 20% probability, the lamp will go out of service before it makes 3.5×10^5 flashes." Note that 3.5×10^5 flashes is a very high value. You will not reach it even if you trigger your lamp every 30 seconds for eight hours a day for a whole year. Nevertheless, the value 80% still looks unclear. What does this really mean? Let us consider the function $F(x)$ defined by the condition $P(\xi < x) = F(x)$, where the random value ξ in our example is the fatal flash that results in the destruction of the lamp. Consider Figure 1.1, in which the function $F(x)$ is plotted. The number of flashes is plotted along the x-axis, whereas the y-axis corresponds to the probability value. In this plot, the point A corresponds to $x = 3.5 \times 10^5$. The value $F(x) = F(3.5 \times 10^5) \cong 0.2$ of the distribution function at this point means that the value of ξ will fall into the interval between 0 and 3.5×10^5 with the probability of 0.2 (20%), which means that with a probability of 20%, the lamp will not survive longer. Until now we have not received much additional information, but we have discussed just one point of the curve presented in Figure 1.1. Let us consider the other points, for instance, the point B ($x = 2 \times 10^4$). Figure 1.1 shows that the probability of lamp failure before 2×10^4 flashes is only 2%. By the way, as soon as we are talking about any given lamp, we can only make some statement about the probability of the lamp to survive after some flashes. However, as soon as a large number of lamps

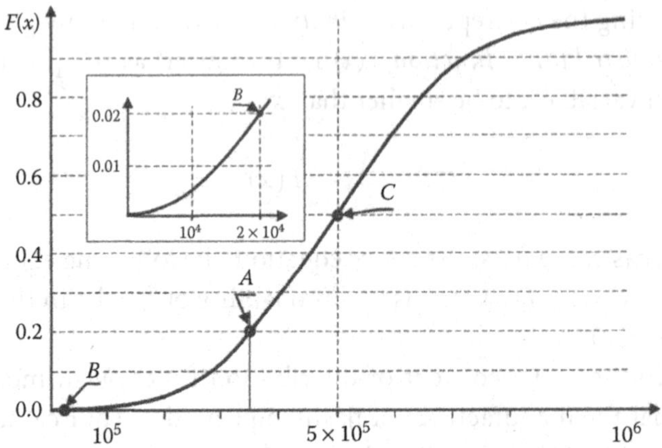

FIGURE 1.1 The distribution function $F(x)$ of the random variable ξ. In this example, the distribution function illustrates the probability that a flashlamp will fail after a certain number of flashes plotted along the x-axis.

are concerned, we can, taking advantage of Equation 1.3, make one more general statement, namely, we can claim that only 2% of the manufactured lamps will not survive before 2×10^4 flashes (point B), half of the lamps will die after 5×10^5 flashes, and so on. Using the curve from Figure 1.1, we can determine the probability of lamp failure between 3.5×10^5 and 5×10^5 flashes. For doing that, it is required to just calculate the difference between the magnitudes of the function $F(x)$ in the points A and B. As one can find from the plot in Figure 1.1, $F(B) - F(A) \approx 0.3$. Speaking more generally, the difference $F(x + \mathrm{d}x) - F(x) = p(x)\mathrm{d}x$ gives the probability of a random value ξ falling into the interval $\mathrm{d}x$. It is also clear that the function $p(x)$, called the *probability density*, represents the derivative of $F(x)$.

Indeed, the probability of ξ to be smaller than x is

$$P(\xi < x) = F(x)$$

and the probability of ξ falling within the interval $[x, \infty]$ is

$$P(\xi \geq x + \mathrm{d}x) = 1 - F(x + \mathrm{d}x)$$

It readily follows from these two equations that the probability of ξ to lie outside the interval $[x, x + \mathrm{d}x]$ is

$$P\left[(\xi < x) \cup (\xi \geq x + \mathrm{d}x)\right] = 1 - F(x + \mathrm{d}x) + F(x)$$

and finally, the probability of ξ to lie in the interval $[x, x + \mathrm{d}x]$ is expressed as follows:

$$P(x \le \xi < x + \mathrm{d}x) = 1 - \left[1 - F(x + \mathrm{d}x) + F(x)\right] = F(x + \mathrm{d}x) - F(x)$$

Consequently, we assume by definition that

$$F'(x)\mathrm{d}x = F(x + \mathrm{d}x) - F(x) = p(x)\mathrm{d}x$$

Now we can provide another definition of the probability distribution function. The probability distribution function $F(x)$ of a random variable ξ is a function of x such that the probability of the variable ξ to be inside the interval $[x, x + \mathrm{d}x]$ is $F(x + \mathrm{d}x) - F(x)$.

The new function $p(x)$ is called the probability density of the random variable.* Let us consider some of the functions that will be used extensively later.

First, let us examine the so-called *normal distribution* (also known as a Gaussian distribution) which is described by the function:

$$p(x) = \frac{1}{\sigma\sqrt{2\pi}} \exp\left[-\frac{(x - \mu)^2}{2\sigma^2}\right] \tag{1.5}$$

In this case, the distribution function $F(x)$ turns out to be the well-known *probability integral*:

$$F(x) = \frac{1}{\sigma\sqrt{2\pi}} \int_{-\infty}^{x} \exp\left[-\frac{(t - \mu)^2}{2\sigma^2}\right] \mathrm{d}t \tag{1.6}$$

This integral cannot be calculated in terms of elementary functions, although its detailed tabulated values can be found in any textbook on the theory of probabilities, for example, the work of Hudson [1]. The functions indicated in Equations 1.5 and 1.6 are plotted in Figure 1.2.†

The second subject of our attention is *uniform distribution* in which all values of the random variables within the interval $[a, b]$ have the same

* Sometimes in the physical literature the term "distribution function" is used for the function $p(x)$ instead of $F(x)$. This is not only nonstrict but also incorrect.

† Note that the sum of the squares of the random normally distributed variables is also a random variable, although its distribution is not Gaussian. This distribution is often used in the processing of the experimental data and has its own name—χ^2 distributions. The tabulated values of this distribution can also be found in the textbooks on the theory of probabilities.

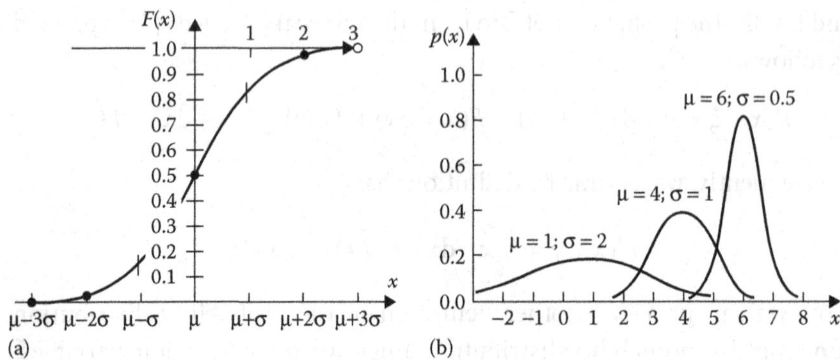

FIGURE 1.2 (a) The distribution function $F(x)$ and (b) the probability density $p(x)$ of the normal distributed random variable.

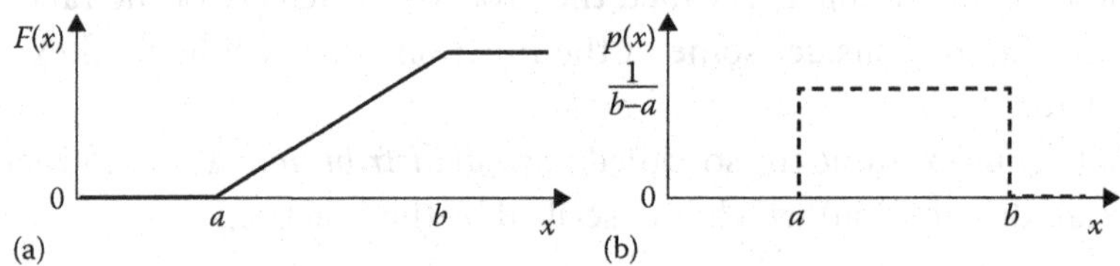

FIGURE 1.3 (a) The distribution function $F(x)$ and (b) the probability density $p(x)$ of the uniform distribution of random variable.

probability. Consequently, one may write the following equations for the probability density and distribution function, which are plotted in Figure 1.3:

$$p(x)=\begin{cases}0 & \text{if } -\infty<x<a\\ \dfrac{1}{b-a} & \text{if } a\le x\le b\\ 0 & \text{if } b<x<\infty\end{cases} \tag{1.7}$$

$$F(x)=\begin{cases}0 & \text{if } -\infty<x<a\\ \dfrac{x-a}{b-a} & \text{if } a\le x\le b\\ 1 & \text{if } b<x<\infty\end{cases} \tag{1.8}$$

Finally, let us consider a *Poisson distribution*. We present two different representations of this probability density function. These very important representations correspond to two principally different processes. As an example of the first one, let us consider the photoelectronic emission from the cathode of an electron optical brightness amplifier also called a light amplifier.* Suppose a photocathode of area S is illuminated with a uniform light flux so that the areal density of photons (number of photons per unit area of a photocathode) is m. The total number of photoelectrons emitted from the photocathode is $N_e = Sm\mu$, and the value of μ is usually called the *quantum efficiency*. Note that this equation is only approximately valid, and the larger the area S and the areal density m, the higher the accuracy. Due to the fact that photoelectron generation is a random process and because the readout of an electronic image is performed using a relatively small window (ΔS is small), the number of electrons counted at different regions of the cathode will not be a constant, $n_e\,(x,y) \neq \Delta Sm\mu$, even for uniform flux $[m(x,y) = \text{const}]$. This number will be a random variable determined by the statistics of photoelectrons. Moreover, repeating the measurements, one would observe a different number of electrons at the same segment of photocathode. Actually, this means that the value of efficiency μ is not a constant. Let us denote this varying value as μx. It was found experimentally that the probability density of electrons emitted at different readings is well approximated by the Poisson distribution:

$$p(x) = \mu x \exp(-\mu x) \tag{1.9}$$

The function introduced in Equation 1.9 is plotted in Figure 1.4. This function is normalized to unity, that is, $\int_0^\infty p(x)\mathrm{d}x = 1$. Consequently, the area under the curve limited by two vertical lines determines the fraction of electrons with respect to their total number. For instance, the dashed interval ($0.5 \leq \mu x \leq 1.5$) in Figure 1.4 contains almost 40% of all emitted electrons, the interval $1.5 \leq \mu x \leq 2.5$ contains about 30%, the interval

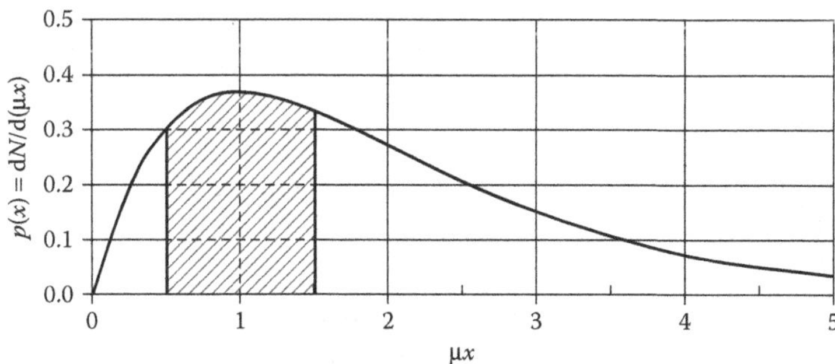

FIGURE 1.4 The probability density of electrons emitted at different readings.

* Light amplifiers will be discussed in detail in Chapter 4.

$2.5 \leq \mu x \leq 3.5$ contains ~15%, and so on. It should be pointed out that a wide range of various physical processes can be described with a Poisson distribution, such as the propagation of quanta in matter or the distribution of photoelectrons emitted by semiconductor photodiodes.

Another form of a Poisson distribution is used for describing a response function* $g(x)$ of electron optical amplifiers in cases when the level of a photocathode illumination is low:

$$p(x) = g(x) = \frac{\mu}{2}\exp(-\mu|x|) \tag{1.10}$$

This even function shows how the density of photoelectrons falls as the observation point is moved away from the maximum at $g(0)$. The distribution function in this case can be written as follows:

$$F(x) = \begin{cases} \frac{1}{2}\exp(\mu x) & \text{if } x < 0 \\ 1 - \frac{1}{2}\exp(-\mu x) & \text{if } x > 0 \end{cases} \tag{1.11}$$

The *expectation* value $\mathcal{M}(\xi)$ (or $\mathcal{M}\xi$) of a random variable ξ is defined as the average value of ξ (taking into account the probability or probability density of this variable). For a discrete distribution of the variable ξ, we have

$$\mathcal{M}(\xi) = \sum_{i=1}^{k} \xi_i P(\xi_i) = \mu \tag{1.12}$$

and for a continuous distribution, we have

$$\mathcal{M}(\xi) = \int_{-\infty}^{\infty} \xi p(\xi) d\xi = \mu \tag{1.13}$$

The *variance* of a random variable ξ is the mathematical expectation of the squared deviation of ξ and $\mathcal{M}(\xi)$:

$$\mathcal{D}(\xi) = \mathcal{M}(\xi - \mathcal{M}\xi)^2 = \sigma^2(\xi) \tag{1.14}$$

* The response function is the intensity distribution at the output screen of the electron optical amplifier when its cathode is illuminated with an extremely small spot (for more details, see Chapter 2).

The quantity σ (called the standard deviation or root mean square [RMS]) is a spread of a variable ξ around its average value $\mathcal{M}(\xi)$. Note that for both the Poisson and Gaussian distributions, we have $\mathcal{M}\xi = \mu$; the variance for Gaussian (normal) and Poisson distributions is $\mathcal{D}\xi = \sigma^2$ and $\mathcal{D}\xi = \mu$, respectively. Note that for a Poisson distribution, the following relation holds: $\sigma = \sqrt{\mu}$. This remarkable feature of a Poisson distribution, which will be widely used later, makes it possible to estimate easily the fundamental value—the signal-to-noise ratio *S*/*N*. Indeed, let our signal contain N_e emitted electrons. Then the dispersion of this signal (the so-called quantum noise) is $\sigma^2 \approx N_e$. Thus, the value of the *S*/*N* is

$$S/N \approx N_e/\sqrt{N_e} = \sqrt{N_e}$$

Let us clarify the physical meaning of these mathematical abstractions by considering the following simple experiment. Suppose that we are going to test a voltmeter designed for measurements in the range from 0 to 10 V. Let the accuracy* of the device be 5%.† This means that the "average error" of the measurements with this device is $\Delta V = 0.05V_{\max}$. (The exact meaning of average error will be explained later.) A reference voltage source with very high stability will be used for testing so that the random voltage fluctuations ΔV_r will not exceed, say, 0.01%. Thus, $\Delta V_r \leq 10^{-4}\,\text{V}$. The procedure of testing will be very simple: After setting an output voltage of the reference source, for instance, at 5 V, we will connect our voltmeter with the source, take a reading of the voltmeter, disconnect it, and repeat this sequence many times. What will we observe? First, we will find that despite the fact that our readings x_i are all in the vicinity of the value 5 (Figure 1.5), an exact reading of 5 happens very

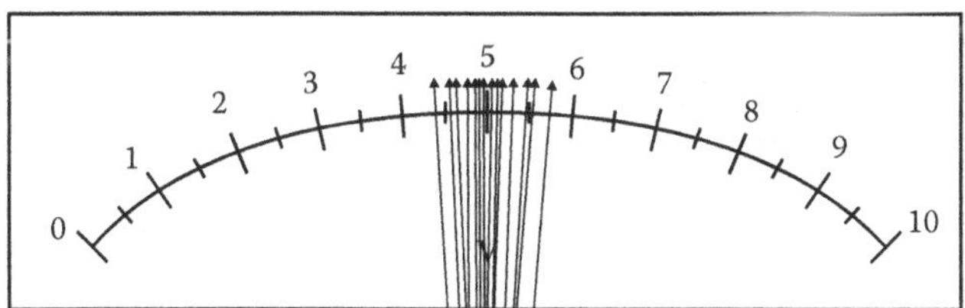

FIGURE 1.5 Scattering of the voltmeter reading.

* The meaning of "accuracy" in this case will be explained later.

† A device with such bad accuracy is very rarely used in real practice, and this extreme example is taken here simply for more vivid demonstration.

rarely or never at all. Second, as the number of readings n increases, we will see that their average value $\bar{x} = (1/n)\sum x_i$ gradually approaches closer and closer to 5 V. (Indeed, we would expect something like this since we have some readings where $x_i > 5$ and some readings where $x_i < 5$; of course, we also expected to observe the readings concentrating in the vicinity of 5 because we knew the reference voltage before the experiment.) However, the above speculations are just the result of our intuitive feelings, so can we put these speculations onto a more formal basis? The answer is yes; we can do this if we know the properties of the random variable x_i—its distribution function (or probability density), expectation value $\mathcal{M}(x_i) = \mu$, and variance $\mathcal{D}(x_i) = \sigma^2$ (or the standard deviation σ). Fortunately, all these data are available—otherwise we could have found them experimentally by analyzing the set of measured data. Indeed, the distribution function of the random variable x_i in our device is determined just by a number of physical reasons including the friction in the bearings of the voltmeter needle, the initial position of the needle before each specific measurement, the transient processes in the electric circuits of the device after it has been connected to the voltage source, and so on. Previous experimental experience has shown that in this situation the measurements will have a normal distribution, which, as we already know, is described by the Equations 1.5 and 1.6. It follows from the previous discussion that for the expectation value in this case, we have $\mathcal{M}(x_i) = \mu = 5\text{ V}$. As far as the variance is concerned, it is known that the scale divisions in an analog voltmeter are spaced by $\Delta V = \sigma$ (it is usual thing). Therefore, in our case, we have $\sigma = \Delta V = 0.05V_{\max} = 0.5\text{ V}$, and consequently, the variance is $\mathcal{D}(x_i) = \sigma^2 = 0.25\ V^2$. Taking advantage of Equation 1.6, we can now calculate the probability of the reading in any specific measurement to fall into the interval $(\mu - \sigma) \le x_i < (\mu + \sigma)$ (the marked region in Figure 1.6a):

$$P[(\mu-\sigma) \le x_i < (\mu+\sigma)] = \int_{\mu-\sigma}^{\mu+\sigma} \frac{1}{\sigma\sqrt{2\pi}} \exp\left[-\frac{(x-\mu)^2}{2\sigma^2}\right] dx = 0.683 \approx 0.7$$

In other words, only about 70% of the measurements will result in readings that fall within the interval between 4.5 and 5.5 V. This probability turns out to be quite moderate, and as one can see from Figure 1.6b, it falls off practically linearly with the decrease of the width of the interval. This is why in any real or virtual experiment, it makes no sense to wait until the voltmeter needle points somewhere very close to the position 5—the

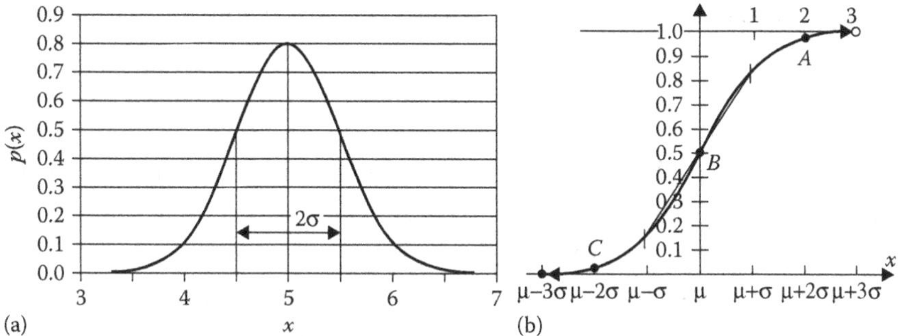

FIGURE 1.6 The probability of obtaining the reading x_i in the interval $\pm\sigma$, where $\sigma = \sqrt{\sigma^2}$ is the standard deviation and σ^2 is the variance of a random variable; (a) is the probability density $p(x)$, (b) is the probability distribution function $F(x)$. The points A and C indicate the values for arguments $\mu + 2\sigma$ and $\mu - 2\sigma$.

probability of this event is negligibly small. For this reason, in all trials when the reading x_i falls into the interval $[(\mu - \sigma), (\mu + \sigma)]$, it is usually known that the magnitude of a measured variable is μ, which is 5 V in our example. More rigorous experimentalists would say that the measured variable is (5 ± 0.5) V, that is, $(\mu \pm \sigma)$, although both expressions are not quite correct. The correct formulation of this result is as follows: The probability that the measured value is in the interval (5 ± 0.5) V is approximately equal to 70%. In practice, in everyday life, nobody talks in such a complicated way, but in fact this is implied when one says that some variable value is equal to $(\mu \pm \sigma)$. The main conclusion from the previous discussion is that *in real experiments we can never exactly determine the magnitude of the measured value but we can definitely establish the probability for this variable to lie within any given interval.*

Let us come back to our example of the voltmeter. The conditions of our testing experiment, as well as the experimental data we are going to acquire, were formulated above. What is the main point of the testing? The point is to compare manufacturer's certification value $\mathcal{M}(x_i) = \mu$ and $\mathcal{D}(x_i) = \sigma^2$ with the experimental data. For doing this, one should treat the acquired data and calculate the *sample mean* and the *standard deviation* of the data x_i from the mean value.

By definition, the sample mean of the random variable x_i is its average value calculated based on a sample of volume n:

$$\bar{x} = \frac{1}{n}\sum_{i=1}^{n} x_i \approx \mu \tag{1.15}$$

The above equation is the experimental average, which is an estimate for the mathematical expectation μ. Comparing Equations 1.12 and 1.15, one can see that the probability $P(x_i)$ does not appear in the Equation 1.15. The reason for this is quite clear: In theoretical calculations, we should take into account the probability of getting any given value x_i, whereas in a real experiment, this probability is realized by nature, since the frequency of various x_i's is different within any given sample.

The second power of a standard deviation of the experimental data from the average value is the estimate of experimental data variance, that is, the data spread around the average value*:

$$\sigma^2(x_i) = \frac{1}{n}\sum_{i=1}^{n}(x_i - \bar{x})^2 \approx \mathcal{D}(x_i) \tag{1.16}$$

The accuracy of estimation of the expectation μ with the experimental average $\bar{x}$ is described by the *variance of sample mean* and can be expressed as follows:

$$\mathcal{D}(\bar{x}_j) \cong \sigma^2(\bar{x}_j) = \frac{1}{n}\sum_{j=1}^{n}(\bar{x}_j - \bar{x})^2 = \frac{1}{n^2}\sum_{i=1}^{n}(x_i - \bar{x})^2$$

From the above equation, we immediately get the following important equation:

$$\sigma(\bar{x}) = \frac{\sigma(x_i)}{\sqrt{n}} \tag{1.17}$$

In other words, the variance of $\bar{x}$ is smaller compared to that of x_i. For this reason, it gives a more accurate estimate for μ than any particular value of x_i, and the estimation accuracy is proportional to $1/\sqrt{n}$. The reason is that in cases where some quantity is measured in n trials, the experimental result is usually presented as follows:

$$\bar{x} + \sigma(\bar{x}) = \bar{x} \pm \frac{\sigma(x_i)}{\sqrt{n}}$$

* It can be shown that for small n the estimate of variance with the Equation 1.16 is biased. For this reason, it is better to use the equation $\sigma^2(x_i) = (1/n-1)\sum_{i=1}^{n}(x_i - \bar{x})^2$ for small-sized samples.

Now we are going to make use of the well-known *Chebyshev's inequality*

$$P(|x-\mu| \geq k\sigma) \leq \frac{1}{k^2}$$

for proving the following important statement: The sample average $\bar{x}$ (with standard deviation $\sigma(\bar{x}) = \sigma/\sqrt{n}$) with high probability converges to μ with an increase in the sample size. Let us rewrite the Chebyshev's inequality using the following $\sqrt{n}$ definitions: $\varepsilon = k\sigma/\sqrt{n}$ and $k = \varepsilon\sqrt{n}/\sigma$. We have $P(|\bar{x}-\mu| \geq \varepsilon) \leq \sigma^2 / n\varepsilon^2$, and taking into account that $P(|\bar{x}-\mu| < \varepsilon) \geq 1-\sigma^2 / n\varepsilon^2$, we readily get

$$P\left[-\varepsilon < (\bar{x}-\mu) < \varepsilon\right] \geq 1-\sigma^2/n\varepsilon^2 \tag{1.18}$$

It follows from the above equation that the probability of $\bar{x} \rightarrow \mu$ tends to unity if $n \rightarrow \infty$.

Well, as a result of the test, we have found the value of σ, compared it with the manufacturer's value ΔV, and established in this way whether the device under test satisfies the certificate of quality. However, we have found only a *random component of the error.* We simply were fortunate that all our readings x_i were in the vicinity of the division 5 V and the value of $\bar{x}$ was approximately 5 V. This would not happen unless the *systematic error*, which by definition is $\mu - \bar{x}$, is equal to 0 as it is in our case. Note that the systematic error is not the random value. Therefore, it is necessary in any way to determine this component of error only once and then to take it into account during experimental data processing. In fact, it is easier to say than to do. In our case, we knew, as the saying goes, "the answer of the exercise" because the highly stable reference voltage source was used for testing. An entirely different situation takes place when we measure the parameters of some unknown object for the first time (for instance, the temperature of a plasma jet) and we have nothing to compare them with the results of our measurements. What can we do in this case? Let us leave this question without an answer for a while and we will come back to it later.

In conclusion, we present the rules for addition and multiplication of the probabilities. Suppose that the events A and B are independent and exclusive. In this case, the probability of one of these events to happen is

$$P(A+B) = P(A) + P(B)$$

If the events A and B are not independent, then

$$P(A+B)=P(A)+P(B)-P(AB)$$

where:

$P(AB)$ is the probability of these two events to happen simultaneously

For independent A and B, the probability is

$$P(AB)=P(A)P(B)$$

In the general case, the following relationship holds:

$$P(AB)=P(A)P_A(B)=P(B)-P_B(A)$$

where:

The *conditional probability* $P_A(B)$ is the probability of event B to happen in conditions when the event A already took place.

This probability is given by the following equation:

$$P_A(B)=\frac{P(AB)}{P(A)}$$

which is completely similar to the following equation:

$$P_B(A)=\frac{P(AB)}{P(B)}$$

REFERENCE

1. Hudson, D. J. 1963. *Lectures on Elementary Statistics and Probability*. Geneva, Switzerland: CERN.

CHAPTER 2

General Properties of Measuring–Recording Systems

IT GOES WITHOUT SAYING that any device has to be tested before it can be used for measurements. Generally speaking, the testing procedure is quite simple: A known or specially generated signal I_{in} is applied to the device input and the output signal I_{out} is compared against its expected value, which is $\boldsymbol{G}I_{in} = U$. Let σ be the measurement error, which is determined by the level of noise in the measuring system. If the condition $\|I_{out} - \boldsymbol{G}I_{in}\| \leq \sigma^2$ is fulfilled, then everything goes well, the device works correctly, and it can be safely used for measurements. This looks really simple, provided we know the operator $\boldsymbol{G}$ and the value of σ, which describe, respectively, the properties of the measuring system and the noise acting in it. As soon as the operator $\boldsymbol{G}$ is known, one can always estimate the output signal $U = I_{out} = \boldsymbol{G}I_{in}$ or reconstruct the input signal from the output signal with the help of the equation $U = \boldsymbol{G}I_{in} + N$, where N represents the noise in the system. The success of this latter procedure depends on the magnitude of the signal-to-noise ratio (*S/N*) $\boldsymbol{G}I_{in}/N$. This equation can also be used to obtain the operator $\boldsymbol{G}$ itself. This option is often much simpler to realize since one can always use the input signals of fairly high amplitude so that the relation $\boldsymbol{G}I_{in} >> N$ is fulfilled. We start our discussion by illustrating these methods and related problems for linear measuring–recording systems.

2.1 LINEAR MEASURING–RECORDING SYSTEMS

As a rule, measurement devices used in experimental physics are linear systems, which means that they can be described by linear equations. To be more precise, such devices are usually designed to be linear within the range over which they are supposed to be used for measurements. The most important feature of a linear system is the validity of the superposition principle: $f(x+y)=f(x)+f(y)$. In particular, this means that in a linear system the noise is of an additive character—it just can be summed up with the signal. This situation makes it possible in linear systems to analyze independently the noise and the signals.

The pulse and spectral approaches are the main methods applied for investigating linear systems. Generally speaking, these methods are equivalent to each other, for which reason in any specific case one may use either the method, depending on accuracy requirements, the convenience, or just the equipment that is available.

In the theory of *linear systems*, the input signal $f_{in}(x)$ [or $f_{in}(t)$] is usually called the *influence*, whereas the output (or recorded) signal $f_{out}(x)$ is called the *response*. Let us first consider the *pulse approach* for linear system study. There are several test signals generally used as the standard influence. One of them is the unit function:

$$\chi(x)=\begin{cases}0 \text{ if } x<0\\ 1 \text{ if } x\geq 0\end{cases} \tag{2.1}$$

Another standard signal is the so-called single or unit pulse (δ-function) determined as follows:

$$\delta(x)=\begin{cases}0 \text{ if } x\neq 0\\ \infty \text{ if } x=0\end{cases} \quad \text{and} \quad \int_{-\infty}^{+\infty}\delta(x)\mathrm{d}x=1 \tag{2.2}$$

(In optics, these functions are used to describe the *illuminated field edge* and the *slit with a width less than the normal.*) Note that the δ-function is called as a *unit pulse* because the relation $\int_{-\infty}^{\infty}\delta(x)\mathrm{d}x=1$ is fulfilled—although the signal amplitude is not equal to unity. It should also be noted that the δ-function is often introduced as the operator realizing the relation $\int_{-\infty}^{\infty}\delta(x)f(x)\mathrm{d}x=f(0)$, so, in particular, the integral $\int_{-\infty}^{\infty}\delta(x)\chi(x)\mathrm{d}x=1$ [since $\chi(0)=1$]. Although very convenient for theoretical modeling, these pure mathematical Abstractions 2.1 and 2.2 cannot be exactly represented by real physical objects. This is why in practical applications these idealized signals are replaced by various approximations, which are more realistic from an experimental point of view and, simultaneously, quite convenient for numerical simulations. For instance, the following two functions are widely used for this purpose:

$$\delta_1(x) = \frac{\alpha}{\sqrt{\pi}} \exp[-(\alpha x)^2] \tag{2.3}$$

$$\chi_1(z) = \frac{1}{2} + \Phi(z) \tag{2.4}$$

where:

$\Phi(z) = \frac{1}{\sqrt{2\pi}} \int_0^z \exp(-u^2/2)\mathrm{d}u$ is the probability integral

Note also that $\int_{-\infty}^{\infty} \delta_1(x)\mathrm{d}x = 1$. (We will discuss later the framework for correct modeling of the χ and δ functions with such functions.) It is clear that the designation of variables (t or x) is selected depending on whether the functions under consideration are temporally or spatially dependent. Note, that both $\delta_1(x)$ and $\delta_1(t)$ functions are not dimensionless. For the time- and space-dependent functions, the dimensions are, respectively, [time^{-1}] and [length^{-1}]. For the same reason, the integrals $\int_{-\infty}^{+\infty} \delta_1(x)\mathrm{d}x$ and $\int_{-\infty}^{+\infty} \delta_1(t)\mathrm{d}t$ are dimensionless. For instance, the device input current can be represented as $I_{\mathrm{in}}(t) = Q_0\delta_1(t)$, since the total electric charge flowing at the input circuit is $\int_{-\infty}^{+\infty} I(t)\mathrm{d}t = \int_{-\infty}^{+\infty} Q_0\delta_1(t)\mathrm{d}t = Q_0$.

If the two-dimensional (2D) case is considered, the 2D δ-function $\delta(x,y)$ can be used. The Gaussian function of two variables is also used in this case:

$$\delta_1(x, y) = \frac{\alpha^2}{\pi} \exp[-\alpha^2(x^2 + y^2)] \tag{2.5}$$

and the so-called *emitting point* function (a representing the point diameter):

$$\delta_2(x, y) = \begin{cases} 1/\pi a^2 & \text{if } (x^2 + y^2) \leq a^2 \\ 0 & \text{if } (x^2 + y^2) > a^2 \end{cases} \tag{2.6}$$

The dimension of the 2D functions $\delta_1(x,y)$ and $\delta_2(x,y)$ is length^{-2} (for instance, cm^{-2}), so that if the unit pulse of duration τ carrying energy $\mathcal{E}_0$ is applied at the input of some optical system, the corresponding intensity distribution is $I_{\mathrm{in}}(x, y) = \mathcal{E}_0\,\delta_2(x, y)/\tau$ [W cm^{-2}]. In the one-dimensional (1D) case, the function $\delta_2(x,y)$ reduces to the function $\delta_2(t)$ that represents the rectangular pulse, which is widely used in electronics. The amplitude of a rectangular pulse of duration τ is equal to A/τ. We should emphasize once more that in all cases the integrals $\int_{-\infty}^{+\infty}\delta_1(x)\mathrm{d}x$, $\int_{-\infty}^{+\infty}\delta_2(t)\mathrm{d}t$, and $\iint\delta_1(x,y)\mathrm{d}x\mathrm{d}y$ are dimensionless and equal to unity.

In the *frequency method* for studying linear systems, input signals of sinusoidal shape are used: $f_{in}(t) = A_0\sin(\omega t)$ or $f_{in}(x) = A_0[1+\sin(\omega x)]/2$, the value of ω being varied within the bandwidth of the investigated measuring equipment. This method will be discussed in more detail in Section 2.3.

2.2 TRANSFER FUNCTION AND CONVOLUTION EQUATION

The response of the system to a single (or unit) pulse $\delta(x)$ is called the *single-pulse response* or the *transfer* (or *spread*) *function*. For this function, we use the designations $g(t)$ or $g(x)$. The response to the unit function is called the transient function. For this function, we use the designation $h(x)$. The above functions can be shown to satisfy the following equation: $\mathrm{d}\left[h(x)\right]/\mathrm{d}x = g(x)$. Let us show that each of these two functions describes completely a noise-free measuring system. To be more rigorous, what we are going to show is that when the input signal is given, the output signal can be determined unambiguously with the help of any of the above functions. But first we consider some general properties of the transfer functions. Suppose that we have a passive radio technical system without losses. Let the input and output signals be $I_{in}(t)$ and $I_{out}(t)$, respectively. The charge flowing into the system during the time interval Δt is $I_{in}(t)\Delta t$. Taking into consideration the conservation law (a net charge cannot accumulate within the system), we obtain the following relation:

$$\int_{-\infty}^{+\infty} I_{in}(t)\mathrm{d}t = \int_{-\infty}^{+\infty} I_{out}(t)\mathrm{d}t \tag{2.7}$$

The amplification or attenuation of the signal is taken into account by introducing a permanent multiplier into the left-hand side of the above equation. Since the transfer function $g(t)$ by definition represents the system response to the single pulse $\delta(t)$, the following relation is valid:

$$Q_0\int_{-\infty}^{+\infty}\delta(t)\mathrm{d}t = Q_0\int_{-\infty}^{+\infty} g(t)\mathrm{d}t$$

In other words, the quantity $Q_0g(t)$ represents the output signal $I_{out}(t)$ for the case when the signal $I_{in}(t) = Q_0\delta(t)$ is applied to the system input. It follows from this last relation that the functions $\delta(t)$ and $g(t)$ have the same dimension. A similar equation is used in optical systems:

$$\mathcal{E}_0\iint\delta(x,y)\mathrm{d}x\mathrm{d}y = \mathcal{E}_0\iint g(x,y)\mathrm{d}x\mathrm{d}y$$

The above equation, along with the two previous equations, just represents the conservation law.

In Figure 2.1, two rectangular pulses $\delta_2(t)$ of different duration and amplitude are shown, the amplitudes and durations being related as $I_1\tau_1 = I_2\tau_2 = Q_0$. The responses $[g(t)]$ to these two signals and the principal electrical circuit illustrating the experimental configuration are also shown in the figure. This figure illustrates the situation that is typical for electrical circuits: The parasitic capacitances and inductances determine the output signal rise time, while the energy stored in these circuit components is responsible for the descending part of the output signal, which is usually fairly long. Note the following circumstances: First, the output signal amplitudes are very small compared to those at the system input. This is clear since the areas under the curves $I_{in}(t)$ and $I_{out}(t)$ should be equal to each other. Second, as one can see in Figure 2.1, the output pulses have almost identical shapes despite the twofold difference in the input pulse duration ($2\tau_1 = \tau_2$). There are two reasons for this fact: the equality $I_1\tau_1 = I_2\tau_2$ and the condition $\beta >> \tau$. The first reason is clear—one can never satisfy the condition (2.7) and, consequently, get equal outputs for $I_1\tau_1 \neq I_2\tau_2$. The physical considerations of the condition $\beta >> \tau$ will be discussed in Section 2.3. Now let us concentrate on the terminology, the history, and the equation relating the input and output signals.

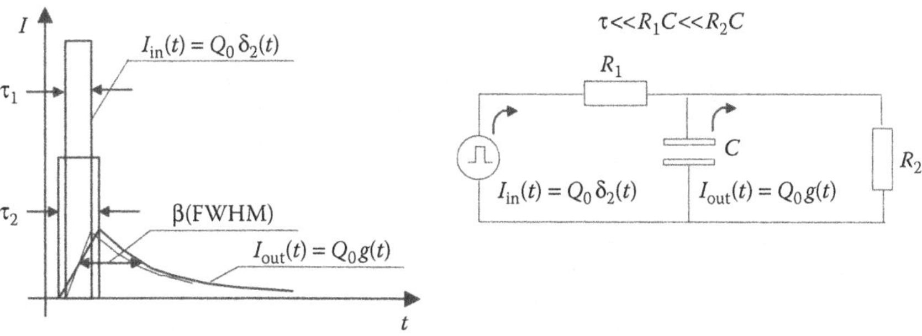

FIGURE 2.1 Influence and response in a linear system.

The parameter β is denoted as FWHM (full width at half maximum), which is a good description of the parameter β. In optical systems, where the transfer functions are usually even, $g(x) = g(-x)$, the parameter β can be introduced by the formal relationship: $2g(\beta/2) = g(0)$. Let us consider the curves in Figure 2.2. In this figure, the situation is considered when the duration of the input rectangular pulse is long compared to the transfer function FWHM. It is quite evident that both the rise and fall times of the output pulse are determined by the transient processes in the circuit, while its central part is related only to the input signal. Now let us consider what happens if we start to decrease the input pulse duration. For sufficiently long pulses, the only significant change will be the reduction in the duration of the flat part of the output pulse. This is easy to understand since the rise and fall times represent the response of the system to the sudden appearance and disappearance of the voltage applied to the system input. However, as the duration of the input pulse is further decreased, there will come a point when there will no longer be a flat part of the output pulse at all—at this point, the output pulse shape becomes quite close to that of the transfer function.

It is interesting to mention that a similar experimental approach for measuring the FWHM of the transfer function was developed many years ago in spectroscopy; for example, the spectral lines at a spectrograph output were found to become sharper as the width of the entrance slit was decreased, but only up to some limit. As soon as the slit width became smaller than this limit (called the *normal width*), a further reduction in the slit width resulted only in the limitation of the total light intensity, and the shape of the line profile did not change.

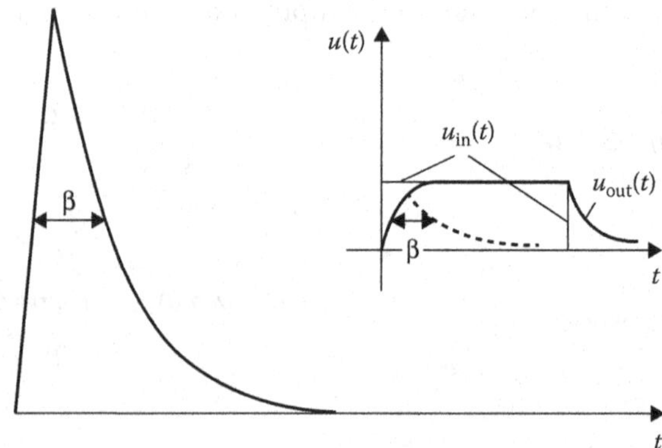

FIGURE 2.2 Transfer function.

Therefore, let us assume that the input signal and the transfer function have been specified. How can the output signal be calculated? The hidden problem in this calculation originates from the fact that the magnitude of the output signal $I_{out}(t)$ at some instant t_k is determined not only by the input signal amplitude $I_{in}(t)$ at this same instant, but also by the time history of the signal over the whole period preceding the instant t_k. For instance, consider Figures 2.1 and 2.2. As one can see from these figures, the charge deposited in the system by the input signal component $\Delta[I_{in}(t)] = I_{in}(t)\Delta t$ supports the corresponding magnitude of the output current $\Delta[I_{out}(t)] = I_{in}(t)\Delta t g(t)$ for a sufficiently long time. In this sense, the transfer function $g(t)$ represents the influence function that determines the effect of the previous input signal components on the output signal at all later moments in time. (Indeed, the longer the time, the smaller the influence.) Taking into account the above consideration along with the linearity of the system (validity of the superposition principle), we can propose the following procedure for calculating the output signal. Let us first divide the time interval $[0, t]$ into a series of subintervals Δt so that $\Delta t << \beta$ (Figure 2.3). Then we calculate the charge $Q_i = I(t_i)\Delta t_i$ flowing into the system during the corresponding interval Δt_i. It follows from the above discussion that the elementary component of the output signal $\Delta[I_{out}(t_j)]$ at the instant $t_j \geq t_i$ can be presented as $\Delta[I_{out}(t_j)] = I_{in}(t_i)\Delta t_i g(t_j - t_i)$. Taking advantage of the superposition principle, we can express the output signal $I_{out}(t_j)$

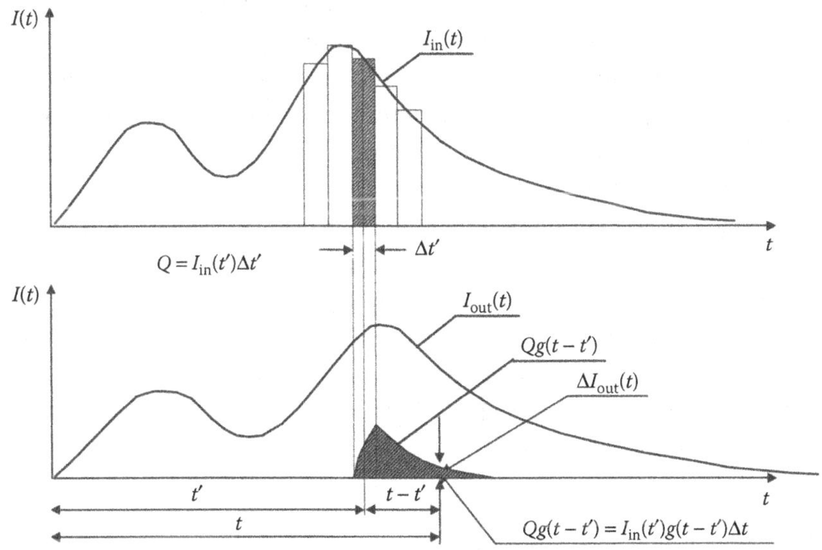

FIGURE 2.3 Input and output signals.

as the sum of its elementary components: $I_{\text{out}}(t_j)=\sum_{i=0}^{i=j} I_{\text{in}}(t_i)g(t_j-t_i)\Delta t$. Using the definitions of Figure 2.3, we can express the elementary component of the output signal as $\Delta[I_{\text{out}}(t)] = I_{\text{in}}(t')g(t-t')\Delta t'$. Finally, taking the limit for infinitely small time subintervals, we arrive at the following integral as the final answer:

$$I_{\text{out}}(t)=\int_{0}^{t} I_{\text{in}}(t')g(t-t')\mathrm{d}t' \tag{2.8}$$

The integral on the right-hand side of the above equation is called the *Duhamel's integral.* Using this equation, one can calculate (in linear systems!) the output signal, in cases when the input signal and the transfer function are known. The same equation can also be used for reconstructing the input signal, provided the output signal and the transfer function are known, as usually is the case in real measurements. It should be emphasized that in real systems with noise, any reconstruction procedure faces a number of problems that are of a fundamental rather than a technical nature. This point is emphasized here because it is encountered many times in Chapters 3 through 12.

The lower limit of the Duhamel's integral is determined by the time domain in which the input signal exists and the upper limit is set by the causality principle: The output signal $I_{\text{out}}(t)$ can be affected only by those input signal components that arrive at the system input before, but not after, the instant t. The situation is different for optical systems, in which the intensity $I(x)$ at some point x in the output image plane can be affected by the input image points located on both sides of the point x, provided we deal with 1D systems. [A conventional spectrogram, with the line intensity constant along the line, $I(y)$ = const, presents the typical example of a 1D image.] In 2D configurations, the intensity $I = f(x,y)$ is a function of two variables, and its magnitude at some given point x,y in the output image plane can be affected by all the input image regions located in the vicinity of the conjugated point in the image plane. With this situation in mind, we can reintroduce the above considerations and get the following equation, which relates the intensity distributions at the input and output of any (linear!) optical system:

$$I_{\text{out}}(x)=\int_{-\infty}^{+\infty} I_{\text{in}}(x')g(x-x')\mathrm{d}x' \tag{2.9}$$

This equation is called a *convolution equation* and is generally used for solving the three following problems: (1) calculation of the output image intensity distribution $I_{out}(x)$ for any given input distribution $I_{in}(x)$ and transfer function $g(x)$; (2) reconstruction of the input distribution using the output image and transfer function $g(x)$ (bearing in mind the above comments regarding problems with noise); and (3) determination of the transfer function $g(x)$ where the input and output images are known.

In the case of 2D images, the convolution equation is written as follows:

$$I_{out}(x,y) = \iint I_{in}(x',y')g(x-x',y-y')\mathrm{d}x'\mathrm{d}y' \tag{2.10}$$

In the literature, the convolution procedure is often designated as $I_{out}(x) = I_{in}(x) \otimes g(x)$. However, we prefer to use the expression introduced previously: $I_{out}(x) = \mathbf{G}I_{in}(x)$, having in mind that in linear systems the operator $\mathbf{G}$ is determined by Equations 2.8 through 2.10. It should be pointed out that it is possible to introduce a similar operator in nonlinear systems, thus formalizing the relationship between the input and output signals.

2.3 TRANSFER RATIO, AMPLITUDE–FREQUENCY AND PHASE–FREQUENCY CHARACTERISTICS, AND RELATIONSHIP BETWEEN INPUT AND OUTPUT SIGNALS IN FOURIER SPACE

As we have mentioned already in Section 2.2, to study linear systems with the frequency method, sinusoidal signals are applied to the system input. Usually, for time-resolved testing, the input signal takes the form $f_{in}(t) = A_0\sin(\omega t)$, whereas for testing spatially resolving devices, the input signals of the form $f_{in}(x) = A_0[1+\sin(\omega x)]/2$ are widely used.

In the frequency method, the transfer ratio is the function that replaces the transfer function. By definition, the transfer ratio function is the ratio of two complex variables—the output signal $\dot{I}_{out} = I_0\exp\{\mathrm{i}[\omega t + \varphi(\omega)]\}$ and the input signal $\dot{I}_{in} = I_0\exp(\mathrm{i}\omega t)$ measured for different frequencies supplied at the system input:

$$K(\omega) = \left.\frac{\dot{I}_{out}}{\dot{I}_{in}}\right|_{\omega} \tag{2.11}$$

Usually, the complex function $K(\omega)$ is represented as

$$K(\omega) = A(\omega)\exp[-\mathrm{i}\varphi(\omega)]$$

The function $A(\omega) = |K(\omega)|$ is called the *amplitude–frequency characteristic*. This function determines the ratio of the amplitudes of the output and input signals. The function $\varphi(\omega)$ is called the *phase–frequency characteristic* and it represents the phase shift so that for the input signal $f_{in}(t) = I_0\sin(\omega t)$ we have the output signal $f_{out}(t) = I_0\sin(\omega t + \varphi)$.

Let us recall that the Fourier transform $\Phi(\omega)$ of the function $f(x)$ is given by the following integral:

$$\Phi(\omega) = \int_{-\infty}^{\infty} f(x)\exp(-i\omega x)dx \tag{2.12}$$

Actually, the function $\Phi(\omega)$, which is the image of $f(x)$ in Fourier space, represents the spectral expansion of the function $f(x)$. The inverse Fourier transform or reconstruction is represented by the following equation:

$$f(x) = \frac{1}{2\pi}\int_{-\infty}^{\infty} \Phi(\omega)\exp(i\omega x)d\omega \tag{2.13}$$

Such operations in Fourier space are widely used in numerical simulations and in the processing of experimental data as well. This is due to the simplification of calculation procedures when operating with Fourier images. An analogous situation takes place when dealing with logarithms: Instead of multiplication we have addition; the derivation of roots is replaced by division, and so on. Similarly, when we deal with function images in Fourier space, the problem of solving the integral equations is reduced to solving the algebraic equations. After this, the solution can be found using the inverse Fourier transform procedure.

It is clear that the input signal $f_{in}(x)$ and its spectral expansion $\Phi_{in}(\omega)$ are related by the inverse Fourier transform:

$$f_{in}(x) = \frac{1}{2\pi}\int_{-\infty}^{\infty} \Phi_{in}(\omega)\exp(i\omega x)d\omega \tag{2.14}$$

The component of the input signal in configuration space corresponding to its spectral component at some given frequency ω^* localized within the frequency interval $\Delta\omega$ can be expressed as follows:

$$\begin{aligned}\Delta[f_{in}(x)]_{\omega^*} &= \frac{1}{2\pi}\int_{\omega^*-\Delta\omega/2}^{\omega^*+\Delta\omega/2} \Phi_{in}(\omega)\exp(i\omega x)d\omega \\ &\cong \frac{1}{2\pi}\Phi_{in}(\omega^*)\exp(i\omega^* x)\Delta\omega\end{aligned} \tag{2.15}$$

Note that the right-hand side of the above equation converges to the exact value of the integral as the magnitude $\Delta\omega$ tends to 0. However, it follows from the definition that the ratio of the spectral components of the output and input signals is the transfer ratio:

$$K(\omega^*) = \left.\frac{\dot{I}_{\text{out}}}{\dot{I}_{\text{in}}}\right|_{\omega^*} = \frac{\Delta[f_{\text{out}}(x)]_{\omega^*}}{\Delta[f_{\text{in}}(x)]_{\omega^*}}$$

It follows from the above equation that

$$\Delta[f_{\text{out}}(x)]_{\omega^*} = \Delta[f_{\text{in}}(x)]_{\omega^*} K(\omega^*) \tag{2.16}$$

Now, taking advantage of Equation 2.15, we find that

$$\Delta[f_{\text{out}}(x)]_{\omega^*} = \Delta[f_{\text{in}}(x)]_{\omega^*} K(\omega^*) = \frac{1}{2\pi}\Phi_{\text{in}}(\omega^*)K(\omega^*)\exp(i\omega^* x)\Delta\omega \tag{2.17}$$

Carrying out the integration over the whole frequency range, we can write the relationship:

$$f_{\text{out}}(x) = \frac{1}{2\pi}\int_{-\infty}^{\infty}\Phi_{\text{in}}(\omega)K(\omega)\exp(i\omega x)d\omega \tag{2.18}$$

Using the definition of the Fourier transform, we can represent the output signal as follows:

$$f_{\text{out}}(x) = \frac{1}{2\pi}\int_{-\infty}^{\infty}\Phi_{\text{out}}(\omega)\exp\left(i\omega x\right)d\omega \tag{2.19}$$

Comparing Equations 2.18 and 2.19, we finally get the important equation:

$$\Phi_{\text{out}}(\omega) = \Phi_{\text{in}}(\omega)K(\omega) \tag{2.20}$$

This relationship between the input and output signals in Fourier space turns out to be so simple because it is the transfer ratio that—just by definition—shows how the spectral components of the input signal are attenuated when passing through the system. Equation 2.20 presents the convolution equation in Fourier space. The obvious solution of the following equation

$$\Phi_{\text{in}}(\omega) = \frac{\Phi_{\text{out}}(\omega)}{K(\omega)} \tag{2.21}$$

can easily be understood: To recover the input signal from the output signal, each spectral component of the output signal has to be divided by its attenuation within the system, that is, has to be multiplied by $1/K(\omega)$. [It follows from these relationships that if in some frequency range the transfer ratio $K(\omega)$ is equal to 1, then the equality $\Phi_{\text{in}}(\omega) = \Phi_{\text{out}}(\omega)$ is valid.]

The above considerations make clear the method for recovering (or reconstructing) the input signal: One should go into Fourier space, find the Fourier transform of the output signal, divide it by the function $K(\omega)$, and then go back into configuration space (perform the inverse Fourier transform). Unfortunately, the situation is much more complicated in real life. When processing real experimental results, we usually deal with the following equation:

$$\Phi_{\text{in}}(\omega) = \frac{\Phi_{\text{out}}(\omega) + N(\omega)}{K(\omega)} \tag{2.22}$$

where:

The function $N(\omega)$ stands for the spectral component of the noise present in the system

Note that as $\omega \to \infty$ all functions on the right-hand side of the above equation tend to 0, but at different rates. The function $N(\omega)$ exhibits the lowest rate. This fact can be explained as follows: Both $\Phi_{\text{out}}(\omega)$ and $K(\omega)$ are smooth functions, whereas the function $N(\omega)$ is not. This is why it converges to 0 very slowly. Having in mind this fact, one can readily see from Equation 2.22 that, starting from some critical frequency ω_c [determined by the requirements that $N(\omega) > \Phi_{\text{out}}(\omega)$ and that $K(\omega)$ is close to 0], the error in the determination of $\Phi_{\text{in}}(\omega)$ may become arbitrarily large. In such cases, mathematicians say *the problem is unstable*. What does this mean from the physical point of view? The case is that the weak dependence of spectral noise on frequency is the result of averaging over a large number of samples, while any given realization of the random process indeed results in a fluctuating function. For this reason, one can always find some frequency ω_j in the region $\omega > \omega_c$ such that the spectral component $N(\omega_j)$ is large, while $K(\omega_j)$ is close to 0. Due to this fact, instead of getting a real (or, at least, approximate) solution, an artificial solution $f_{\text{in}}(x) = A_j \sin(\omega_j x)$ will be obtained since the spectral component $\Phi_{\text{in}}(\omega_j)$ would dominate in the integral (2.19). It should be emphasized that the magnitude of A_j can be arbitrarily large. We have considered the unstable behavior of solutions of the convolution equation in Fourier space because in this case the effect is better illustrated. Because of the same physical nature, the unstable behavior is also observed when solving similar problems in configuration space. These problems are well known as ill-posed problems. Nowadays, after the pioneering work of Tikhonov, there are a number of methods for solving such ill-posed problems. These methods will be discussed later in Chapter 12.

It follows from definition that the transfer function $g(x)$ is the signal at the output of the system to the input of which a signal $f_{in}(x) = \delta(x)$ is applied. Consequently, according to Equation 2.18, we can write the following equation:

$$g(x) = \frac{1}{2\pi}\int_{-\infty}^{\infty} \Phi_\delta(\omega)K(\omega)\exp(i\omega x)d\omega \tag{2.23}$$

Taking into account the fact that $\Phi_\delta(\omega) \equiv 1$, we readily obtain the important relationship

$$g(x) = \frac{1}{2\pi}\int_{-\infty}^{\infty} K(\omega)\exp(i\omega x)d\omega \tag{2.24}$$

that the transfer ratio represents the spectrum of the transfer function. Indeed, the inverse equation is also fulfilled:

$$K(\omega) = \int_{-\infty} g(x)\exp(-i\omega x)dx \tag{2.25}$$

Thus, the transfer ratio and transfer functions are coupled by a pair of Fourier transforms.

2.4 SOME CONSEQUENCES

Let us consider some important and useful transfer ratio and transfer function properties.

1. Suppose a measuring system is given such that $A(\omega) = |K(\omega)| = C = \text{const}$ and $\varphi(\omega) = a\omega$, that is, $K(\omega) = C\exp(-ia\omega)$. Let the input signal be $I_{in} = f_{in}(t)$. Let us find the output signal $I_{out} = f_{out}(t)$.

$$I_{out}(t) = \frac{C}{2\pi}\int_0^t \Phi_{in}(\omega)\exp(-ia\omega)\exp(i\omega t)d\omega$$

$$= \frac{C}{2\pi}\int_0^t \Phi_{in}(\omega)\exp[i\omega(t-a)]d\omega = Cf_{in}(t-a)$$

Consequently, for $A(\omega) = C = \text{const}$ and $\varphi(\omega) = a\omega$, the shape of the signal is not changed by the system, although the signal is delayed with respect to the input by the time a. This time is called the *delay time*.

2. It is worth mentioning that as a rule the transfer functions of optical systems exhibit a symmetry with respect to their maxima—that is, they are even functions. Let us show that $\varphi(\omega) = 0$ for the even transfer functions $g(x) = g(-x)$. In the following equations, we will sequentially substitute x by $-x$, and then $g(x)$ by $g(-x)$:

$$K(\omega) = \int_{-\infty}^{\infty} g(x)\exp(-i\omega x)dx = \int_{-\infty}^{\infty} g(-x)\exp(i\omega x)dx$$

$$= \int_{-\infty}^{\infty} g(x)\exp(i\omega x)dx = K(-\omega)$$

Thus, for even transfer functions, the transfer ratio is proved to be an even function:

$$K(\omega) = K(-\omega) \tag{2.26}$$

Bearing in mind that $g(x)$ is a function of a real variable and $K(\omega) = K(-\omega)$, we can write the following equation for the function that is the complex conjugate of $K(\omega)$:

$$K^*(\omega) = \int_{-\infty}^{\infty} g(x)\exp(i\omega x)dx = K(-\omega)$$

In other words, we have proved that the following relation is valid for $g(x)$ and, by the way, also for every even function of a real variable:

$$K^*(\omega) = K(-\omega) \tag{2.27}$$

Let us rewrite this equation as

$$A(\omega)\exp[i\varphi(\omega)] = A(-\omega)\exp[-i\varphi(-\omega)]$$

or equivalently as

$$A(\omega)[\cos\varphi(\omega) + i\sin\varphi(\omega)] = A(-\omega)[\cos\varphi(-\omega) - i\sin\varphi(-\omega)]$$

These equations can be valid only if $A(\omega) = A(-\omega)$ and $\varphi(\omega) = -\varphi(-\omega)$. Writing Equation 2.26 as

$$A(\omega)\exp[-i\varphi(\omega)] = A(-\omega)\exp[-i\varphi(-\omega)]$$

and making the substitutions $A(-\omega) = A(\omega)$ and $-\varphi(-\omega) = \varphi(\omega)$ on the right-hand side, we get the following equation:

$$A(\omega)\exp[-i\varphi(\omega)] = A(\omega)\exp[i\varphi(\omega)]$$

which can be satisfied only by assuming that $\varphi(\omega) \equiv 0$, or equivalently that $K(\omega) = A(\omega)$. This corresponds exactly to what we set out to prove.

3. In optics, the transfer ratio is called the *frequency-contrast function* or the *frequency-contrast characteristic*. It is also referred to as the *modulation transfer function* (MTF). Experimentally, the MTF is determined using the output image as follows:

$$A(\omega) = \frac{E_{max} - E_{min}}{E_{max} + E_{min}} \tag{2.28}$$

where the signal $f_{in}(x) = E_0[1 + \sin(\omega x)]$ is assumed to be applied to the system input. It is clear that in this case the output signal can be expressed as $f_{out}(x) = E_0[1 + A(\omega)\sin(\omega x)]$, with the maximum and minimum values of the output signal being realized for $\omega x = \pi/2 + k\pi$ and $\omega x = 3\pi/2 + k\pi$, respectively (here k are even numbers). Note that in optical systems without losses, the condition $(E_{max} + E_{min})/2 = E_0$ is valid for all frequencies ω in spite of the fact that both of the amplitudes E_{max} and E_{min} are frequency dependent. Due to this fact, one can find the MTF using only the output image.

4. It is well known from the spectral theory that the product of the pulse bandwidth and pulse duration is approximately equal to unity: $\Delta t \times \Delta\nu \approx 1$ ($\Delta\nu = \Delta\omega/2\pi$). Indeed, this is a rough estimate. In addition, it is not quite clear how we can determine the pulse duration for pulses of an arbitrary shape. Indeed, even more questions arise so far as $\Delta\omega$ or $\Delta\nu$ are concerned. However, bearing in mind that we are trying to get no more than a rough estimate, let us assume that Δt, $\Delta\omega$, and $\Delta\nu$ are determined by the FWHM of the corresponding functions $f(t)$, $\Phi(\omega)$, and $\Phi(\nu)$, respectively: $\Delta t \approx \beta_t$ and $\Delta\nu \approx \beta_\nu$. Let us also recall that the FWHM for even functions can be calculated using the relationship $2f(\beta/2) = f(0)$. After these remarks, we can calculate the values β_t and β_ν for the function $\delta_1(t) = (\alpha/\sqrt{\pi})\exp[-(\alpha t)^2]$ (see Equation 2.3), which we had already used for modeling the δ-function

in the pulse method for studying linear systems. With the help of the equation $(2\alpha/\sqrt{\pi})\exp[-(\alpha\beta/2)^2] = (2\alpha/\sqrt{\pi})\exp[-(0)^2]$, we can easily find that $\beta_t = 2\sqrt{\ln 2}/\alpha$. The Fourier transform of Equation 2.3 can be written as

$$\Phi(\omega) = \frac{\alpha}{\sqrt{\pi}} \int_{-\infty}^{\infty} \exp[-(\alpha t)^2]\exp[-i\omega t]dt = \frac{\sqrt{\pi}}{\alpha}\exp\left[-\left(\frac{\omega}{2\alpha}\right)^2\right]$$

and, consequently, $\Phi(\nu) = (\sqrt{\pi}/\alpha)\exp[-(\pi\nu/\alpha)^2]$ and $\beta_\nu = (2\sqrt{\ln 2}/\pi)\alpha$. Using the resulting expressions, we have immediately $\Delta t \times \Delta\nu \approx \beta_t \times \beta_\nu = (4\ln 2/\pi) \approx 0.9$. In other words, it turns out that the estimation $\Delta t \times \Delta\nu \approx 1$ is not so bad.

There are at least two reasons for carrying out these exercises. First, we have found out that for the given characteristic time Δt (or the spatial scale Δx) describing the evolution of a function in the time (or the space) domains, using the equation $\Delta\nu \approx 1/\Delta t$ or $\Delta\nu \approx 1/\Delta x$ we can estimate (though not calculate precisely) the spectral width of the signal. (For the functions possessing a single maximum, the FWHM can be used as the characteristic scale.) Second, we are interested in finding the criteria for the correct use of regular functions as replacements for the purely singular δ-function. Now let us compare the frequency and pulse methods. In the frequency method, we measure the function $K(\omega)$ gradually changing the frequency of the input signal in the range from $\omega = 0$ to ω_{max}, where the frequency ω_{max} is determined by the condition $\Phi_{out}(\omega_{max}) \geq N(\omega_{max})$ [here $N(\omega_{max})$ is the spectral component of the noise at the frequency ω_{max}]. For measuring the transfer function in the pulse method, we apply a signal of short duration to the system input. Indeed, all the frequencies within the range $[0,\omega_{max}]$ have to be represented in the spectrum of this signal. However, in the first case the value of ω_{max} can be easily determined by the output signal—we just have to increase the frequency until the output signal submerges in the noise. The situation is quite different in the second case—it is not so easy to decide just by the shape of the transfer function if the signal at the system input is short enough to represent all the frequencies in the range $[0,\omega_{max}]$. However, as shown above, the system bandwidth $\Delta\omega$ can be estimated as $\Delta\omega \sim 2\pi/\beta$. The spectral width of an input pulse with duration δ can be written as $2\pi/\delta$. Undoubtedly, the frequency range of the input signal should be large compared to the system bandwidth $\Delta\omega$, that is, the following inequalities have to be fulfilled: $2\pi/\delta >> 2\pi/\beta$ or $\delta << \beta$. How large should be the difference? The calculations show that if the value of δ is as small as $\sim\beta/3$, the inaccuracy in the determination of β will not exceed several percents.

5. Now let us discuss how the technique that we have developed can be applied to characterize the quality of different measuring–recording systems. It should be emphasized that in all systems with a low level of noise, we can immediately determine the output signal from the input by using the transfer function or the transfer ratio of the system. In this sense, these functions are very useful. Moreover, these functions are also convenient at the stage of designing the measuring system, when the system is not experimentally tested, but the parameters of the system components are available. For instance, let us assume that our system consists of a number of sequentially installed subsystems [an optical system, a charge-coupled device (CCD) camera, a data transfer line, etc.], and that we know the transfer functions $g_1(x)$, $g_2(x)$, $g_3(x)$,... or the transfer ratios $K_1(\omega)$, $K_2(\omega)$, $K_3(\omega)$,... of each of these subsystems. In this case, the transfer function and transfer ratio of the whole system can be represented by the following equations:

$$g_\Sigma(x) = \iiint_{-\infty}^{\infty} g_n(x - x_{n-1})g_{n-1}(x_{n-1} - x_{n-2})\ldots$$

$$g_2(x_2 - x_1)g_1(x_1)\mathrm{d}x_1\ldots\mathrm{d}x_{n-1}$$

$$K_\Sigma(\omega) = K_1(\omega)K_2(\omega)K_3(\omega)K_4(\omega)\ldots$$

Despite all these facts, we still do not have any *criteria* for the evaluation of the output signal. Consequently, we cannot describe the different systems quantitatively and compare one against the other. The point is that the functions that have been discussed represent the aberration properties of the systems. This is an important issue, but not the only one as any real system is always a system with noise. In a system with noise, the final quality of the output signal is very much dependent on the *S/N*. Now let us pay attention to this aspect of the problem.

First of all, we note that the output signal of the system under consideration is supposed to be processed as the final stage. It is also important that any real physical system is not frequency limited because the function $K(\omega)$ tends to 0 only asymptotically at $\omega \to \infty$. This implies that for those cases where $S/N \to \infty$, the input signal can be reconstructed from the output signal with any accuracy for any transfer function. This can be done, for instance, with the help of the procedure $\Phi_{in}(x) = \Phi_{out}(\omega)/K(\omega)$. In this sense, all systems are identical. However, as shown previously, real situations are much more complicated, and the successful reconstruction of the input signal is extremely dependent on the noise present in the system. This is why noisy systems cannot be quantitatively characterized solely by the functions $g(x)$ and $K(\omega)$.

2.5 DISCRETIZATION

As a rule, the quantities measured in experiments represent continuous functions of some variables. However, during data processing, both the functions and the variables are replaced with sets of discrete values—in this process, an infinite set of function values and an infinite set of argument values are replaced by a finite number of values. This procedure can be performed correctly because the functions under consideration are limited in configuration space as well as in Fourier space.* If the discrete values of the function are spaced uniformly along the ordinate axis so that $m_k = k\Delta m$, then the discretization (quantification) procedure can be expressed by the following equation:

$$m_k = \left[\frac{m}{\Delta m} + \frac{1}{2}\right]\Delta m \tag{2.29}$$

where the part of the equation within the square brackets stands for the "integer part" and Δm represents the discretization step. It goes without saying that the discretization procedure produces some error $\varepsilon = m - mk$ since the real magnitude of the function m is replaced by its approximate magnitude m_k. This error is referred to as the quantization error or the quantization noise. It readily follows from Equation 2.29 that $\varepsilon \le \Delta m/2$. Usually, the value of Δm is selected in such a way that the value of ε is of the order of the errors in the recording devices and amplification–transmission lines. For normally distributed noise, the discretization step is selected so that $\overline{\varepsilon^2} = \sum_{j=1}^{n} \sigma_j^2 = \sigma^2$, where the summation is carried out taking into account all independent noise-generating processes. What

* The assertion that "the functions are limited in both configuration and Fourier spaces" is incorrect from a mathematical point of view. However, as will be shown later, from the experimental data accuracy point of view, these functions could be represented as *practically limited* in both the spaces.

will be the result when we select the discretization step to be $\varepsilon = \bar{\sigma}$? For a normal noise distribution in the recording–transmission line, the condition $\Delta m = 2\sigma$ means that the probability of an error of *one-scale division* ($\pm\Delta m$) is ~30%. At the same time, the probability of getting a two-scale division error is only ~5% and the probability of getting a three-scale division error is just no more than ~0.5%.

The theoretical basis for the argument value discretization is given by the theorem of Kotelnikov–Shannon. We state one of the versions of this well-known theorem as follows: *For unambiguous determination of the function f(x), which is strictly limited in Fourier space by the frequency* ν_c*, it is necessary and sufficient to determine the magnitude of this function at the discrete points separated by* $\Delta x \leq 1/2\nu_c$. Taking advantage of this theorem, we can give another representation of the relationships indicated by Equations 2.12 and 2.13. Following the theorem, let us assume that the quantization step is $\Delta x = 1/2\nu_c$. The discrete values of x are designated as $x_k = \Delta xk = k/2\nu_c$, the total number of members of the series is given as $N = T/\Delta x = 2T\nu_c$, where $T = x_{max} - x_{min}$ is the variation interval of the function argument. The magnitude of $\Delta\omega$ is determined by the smallest frequency $\nu_{min} = 1/T$ using the relation $\Delta\omega = 2\pi/T$, and, as usual, $\omega_j = j\Delta\omega$. (In the above equations, the values k and j indicate integers.) By substituting into Equation 2.12 the values of x, ω, and dx by x_k, ω_j, and $\Delta x = 1/2\nu_c$, respectively, and carrying out the summation over all the values of $f(x)$ from $f(x_{min})$ to $f(x_{max})$, we obtain the following important equation:

$$\Phi(\omega_j) = \frac{1}{2\nu_c} \sum_{k=-T\nu_c}^{T\nu_c} f(x_k)\exp(-i\omega_j x_k) \tag{2.30}$$

Making similar substitutions in Equation 2.13, one can obtain the following equation after summation over all the values of $\Phi(\omega_j)$ from $\Phi(-\omega_{max})$ to $\Phi(\omega_{max})$:

$$f(x_k) = \frac{1}{T} \sum_{j=-T\nu_c}^{T\nu_c} \Phi(\omega_j)\exp(i\omega_j x_k) \tag{2.31}$$

where:

$\omega_{max} = 2\pi\nu_c$

Generally speaking, Equations 2.30 and 2.31 are incorrect because they can be applied to functions that are limited in both configuration and Fourier spaces, which, indeed, is impossible. However, these functions, which are of practical importance in experimental data processing, are *practically limited* in both spaces. This means that for $|\omega_j| > 2\pi\nu_c$ and $|x_k| > T/2$, the contribution of the corresponding terms in the above equations becomes negligible compared to $\Phi(\omega_j)$ and $f(x_k)$, respectively, or these terms become small compared to the noise level. If, as often happens in practice, the function $f(x)$ is determined at the points x_k, $\Delta x = |x_k - x_{k-1}| \leq 1/2\nu_c$, and the total number of points x_k within the interval $[x_n, x_{-n}]$ is $2n$, then the Fourier transform of this function can be calculated using the following equation:

$$\Phi(\omega_j) = \Delta x \sum_{k=-n}^{n} f(x_k) \exp\left(-i\frac{\pi jk}{n}\right) \tag{2.32}$$

2.6 COMMUNICATION THEORY APPROACH

In this section we will discuss the data transfer rate in optical and electronic channels, information capacity of memory storage, energy equivalence of recorded information (this is the input signal energy that is necessary to get one bit of output information [J/bit]) because just these parameters determine optimal data processing, quality of measuring-recording systems, and right way to experimental tools certification.

Before proceeding to experimental data processing, we enter into the final stage of experiment—the creation of the numerical array carrying the information about the signal at the input of the measuring system. The requirements for this array are determined by the accuracy that we need to provide for precise characterization of the set of space- and time-dependent functions that describe the object under investigation and which are the final goal of the measurements. The possibility that these requirements can be satisfied depends on the specific features of the equipment used for measurements. Usually, this equipment consists of a set of devices that transform the information carried by the input signal. In consequence, the output (recorded) signals can be appreciably different compared to those at the input of the system. (This fact is the main reason for the loss of accuracy.) Definitely, some general method is required for analyzing measuring systems that could contain many different subunits, including optical devices, electronic equipment, analog–digital converters, transmission lines, and so on. Communication theory seems to be the most adequate tool for addressing this problem since this theory was developed specifically for the analysis and optimization of the transmission rates of communication channels and the information capacity of the data storage. Indeed, the devices listed above do represent the equipment for receiving, transmitting, processing, and recording the information. The information can be contained in the temporal or spatial structure of the signals—this difference is not essentially of principal importance. The really important fact is that all the devices mentioned above satisfy a number of fundamental requirements. Among these requirements, the most important are linearity and invariance (the independence of system properties on a temporal or spatial shift). This fact makes it possible to study the systems using the same mathematical methods. In particular, the well-developed methods of communication theory can be successfully applied for analyzing a variety of measuring systems, including optical ones.

It is known that the measure of uncertainty, or, more precisely, the measure of information about this uncertainty contained in the communication, is called the information entropy. The information entropy is determined by the following equation:

$$H(A) = -\sum_{i=1}^{n} P(A_i)\log_2 P(A_i) \tag{2.33}$$

where:

$P(A_i)$ is the probability of the elementary event A_i (in our case, A_i is the probability of getting some given value of the investigated function)

This equation determines the entropy of some finite probability space $\{A\}$. For the continuous random variable y characterized by the probability density $p(y)$, the information entropy is introduced as follows:

$$H(y) = -\int_{-\infty}^{\infty} p(y)\log_2[p(y)]\mathrm{d}y \tag{2.34}$$

Note that in communication theory, $P(A_i) \neq$ const, since the probabilities of transmitting symbols (letters, words, etc.) are very different. The situation is quite different when we receive and record communications related to the physical processes that have been "encoded" by nature. Oscillograms, interferograms, conoscopic images, and so on represent the communications of this type. Another example is the analysis of halftone images when we have no *a priori* information and the probability of observing some given intensity is approximately constant for any point in the image plane. Let us assume that the magnitude of the measured function is limited by the value m_{max}. Suppose also that the quantization step as defined by the measurement accuracy is equal to Δm and the probability of observing any count within the range from 0 to $m_{max}/\Delta m$ is constant for each point (x, y) of the image. In this case, the probability of recording any given value of the function at some given image point is $1/L$, where $L = (m_{max}/\Delta m) + 1$. Consequently, the probability of recording any given set of independent values of the function at k points is $1/L^k$. Finally, taking into account the conditions $\sum_{i-1}^{n} P(A_i) = 1$ and $P(A_i) \equiv 1/L^k$, we can write the following equation for the information entropy:

$$H(A) = -\sum_{i-1}^{n} P(A_i)\log_2 P(A_i) = (-k\log_2 L)\left(-\sum_{i-1}^{n} P(A_i)\right) = k\log_2 L \quad (2.35)$$

Thus, if the discrete independent values of the investigated function are uniformly distributed over the range L, the entropy of the communication of length k (k samples are taken) can be calculated using Equation 2.35. Accordingly, if we can perform k independent measurements of intensity (brightness) per unit area of the image, and if the measurement range (as it is introduced above) equals L, then the entropy per unit area of the image is $H = k\log_2 L$. This means that the maximum amount of information that can be stored (and readout) at the unit area of the image field is determined only by the values of k and L. It should be emphasized that this simple and clear relationship determines just the entropy of the aggregate of discrete random variables distributed so that $P(A_i) = 1/L^k$. For the case of real optical images, this equation provides just a rough estimate of the image entropy. For instance, if the transfer function FWHM satisfies the condition $\beta << \Delta$, $\nu_c << 1/\Delta$, where Δ is the readout step ($1/\Delta^2 = k$), and $L >> 1$, the information content of the image calculated using Equation 2.35 exceeds the correct value by 20%–30%. However, the magnitude of this error is not so important—the case is that the above-stated conditions ($\beta << \Delta << 1/\nu_c$) are very far from optimum. In the optimum situations, this simple equation is not valid, that is, it cannot be used whenever it is exactly needed. The correct equations for entropy calculations (including optical images) will be presented later. Now let us concentrate on the following issue. It is known that the information entropy of the random variable y with normal distribution and zero expectation is given by the following equation:

$$H = \frac{1}{2}\log_2(2\pi e \sigma_y^2) \tag{2.36}$$

where:

σ_y^2 is the variance of the variable y

Suppose we consider some variable u that is not normally distributed. For describing such variable, two quantities are used in information theory—the variance σ_u^2 and the quantity $\overline{\sigma_u^2} \equiv 2^{2H(y)}/2\pi e$. This latter quantity is called (after C. Shannon) the entropy power. It follows from this definition that the entropy power is equal to the variance of some normally distributed random value with the entropy equal to that of the variable u. In communication theory, the transmission capacity C of the channel is determined by the maximum entropy of communication that can be transmitted through the channel in a unit time with a negligibly small error. Let us consider the channel with additive noises characterized by the variance σ_u^2 and the entropy power $\overline{\sigma_u^2}$. If the variance of the input signal does not exceed some given value $\overline{u^2}$, then the following estimate can be written for the transmission capacity of the channel:

$$\frac{1}{2}\log_2\left(1+\frac{\overline{u^2}}{\overline{\sigma_u^2}}\right) \le C \le \frac{1}{2}\log_2\left(\frac{\sigma_u^2+\overline{u^2}}{\overline{\sigma_u^2}}\right) \tag{2.37}$$

In the case of Guassian noise, the inequality signs should be replaced by "=." For Gaussian noise, the condition $\overline{\sigma_u^2} = \sigma_u^2$ holds; thus, the information that can be transmitted through the channel can be presented as follows:

$$H = \frac{1}{2}\log_2\left(1+\frac{\overline{u^2}}{\sigma_u^2}\right) \tag{2.38}$$

It is proved in communication theory that $\sigma_u^2 \ge \overline{\sigma_u^2}$. Due to this fact and taking into account the inequality (2.37), we can conclude that Gaussian noise channels have the lowest possible transmission capacity, that is, the noise background has the maximum effect for Gaussian noise. Consequently, Equation 2.38 provides the lower estimate of the information that can be transmitted through a noisy channel. Taking advantage of the Kotelnikov–Shannon theorem, we can extend the above considerations to the case of continuous channels. For instance, the transmission capacity of the bandwidth-limited channel with additive white noise can be written as follows:

$$C = \Delta\nu \log_2\left(1 + \frac{\overline{P}}{N_0 \Delta\nu}\right) \tag{2.39}$$

where:

$\Delta\nu$ stands for the channel bandwidth
$\overline{P}$ is the average power at the channel input
N_0 is the power of noise per unit bandwidth

[Note that the average power of the function $u(t)$ is determined as $\overline{P} = (1/T)\int_{-T/2}^{T/2} u^2(t)\mathrm{d}t$. For real physical channels, this quantity is proportional to the power as it is usually introduced, if $u(t)$ represents the signal amplitude.] According to Equation 2.39, the following amount of information can be transmitted during the time interval Δt through the bandwidth-limited channel with additive white noise:

$$H = \Delta\nu\Delta t \log_2\left(1 + \frac{\overline{P}}{N_0 \Delta\nu}\right) = \Delta f \Delta t \log_2\left(1 + \frac{\overline{P}\Delta t}{N_0 \Delta f \Delta t}\right) \tag{2.40}$$

Assume that $\Delta\nu \times \Delta t \sim 1$. Consequently, as one can readily see from Equation 2.40, the amount of information transmitted through a noisy channel is determined by the ratio of the received signal energy $\overline{P}\Delta t$ to the energy of the noise $N_0\Delta\nu\Delta t$. In other words, for successful transmission of the unit information, the energy received in some physical channel has to be at least larger than the noise energy acquired during the communication time. As one can see from Equation 2.39, the transmission capacity of the channel depends on the bandwidth and the ratio of the average input signal power $\overline{P}$ to the noise power $N_0\Delta\nu$ within the bandwidth $\Delta\nu$. For given $\overline{P}$, N_0, and small $\Delta\nu$, the magnitude of C grows very rapidly with any increase in $\Delta\nu$. For large $\Delta\nu$, the magnitude of C tends to some value C_∞, which can be easily calculated:

$$C_\infty = \frac{\overline{P}}{N_0 \ln 2} \approx \frac{\overline{P}}{0.7 N_0} \tag{2.41}$$

The above equation describes the transmission capacity of the channel, provided no limitations are imposed onto the channel bandwidth.

Comparing Equations 2.39 and 2.40, one can readily find that $C = 0.7C_\infty$ even for $\Delta\nu = \overline{P}/N_0$.

The *rate of data generation*, the *rate of data transfer*, and the *transmission capacity of the communication channel* represent the concepts widely used in communication theory. Let us consider these concepts in more detail with the final goal of calculating the entropy of transmitted signals and the energy required for transmitting some given amount of information through a noisy channel. We should also keep in mind that our primary goal is to find an efficient method for analyzing measuring–recording systems. Moreover, we are interested in the method that would make possible the optimum design of these systems. We determine the optimum method as that which allows reconstruction of the input signal from the recorded output with the minimum energy expense. This point is of particular importance as an input signal can be reconstructed with any accuracy, provided the energy at the channel input and its dynamic range are not limited. Let us also note many terms of communication theory such as the *rate of data generation*, the *rate of data transfer*, and the *communication transfer rate*, apropos, which are not the best terminology when we deal with optical images, since in these cases the information is contained not in the temporal but in the spatial structure of signal so the amount of information is applied not to the unit time but to the unit length or area. Therefore, it seems reasonable when we deal with optical images to replace, for instance, the term the *communication transfer rate* by the *density of the transmitted information flux*, which we will use in the following.

Let us consider, for instance, an input signal that is a 2D image given as the input intensity distribution $E_{in}(x,y)$ or the input exposure distribution $\mathcal{E}_{in}(x,y) = E_{in}(x,y)\tau$ (where τ is the exposure time). First we consider the rate of data generation. Note that the continuous function $E(x,y)$ has an infinite number of values. This means that for perfect reconstruction of this function an infinite amount of information has to be transmitted, or in other words, unlimited energy storage is required. This also implies that we would require a channel with infinite transmission capacity or unlimited communication time. Indeed, all this is impossible. However, if we allow some discrepancy between the exact values $E_{in}(x,y)$ and those reconstructed from the output image, the amount of information required to be transmitted through the channel will be finite. In communication theory, *the minimum amount of information required to be transmitted per unit time* (or, as in our case, extracted from the unit length or area of the optical image) *that would allow for reconstruction of the transmitted function with the given accuracy is referred to as the rate of data generation.* This quantity, which depends on the nature and value of the allowed discrepancy, was introduced by Shannon as the "*rate of information generation with respect to the accuracy criterion.*" (Generally speaking, one could consider a number of different criteria; the most frequently used is the root-mean-square [RMS] criterion.) It goes without saying that the channel transmission rate should allow for communication with a rate that is at least not smaller than the rate of data generation at the channel input.

The rate of data transfer through the communication channel is determined by the following Shannon relation [1]:

$$R = H(X) - H_Y(X) = H(Y) - H_X(Y) \tag{2.42}$$

Here X and Y represent the input and output signals, that is, $X = I_{in} = f_{in}(t)$ or $X = I_{in} = f_{in}(x,y)$ and $Y = I_{out} = f_{out}(t)$ or $Y = I_{out} = f_{out}(x,y)$, respectively, no matter which type of signal—either temporal $I(t)$ or spatial $E(x,y)$—carries the information; $H(X)$ and $H(Y)$ denote the entropy of the input and output signals. Let us consider the situation when the input signal $E_{in}(x,y)$ represents the distribution of intensity at the input pupil. In this case, we have the following:

- $H(X) = H[E_{in}(x,y)]$—the entropy of the source that is equal to the entropy at the channel input.
- $H(Y) = H[E_{out}(x,y)]$—the entropy at the channel output that is the recorded image entropy.
- $H_Y(X) = \iint_{-\infty}^{+\infty} p(X,Y)\log_2 p_Y(X)\mathrm{d}X\mathrm{d}Y$—the conditional input entropy (*unreliability*), which is the part of the input information (the part of information contained in the input image) and not present in the recorded image, or the uncertainty in the transmitted image for the cases when the received image is known.
- $H_X(Y) = \iint_{-\infty}^{+\infty} p(X,Y)\log_2 p_X(Y)\mathrm{d}X\mathrm{d}Y$—the conditional output entropy, which is the uncertainty in the received image, provided the transmitted image is known.

It is clear that the specific meaning of the above definitions and designations such as $H(X) = H[\mathrm{E}_{in}(x,y)]$, $H(Y) = H[\mathrm{E}_{out}(x,y)]$, $H_Y(X)$, and $H_X(Y)$ is determined by the fact that we consider for explanation the situation with transmission of 2D optical images. Just a simple substitution of the word "communication" instead of the word "image" will make the above definitions quite general.

Thus, it is clear that the quantity R represents the amount of information about the input signal that is contained in the output signal. Usually, in communication theory, instead of the functions R, $H(X)$, $H(Y)$,…, the quantities R_1, $H_1(X)$, $H_1(Y)$,… are introduced which represent the entropy or information recalculated per *one degree* of freedom [in our case this means "per single count of the functions $\mathrm{E}_{\mathrm{in}}(x,y)$ or $\mathrm{E}_{\mathrm{out}}(x,y)$"]. Let us note the following fact: If the signal and the noise are statistically independent, the conditional entropy at the output is equal to the noise entropy, $H_X(Y) = H(N)$. Moreover, for additive noise, the condition $N = Y–X$ holds true. In the absence of noise, the equality $H(X) = H(Y)$ is fulfilled, provided the transfer ratio is equal to unity within the transmitted bandwidth.

Let us calculate the data transfer rate in a communication channel with Gaussian noise $[p(N)=(\sigma\sqrt{2\pi})^{-1}\exp(-N^2/2\sigma^2)]$. We are going to transmit the information about the aggregate of functions characterized by the probability density $p(X) = 1/E_{max} = \text{const}$ for $0 \le E \le E_{max}$ and $p(X) \equiv 0$ for $E < 0$ or for $E > E_{max}$. It will be shown later that in the case of limited amplitude signals, this aggregate can be transmitted with the maximum rate through the channel, provided the condition $E > N$ is fulfilled. In other words, in this case, the full transmission capability of the channel can be realized. According to Equation 2.42, the information transfer rate can be written as $R = H(Y) - H_X(Y)$. Usually, there is no problem with the second term on the right-hand side of this equation. For instance, it follows from Equation 2.36 that for Gaussian noise this term can be written as $H_X(Y) = H(N) = (1/2)\log_2(2\pi e\sigma_y^2)$. The situation is not so straightforward as far as the first term is concerned since one cannot directly determine the distribution $p(Y)$. However, taking advantage of the well-known equations of probability theory, we can in turn calculate the functions $p(X,Y)$, then $p(Y)$, and finally H_1. Consider the following probabilistic equations:

$$p(X,Y) = p_X(X)p(Y) = p(X)p_X(Y) \tag{2.43}$$

$$p(Y) = \int_{-\infty}^{\infty} p(X,Y)\mathrm{d}X \tag{2.44}$$

In our case, we have

$$p(X,Y) = \left(E_{\mathrm{max}}\sigma\sqrt{2\pi}\right)^{-1} \exp\left[-\frac{(Y-X)^2}{2\sigma^2}\right] \tag{2.45}$$

It follows from the above equation that

$$p(Y)=\int_{0}^{E_{max}}\left(E_{max}\sigma\sqrt{2\pi}\right)^{-1}\exp\left[-\frac{(Y-X)^2}{2\sigma^2}\right]$$
$$=E_{max}^{-1}\left[\Phi\left(\frac{Y+E_{max}}{\sigma}\right)-\Phi\left(\frac{Y}{\sigma}\right)\right] \tag{2.46}$$

where:

$\Phi(z)=(2\pi)^{-1/2}\int_0^z \exp(-u^2/2)du$ is the probability integral

Now we can arrive at the equations for $H_1(Y)$ and R_1:

$$R_1=-\int_{-\infty}^{+\infty}E_{max}^{-1}\left[\Phi\left(\frac{Y+E_{max}}{\sigma}\right)-\Phi\left(\frac{Y}{\sigma}\right)\right]\log_2$$
$$\left\{E_{max}^{-1}\left[\Phi\left(\frac{Y+E_{max}}{\sigma}\right)-\Phi\left(\frac{Y}{\sigma}\right)\right]\right\}dY-\frac{1}{2}\log_2(2\pi e\sigma^2) \tag{2.47}$$

Unfortunately, the integral in the above equation cannot be calculated analytically. Nevertheless, the function $R_1 = H(Y) - H_x(Y)$ can be well approximated by the following equation [2] that actually represents the main part of an asymptotic expansion of R_1, the accuracy of which increases with L:

$$R_1 \approx \log_2 L + AL^{-1} - B \tag{2.48}$$

where:

$L = E_{max}/2\sigma = \overline{E}/\sigma$ and σ^2 is the noise variance

This result can be presented in a more familiar manner:

$$R_1 = Q_1\log_2 L \tag{2.49}$$

For $L > 1.2$, the slowly varying function $Q(L)$ can be well approximated with the following equation*:

$$Q(L)=1-\frac{1}{\log_2 L}\left(B-\frac{A}{L}\right) \tag{2.50}$$

* For instance, if we assume $A = 1.286$ and $B = 1.047$, then $Q(L)$ will be calculated with the accuracy of ~1% even for $L = 1.5$. For $L = 10$, Equation 2.50 provides the accuracy of $\sim 10^{-6}$.

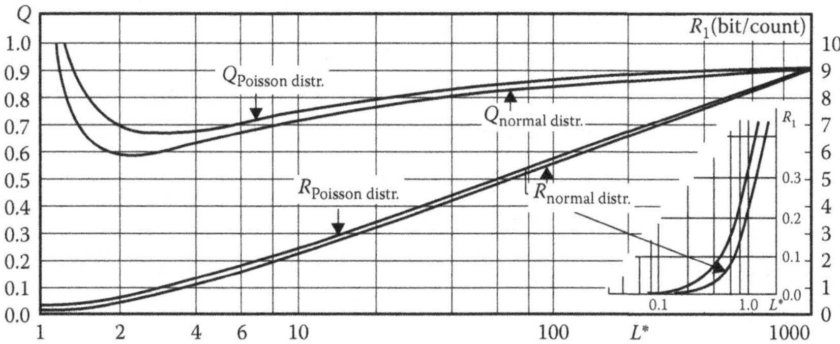

FIGURE 2.4 Data transfer rate in the channel with noise versus the effective *S*/*N*.

These functions are plotted in Figure 2.4. As one can see in this figure, for $L > 0.5$, the function $Q(L)$ is always of the order of 1. In the region of most practical importance $10 < L < 100$, the function $Q(L) \approx 0.8$. As for the function $R(L)$, it gradually increases with an increase in L and it follows just from common sense that the transmitted information should decrease monotonically with a decrease in the *S*/*N*. Nonmonotonic behavior of the function $Q(L)$ can be accounted for by the fact that $\log_2 L \to 0$ for $L \to 1$ [see Equation 2.49, while for $L = 1$ the condition $R(L) = 0$ is not fulfilled, whereas $R(L) \neq 0$ when $L = 1$].

Indeed, for other input signal aggregates and other noise distributions, acting in the communication channel, the data transfer rate (for an image this is the information about the input image contained in the unit length or unit area of the output image) will be different.

Let us consider again the aggregate of the input signals determined in the interval $[0, E_{max}]$ by the probability density $p(X) = 1/E_{max} = \text{const}$. In contrast to the previous discussion, now we assume that a Poisson noise $p(N) = (2\alpha)^{-1}\exp(-N/\alpha)$ distribution acts in the channel. (This is the standard situation for those cases where images are obtained with the help of electron optical converters at a low level of photocathode illumination.) The method for calculating the data transfer rate $R_1(L)$ remains the same: We calculate sequentially $p(Xv, Y)$, $p(Y)$, and then $H(Y)$. Finally, we obtain $H_X(Y)$ using the relationship $H_X(Y) = H(N)$. Omitting the detailed steps in the calculation and taking advantage of Equation 2.42, we can arrive at the following equation for R_1:

$$R_1 \approx \log_2 L + 1.37/L - 0.94 \tag{2.51}$$

At the intermediate stage of calculations, one could obtain the following equations: $H(Y) = \log_2 L + \alpha E_{max}^{-1}(1 + 2\log_2 e)$ and $H_X(Y) = H(N) = \log_2 2\alpha$, where $\alpha = \sqrt{\sigma/2}$ and σ is the noise variance. Taking into account the above results, we can rewrite this equation in a form that is quite similar to Equation 2.49:

$$R_1 = Q_2 \log_2 L \tag{2.52}$$

where:

$Q_2(L) = (0.94 - 1.37/L)/\log_2 L.$

As one can see from the plots in Figure 2.4, the functions $Q_1(L)$ (for Gaussian noise) and $Q_2(L)$ (for Poisson noise) are very close to each other. Thus, the slowly varying function $Q(L)$ represents some correction factor, which, depending on the distribution functions of signals and noise, corrects the value of the data transfer rate for a given value of the S/N.

In communication theory, the transmission capacity C_1 of the channel is introduced as the maximal value R_1, which is selected from all possible distributions $p(X)$ for given $p_X(Y) = (N)$. For amplitude-limited signals, the maximum R_1 is obtained in the case of uniformly distributed function: $p(Y) = 1/E_{max}$ = const. This is why the magnitudes of R_1, calculated using Equations 2.48, 2.49, 2.51, and 2.52 for $L > 2$ (Figure 2.4), actually represent the transmission capacities of the channels with Gaussian and Poisson noise. The reason why communication theory seeks to reach as its transmission capacity the maximum data transfer rate in some given channel is clear: This is exactly one of the main goals of communication theory—to develop efficient codes [i.e., distributions $p(X)$] allowing for $R \to R_{max} = C$. What is our interest in the channel transmission capacity when we transmit decoded signals or signals "coded" by nature such as oscillograms, spectrograms, and X-ray images? This capacity gives us the upper estimate allowing for comparison of the different channels. The exact amount of information about the input signal contained in the output signal can be calculated only on the basis of the statistical properties of the input signal, that is, for known functions $p(X)$. This is described by the concept of communication transfer rate, which, as we have already discussed, is not the best terminology when we deal with optical images and has to be replaced in this case by the density of the transmitted information flux.

Thus, we have established that for $L >> 1$ using the aggregate of the uniformly distributed input signals [$p(E) = 1/E_{max}$ = const], we can get the upper estimate for the communication transfer rate or, in the case of optical images, the density of the transmitted information flux. Let us discuss a particular example illustrating that this estimate is not so bad. Suppose an optical channel with Gaussian noise is used for recording the interferometric pattern of a plane probing wave disturbed by the investigated object. For relatively large and irregular phase delays, the output illumination can be represented as $E(x) = E_0 + E_0\cos(\omega x + \varphi)$, where $E_0 = E_{max}/2$ and φ is the random variable described in the interval $[0, 2\pi]$ by a probability density $p(\varphi) = 1/2\pi$. It is evident that for a sinusoidal input signal the different magnitudes of intensity at the output image plane do not have the same density probability (as in the previous discussion): Most readings will be concentrated in the vicinity of E_{max} and 0. In this case, the probability density can be written as $p(E) = 2(\pi E_{max})^{-1}[1–2(E/E_{max}–1)^2]^{1/2}$, which differs significantly from the situation $p(E) = 1/E_{max} = \text{const}$ (Figure 2.5). At the same time, the corresponding dependencies $R_1(L)$ for $L > 10$ are almost indistinguishable as can be seen in Figure 2.5. Indeed, the situation can be different. For instance, in spectroscopic measurements, the function $p(E)$ has a parabolic shape, which can result in an order of magnitude difference in the values of $R_1(L)$.

The concept of the *data generation rate* R is introduced in communication theory as a minimal value R_1 that is selected from all possible distributions $p(N)$, provided the probability $p(X)$ and the approximation error δ are given. According to Shannon, this means that "all possible channels are considered; the magnitude of R_1 is calculated for each channel, and the minimum value of all those calculated is selected." Since the minimum is obtained for the channel with Gaussian noise, the magnitude of R_1 calculated using Equations 2.51 and 2.53 represents the data generation rate

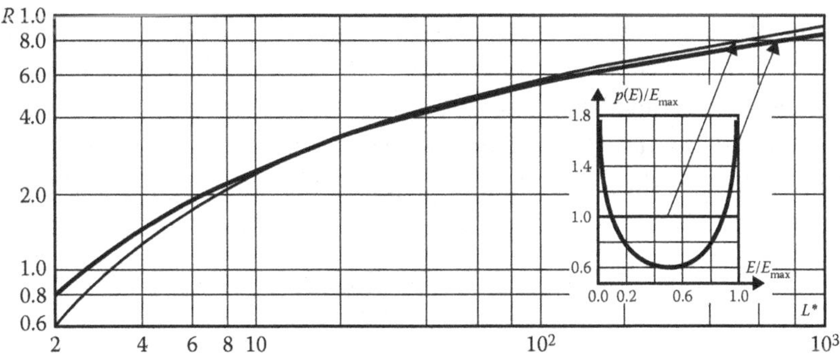

FIGURE 2.5 Probability density of the input illumination $p(E)$ and its corresponding data transfer rates.

for signals characterized by the probability density $p(E) = 1/E_{max} = \text{const}$, provided the variance satisfies the condition $\sigma^2 = \delta^2$. In other words, the data generation rate is the minimum amount of information required to be applied to the system input per unit time (per unit length or area in case of optical images) for providing the reconstruction of the function with the given accuracy δ.

If we can get k independent counts per unit time, length, or area of the optical image, the transmission capacity of the channel is

$$C = kR_1 = kQ\log_2 L \tag{2.53}$$

This value is the upper estimate for the amount of information about the input signal carried by the unit time, the unit length, or the unit area of the output signal. According to Kotelnikov–Shannon's theorem, the sampling of the functions E limited by the value ν_c has to be performed with the step $\Delta \leq 1/2\nu_c$. Let us assume that $\Delta = 1/2\nu_c$. Then, for 1D and 2D signals, we have, respectively, the following equations for the transmission capacity:

$$C = 2\nu_c QL = 2\nu_c Q\log_2 \frac{\overline{E}}{\sigma} \text{[bit/s or bit/cm]}$$

$$C = 4\nu_c^2 QL = 4\nu_c^2 Q\log_2 \frac{\overline{E}}{\sigma} \text{[bit cm}^{-2}\text{]} \tag{2.54}$$

Note that the above equations are valid for the case when the ratio $\overline{E}/\sigma$ does not depend on ω.

So far in our discussion, we have assumed that the transfer ratio of the channel with noise is constant over the whole bandwidth up to the boundary frequency ν_c, that is, $K(\omega) \equiv 1$. Now let us pay attention to the opposite situation—that of a noiseless channel with a variable transfer function $K(\nu) \neq 1$. For noiseless channels with $K(\nu) \equiv 1$, we would have the following condition fulfilled $H(X) = H(Y)$, which means that the information is not lost. If the transfer function varies within the bandwidth, that is, $K(\nu) \neq 1$, then, as shown by Shannon [1], the following equation follows:

$$H_1(Y) = H_1(X) + \frac{1}{\nu_c}\int_0^{\nu_c} \log_2 |K(\nu)| d\nu \tag{2.55}$$

The last term on the right-hand side of the above equation is always negative. Consequently, for those cases when $K(\nu) \neq 1$, the values of R and C

calculated using the above equations have to be reduced by the amount $\left|1/\nu_c \int_0^{\nu_c} \log_2 |K(\nu)| d\nu\right|$. This correction takes into account the information losses caused by the finite width of the transfer function, that is, the energy losses in the high-frequency band. It is very convenient to take into account the information losses caused by the transfer function width using the following value L^* instead of L:

$$L^* = L\exp\left[\frac{1}{\nu_c}\int_0^{\nu_c} \ln|K(\nu)| d\nu\right] = L\mathcal{K} \tag{2.56}$$

By direct analogy with Shannon's terminology, the quantity

$$\mathcal{K} = \exp\left[\frac{1}{\nu_c}\int_0^{\nu_c} \ln|K(\nu)| d\nu\right] \tag{2.57}$$

is called the *power loss factor.*

Finally, for a real channel with noise and $K(\nu) \neq 1$, the data transfer rate can be represented by the following equation:

$$R = kQ(L)\log_2\left\{\frac{E_{max}}{2\sigma}\exp\left[\frac{1}{\nu_c}\int_0^{\nu_c} \ln|K(\nu)| d\nu\right]\right\} \tag{2.58}$$

Let us remark that the information transmission capacity C of the channel is determined as the maximal value R_1 for given $p_X(Y) = (N)$. For signals with limited amplitude, this maximum is obtained for the case when the condition $p(Y) = 1/E_{max} = \text{const}$ is fulfilled. Consequently, the value of R calculated using Equation 2.58 represents the transmission capacity of channels with noise, the statistical properties of noise being accounted by the slowly varying function $Q(L^*)$. Note that Equation 2.58 can also be used for calculating the information capacity of optical storage devices, such as photographic films or CCD matrices.

Now we can reformulate the well-known theorem of communication theory treating the relation between the data generation rate $\mathcal{R}$ and the transmission capacity of the communication channel C as follows: *If the spectrum of the input signal is limited by the value* ν_c, *the channel is characterized by the transfer ratio* $K(\omega)$ *[or the MTF* $K(\nu)$*], and the variance of channel noise is* σ^2, *then the input signal can be reconstructed by the* *output signal with an accuracy δ in those and only in those cases when* $\mathcal{R} \leq C$, *that is,*

$$k_{in} Q \log_2 \frac{\overline{E}_{in}}{\delta} \leq k_{out} Q \log_2 \frac{\overline{E}_{out}}{\delta} \mathcal{K} \tag{2.59}$$

This inequality makes it possible to estimate the transmission capacity of the channel for given approximation accuracy. It can also be used if it is necessary to estimate the input signal approximation error, provided the channel parameters are given. For instance, if the number of independent samples for the input and output signals is selected so that $k_{in} = k_{out} = 2\nu_c$ or, for the case of 2D images, $k_{in} = k_{out} = 4\nu_c^2$, then the relative reconstruction error $\delta/\overline{E}_{in}$ can be estimated as follows:

$$\frac{\delta}{\overline{E}_{in}} \geq \frac{\sigma}{\overline{E}_{out}\mathcal{K}} = \frac{\sigma}{\overline{E}_{out}} \left\{ \exp\left[\frac{1}{\nu_c} \int_0^{\nu_c} \ln|K(\nu)| d\nu \right] \right\}^{-1} \tag{2.60}$$

Thus, we obtained the lower estimate of the reconstruction error, that is, the minimum possible value of the error. Indeed, this minimum value can be achieved only in the case when k independent samples can be taken per unit time, length, or area of the optical image, and the transmission capacity of the channel C is not smaller than the information flowing into the channel input. Dividing the energy $\overline{E}_{in}$ supplied to the channel input per unit time (or per unit length or area of the input entrance pupil of the channel) by the amount of information H transmitted per unit time (contained in a unit length or area of the output image), we get the energy equivalence of recorded information for the unit information transmitted through the channel:

$$\frac{\overline{E}_{in}}{H} \geq \frac{\overline{E}_{in}}{kQ} \left(\log_2 \left\{ \frac{\overline{E}_{out}}{\sigma} \exp\left[\frac{1}{\nu_c} \int_0^{\nu_c} \ln|K(\nu)| d\nu \right] \right\} \right)^{-1} \quad \text{[J per bit]} \tag{2.61}$$

We will see later to what extent (or how close) can one approach to this limit in any specific case. Now we just emphasize that Equations 2.58, 2.60, and 2.61 completely describe any measuring–recording system: These (and only these!) equations make it possible to determine the maximum amount of information that can be received, transmitted, and recorded per unit time (unit length or area of the optical image). These equations give the estimate for the reconstruction error and energy equivalence of recorded information as well. Only with the help of the above equations, one can establish the accuracy attainable when measuring with different devices the time- and space-dependent functions describing the objects under investigation. The quantities calculated using these equations provide the only basis for adequate comparison of different measuring systems.

Let us make several additional remarks. Usually, in real measuring systems, an increase in bandwidth $\Delta\nu$ results in a decrease of $\mathcal{K}$ and an increase of σ (for the case of white noise $\sigma^2 = \Delta\nu\sigma_0^2$, where σ_0^2 is the spectral density of noise). Consequently, the magnitude of the *S/N* is reduced following an increase in the bandwidth. However, as one can see from Equation 2.54, it is beneficial from the energetic point of view to record and transmit the information in a broad bandwidth and with a low *S/N* $\bar{E}/\sigma = L$, that is, with higher spatial/temporal resolution but with lower dynamic range. In this same case, the maximum transmission capacity, the maximum information density, and so on will be provided for a given signal energy, (By the way, this is the main reason why the information capacity of holograms exceeds dramatically the capacity of conventional photographic images.) Consequently, if we were interested in the full realization of the transmission capacity of the channel (trying to satisfy the relation $\mathcal{R} = C$), we should transform the input signal in some optimum manner. These transformations are possible, as has been strictly proven in communication theory. However, in spite of the possibility that exists in principle for designing such coding devices, the engineering problems that would be involved make this approach senseless. Under the circumstances when the input signals are inefficiently "coded" by nature, it seems that we cannot change the situation. Nevertheless, it is not always true. To clarify the problem, let us consider the available options.

Consider, for instance, the case when the input signal is the image represented by the illumination distribution $E_{in}(x,y)$. Let us also assume that the signal $E(x,y)$ is limited by the value E_{max} and that the spectrum of the signal is limited by the value ν_c. Suppose the image is transmitted through a channel described by the transfer ratio $K(\nu_x,\nu_y)$ or by the transfer function $g(x,y)$. Let the noise acting in the channel be the normal distributed noise with variance σ^2. The problem is to reconstruct the input image $E_{in}(x,y)$ with the RMS accuracy δ for a given output image $E_{out}(x,y)$. In other words, we know the data generation rate, that is, the amount of information per unit input area of the channel $\mathcal{R} = k_{in} Q \log_2 \bar{E}/\delta$. We also know the transmission capacity of the channel that can be written as follows:

$$C = k_{out} Q \log_2 \left\{ \frac{E_{max}}{2\sigma} \exp\left[\frac{1}{\nu_c} \int_0^{\nu_{cx}} \int_0^{\nu_{cy}} \ln\left|K(\nu_x,\nu_y)\right| d\nu_x d\nu_y \right] \right\}$$

where:

k_{out} is the number of independent samples per unit area of the output image

(The problem of optimum choice for the value of k_{out} will be discussed later; for the moment we just assume that $\beta\sqrt{k_{out}} < 1$, where β is the FWHM of the transfer function so that the relation $2g(\beta/2) = g(0)$ is fulfilled.) In the case when the inequality $\mathcal{R} > C$ is valid, the incoming information cannot be transmitted through the channel. Consequently, the input image has to be transformed in order to match $\mathcal{R}$ and C. In optical systems, this can be done simply by magnifying the image. If the magnification value is equal to M, then the spatial spectrum is compressed by a factor of M, the illumination is reduced as M^2, and the density of information at the system input is reduced by even more than M^2, taking into account the function $Q(L^*)$.

Now, having in mind our reconstruction problem [the need to reconstruct $E_{in}(x,y)$ with RMS δ], let us consider the matching procedure from the point of view of the *minimum input energy* $\mathcal{E}_{in}(x,y) = E_{in}(x,y)\Delta t$ [i.e., from the point of view of minimizing the exposure time Δt since the illumination $E_{in}(x,y)$ is given] or the *minimum input area* of the channel required for image transmission. First consider the effect of the input area.* In order to do this, we have to come back to the relationship indicated by Equation 2.60. Remember that this relationship was obtained, provided the condition $k_{in} = k_{out} = 4\nu_c^2$ is fulfilled. The relationship indicated by Equation 2.60 means that the relative error of the reconstructed image $\delta/\overline{E}_{in}$ cannot be smaller than the effective signal-to-noise ratio at the system output $\sigma/\overline{E}_{out}\mathcal{K}$—one could call it the *relative accuracy of the output image.* Having in mind the inequality discussed, we can conclude that the relative reconstruction error $\delta/\overline{E}_{in}$ (which is of primary importance, not the absolute value of δ) can be decreased only by increasing the input energy. Thus, if we have plenty of energy, and the main problem is to reduce the

* Matching with the input area is very important because the input windows of many optical devices are limited in size.

input area, the magnification magnitude M has to be properly selected so that $k = 4\nu_c^2 \approx 1/\beta^2$.* For $\beta^2 \rhd 1/k = 1/4\nu_c^2$, the magnitude of $\mathcal{K}$ falls exponentially, and in order to provide the reasonable values of $\delta/\bar{E}_{\text{in}}$, one needs an unrealistically large amount of energy.

Consider the opposite situation where the input area of the optical channel is not restricted, but the input energy is limited. The question is whether we can improve the reconstruction quality without coding the input image just by varying the magnification M, which would mean projecting the investigated image onto the system input with different magnifications? Let us consider the values of M that allow the condition $k_{\text{in}} = 4\nu_c^2 \lhd 1/\beta^2$ to be fulfilled. In this case, for k_{in} elements in the input image, we can provide $k_{\text{out}} \approx 1/\beta^2$ statistically independent elements in the output image. (Indeed, $k_{\text{out}} \rhd k_{\text{in}}$.) This is equivalent to the situation when we transmit the image using $n = k_{\text{out}}/k_{\text{in}}$ independent channels, as if we had n independent output images. As soon as the input energy is limited, the higher the number of images, the lower the accuracy of each image. At a first glance, the only benefit comes from the fact that $\mathcal{K} \to 1$ when the magnification M is increased. It is clear that by using the simple averaging over n images during the reconstruction of $E_{\text{in}}(x,y)$, we decrease the relative reconstruction error by a factor of $\sqrt{n}$. However, the relative error of any of those n images increases by the same $\sqrt{n}$ factor, just due to the fact that the illumination decreases as M^2, and the spectrum compresses by M times, so that the variance $\sigma = \sqrt{\Delta\nu_x \Delta\nu_y \sigma_0^2}$ decreases by M times. Thus, the S/N also decreases by $M = \sqrt{n}$ times. Nevertheless, this situation is advantageous in many additional respects. First, by comparing the "identical" columns and rows in all n images, we can get rid of the "rough" points that fall out of the confidence interval of 2σ or 3σ. The function value at these points can be replaced by half the sum of the neighboring points. This procedure effectively cleans up the image from imperfections caused by factors such as photographic emulsion scratches, multielectron noise components in electron optical converters, and by the structure of microchannel plate intensifiers. Second, by using the local filtering procedure (see Chapter 12) instead of simple averaging, it is possible to reduce significantly the systematic errors. Third, the positioning accuracy can be significantly increased when using the magnified images. A similar approach when applied to stellar navigation systems allows an increase of

* In Chapters 4 and 12, we will discuss the methods for optimal selection of k, the amount of samples per unit area (length) of the image.

more than an order of magnitude in the angular resolution. This list can be continued, but we will postpone this discussion to the following chapters where the communication approach will be applied to specific optical systems and reconstruction problem analysis.

2.7 DETERMINATION OF THE MEASURING–RECORDING SYSTEM PARAMETERS

As shown in the previous sections, this is the maximum density of the information flow through the optical system (i.e., the transmission capacity of the communication channel $C = \mathcal{R}_{max}$) that, together with the energy equivalence of recorded information $\bar{\mathcal{E}}/H$, characterize completely the measuring–recording system. Knowing these parameters, one can compare different systems; calculate the data transfer rates; find the optimum condition for receiving, transmitting, and recording the data; and estimate the measurement accuracy of the input images (i.e., the image reconstruction accuracy). To find these quantities for some given device, specific information is required concerning the transfer function $g(t)$ or $g(x)$, the transfer ratio $K(\omega)$, input sensitivity (quantum efficiency of the receiving unit), maximum adaptable signal E_{max}, and noise. All these data can be obtained only from detailed measurements. It should be mentioned, however, that the measurement procedure can be significantly reduced for systems in which the shape of the transfer function is known and the properties of the noise are given. In this case, there are only four quantities, namely β, μ, E_{max}, and σ, that determine completely the system properties and allow for calculation of the transmission capacity C and energy equivalence of recorded information $\bar{\mathcal{E}}/H$. Below we consider the measurement methods used in such cases. Let us start from the noise considerations.

1. The statistical properties of noise are determined by the probability density function $p[I_{out}(t)]$, which can be determined by applying a steady or continuous signal $I_{in}(t) = \text{const}$ to the system input. [If the variance σ^2 is independent of I_{in}, one can carry out the measurements in the absence of signal, that is, for $I_{in}(t) = 0$.] Thus, the measurement procedure is obvious: The signal at the system input is fixed, $I_{in}(t) = \text{const}$, and the aggregate of the readout data is analyzed with the help of an amplitude analyzer. Standard multichannel amplitude analyzers are most suitable for this purpose, since these devices are specially designed for measuring the probability density

function $p(A_k)$ of the signals A_k applied to their input. These commercially available devices provide amplitude discrimination at the input of each channel, thus transmitting only signals that are in the range from A_k to $A_k + \Delta A$, where the quantity A_k ranges from 0 to A_{max} gradually increasing from channel to channel by the value ΔA. Modern multichannel amplitude analyzers contain thousands of channels, so the probability density (and also the distribution function) of the random variable A_k can be measured with high accuracy, since the value $\Delta A = A_{max}/k_{max}$ is very small.

For studying the spectral and statistical properties of noise in optical images, a uniform signal $\mathcal{E}_{in}(x) = \text{const}$ is applied to the system input and the output signal $\mathcal{E}_{out}(x_i)$ is read out using a sufficiently detailed grid, so that the sampling step $\Delta = x_i - x_{i-1} \ll \beta$. After this, we can calculate the following:

$$\overline{\mathcal{E}_{out}}(x) = \frac{1}{N}\sum_{i=1}^{N}\mathcal{E}_{out}(x_i) \tag{2.62}$$

and the square of RMS

$$\sigma^2 = \frac{1}{N-1}\sum_{i=1}^{N}\left[\mathcal{E}_{out}(x_i) - \overline{\mathcal{E}_{out}(x)}\right]^2 \tag{2.63}$$

that represents the estimate for variance, when $\sigma^2 \to \mathcal{D}[\mathcal{E}_{out}(x)]$ for $N \to \infty$. From these equations, the spectral density of the white noise can be obtained: $\sigma_0^2 = \sigma\Delta$. If the variance depends on the signal amplitude, $\sigma(\mathcal{E})$, the procedure described above should be repeated for different signals $\overline{\mathcal{E}_{in}}$. This procedure is illustrated in Figure 2.6 where the function $N_i(x)$ is plotted. As one can see from the figure, the difference between $\overline{N}(x) \approx 220$ and certain counts $N_i(x)$ comes up to 30 units. For given spectral and statistical noise properties—as in the case of image converters or photographic emulsions—no additional measurements

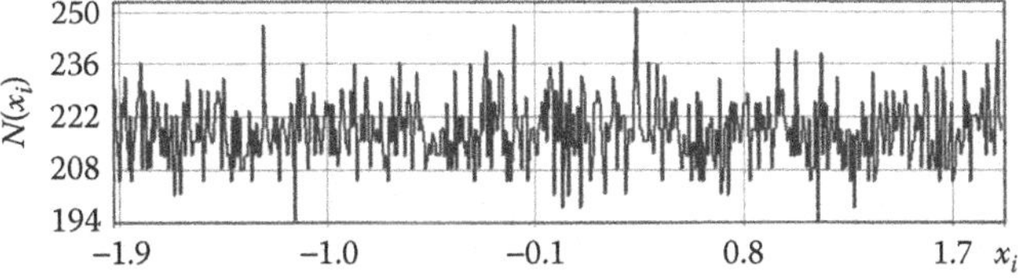

FIGURE 2.6 Noise of CCD matrix.

are required. In the opposite case, one should investigate these properties of noise using the same procedure as we have discussed above: The input signal is fixed, $\mathcal{E}_{\text{in}}(x) = \text{const}$, and the array of output signal data $\mathcal{E}_{\text{out}}(x_i)$ is recorded with a multichannel amplitude analyzer.

The spectral characteristics of noise $n(\omega)$ or $\overline{n^2}(\omega) = N(\omega)$ can be found by averaging the Fourier transforms of the aggregates of particular realizations: A uniform signal $\mathcal{E}_{\text{in}}(x) = \text{const}$ is applied to the input many times, and in each case, the noise component $\mathcal{E}_N^k = \mathcal{E}_{\text{out}}(x) - \overline{\mathcal{E}}_{\text{out}}(x)$ of the output signal $\mathcal{E}_{\text{out}}(x)$ is used to calculate the corresponding Fourier transforms:

$$\mathcal{E}_N^k(\omega) = \frac{1}{X}\int_{-X/2}^{X/2} \mathcal{E}_N^k(x)e^{-i\omega x}dx \tag{2.64}$$

After this the operation of averaging is performed:

$$\frac{1}{K}\sum_{k=1}^{K}\left|\mathcal{E}_N^k(\omega)\right|^2 = \frac{N(\omega)}{X} \tag{2.65}$$

where:

X stands for the realization length*

2. In electronics, it is usual to measure the transient function $h(t)$ rather than the transfer function $g(t)$. [We recall that by definition the transient function $h(t)$ is the system response to the unit function $\chi(t)$; note that the following equation holds: $d[h(t)]/dt = g(t)$.] Standard commercially available devices are used for measuring $h(t)$. These devices include a generator of steplike signals $\chi_1(t)$ (modeling the unit function), which is applied to the input of the system under investigation and the recorder of the output system signal $h(t)$. When using such devices, it is very important to be sure that the rise time Δt of the signal $\chi_1(t)$ from the level $0.1\chi_1(t)_{\max}$ up to level $0.9\chi_1(t)_{\max}$ (this is the standard definition of the rise time in electronics) is significantly shorter than the transfer function FWHM, that is, $\Delta t \ll \beta$.

---

* The frequently used method for calculating $N(\omega)$ based on an estimation of the autocorrelation function $B(\tau) = (1/X)\int_{-X/2}^{X/2}\mathcal{E}(x+\tau)\mathcal{E}(x)dx$ and following Fourier transformation $\int_{-\infty}^{\infty} B(\tau)e^{-i\omega\tau}d\tau = N(\omega)$ is not a correct procedure in our case, since $N(\omega)$ obtained in this way represents the spectral density of the given realization instead of averaging over the aggregate of realizations.

Analogous devices are not available in optics. It seems that there is not any problem here. Really, since the transfer function represents the system response to $\delta(x)$-like signals, it is quite enough to project the image of a narrow slit onto the system input, and if the slit width $d \ll \beta$, then one could expect that $\mathcal{E}_{out}(x) \approx g(x)$. Unfortunately, this method turns out to be extremely inaccurate. There are three reasons for this: (1) inadequate dynamic range of the output signal measurements, (2) imperfection of the projecting optics, and (3) insufficient *S/N*. Let us consider these difficulties in more detail. The first problem originates from the requirement to measure the transfer function with sufficiently high and almost continuous relative accuracy over the whole range of the transfer function definition—in the vicinity of its maximum and in the far wings as well. Only in this case can one expect to get the converging solution when solving the high-accuracy reconstruction problem. The dynamic range of conventional recording systems is usually not high enough to realize such measurements. However, even in this case, there is a trick to bypass this problem. This standard approach consists in placing a glass wedge just behind the collimating objective, the slit being installed in the focal plane of the objective. Due to multiple reflections from the wedge surfaces, a number of slit images with gradually decreasing brightness are projected onto the system input. Now one can measure the maximum and the wings of the transfer function using the different images with subsequent matching of the results. Indeed, this is additional but inevitable work. The second problem results in the strict requirements for the resolution of the projecting optics: Its transfer function FWHM has to be significantly smaller than what we expect to have in the system that is being investigated. This requirement could be satisfied only by using a microscope objective. It is possible but not very convenient. The situation becomes more unpleasant as soon as noise considerations are considered. The case is that the width of the slit image at the system input has to be very small compared to β. As follows from the above discussions, in this case the *S/N*, as well as the corresponding inaccuracy of the transfer function measurements, should be extremely low. Indeed, one can try to increase the accuracy by statistical averaging over multiple repetitive measurements or by increasing the length of the slit, but it is clear that this is not the way for substantial improvement of the situation. Another method consists in applying the unit function $\chi(t)$ to the system input (so-called illuminated field edge). Unfortunately, the situation cannot be improved appreciably in this way. It is reasonable to project an image of a narrow slit onto the system input only when the shape $g(x)$ is known, since all the input samples will be used for calculating only one parameter, namely, the transfer function FWHM, and the acceptable accuracy can be obtained. However, in the general case, this method does not lead to good results.

Therefore, what is the alternative? The answer is clear: We should concentrate on the function $K(\omega)$. Then the function $g(x)$ can be calculated by applying the inverse Fourier transformation. To find $K(\omega)$, a sinusoidal signal $\mathcal{E}_{in}(x) = \bar{\mathcal{E}}(1+\sin\omega x)/2$ is applied to the input, and all the output readings are used for the determination a single parameter, namely, the amplitude $A(\omega)$ of the variable part of the output signal $\mathcal{E}_{out}(x) = \bar{\mathcal{E}}[1+A(\omega)\sin\omega x]/2$. It is worth mentioning that for real even functions $g(x)$, as often occur in optics, the equality $K(\omega) \equiv A(\omega)$ holds true. Therefore, by varying the spatial frequency ω, one can easily find the function $K(\omega) \equiv A(\omega)$. In such measurements, conventional interferometers, for instance, the Michelson interferometer, can be used for generating the proper input signal. The MTF $A(\omega)$ is determined from the experimental data using the following equation:

$$A(\omega) = \frac{\mathcal{E}_{max} - \mathcal{E}_{min}}{\mathcal{E}_{max} + \mathcal{E}_{min}}$$

So far we have discussed the situations where the 1D transfer function $g(x)$ can be used. In those cases, $g(x)$ can be measured by imaging a narrow slit with width $\delta \ll \beta$ onto the system input. In 1D images, there is a defined direction along which the illumination does not change. The standard spectrogram presents the typical example of a 1D image. Images can be treated as 1D also in cases where the spectrum in one direction is significantly broader than in the other, say, $\nu_{cx} \gg \nu_{cy}$. This happens, for instance, in interferograms with low optical path difference so that the interferometric lines are just slightly curved.* In the 2D cases, it is necessary to have the 2D transfer function $g(x,y)$. Generally speaking, this function can be found in individual measurements of $g(x)$ and $g(y)$. In the

* Here the question arises concerning the readout direction. This direction is usually selected to be coincident with one of the coordinate axes. In the data array that is read out from a 1D image, the spectral components ν_x (or ν_y) will be absent only if the axes are selected in such a manner that $E(x)$ = const or $E(y)$ = const. Of course, one cannot pay attention to axis directions at the readout stage of the data processing. The proper axes can be found at the later stage by analysis of the data array. Finding the directions of permanent illumination (or optical density), one can perform the 2D–1D array transformation if the 1D image is processed.

absence of astigmatic distortions, the 2D transfer function can be found using the well-known Abel equation:

$$g(x)=2\int_{x}^{\infty}\frac{g(r)rdr}{\sqrt{r^2-x^2}} \tag{2.66}$$

The solution of the above equation is given by the following equation:

$$g(r)=\frac{1}{\pi}\int_{r}^{\infty}\frac{g'(x)dx}{\sqrt{x^2-r^2}} \tag{2.67}$$

where:

$g(r)=g\left(\sqrt{x^2+y^2}\right)$ is the 2D transfer function $g(x,y)$

Indeed, in this case, the transfer ratio is also a 2D function $K(\omega_x,\omega_y)$. In astigmatic-free systems, $K(\omega_x,\omega_y)=K(\omega_r)=K\left(\sqrt{\omega_x^2+\omega_y^2}\right)$. Actually, this means just a transform to polar coordinates in Fourier space.

In electronics, the frequency–amplitude characteristic $A(\omega)$ is measured with the help of standard signal generators (SSGs). The word "standard" implies that the amplitude of the generator signal $A_0\sin(\omega t)$ remains constant over the whole range of the frequency variation. Therefore, when using these devices, it is sufficient to apply the generator signal to the input and to get $A(\omega) = A_\omega/A_0$ just by varying the frequency ω (A_ω is the output signal amplitude measured at frequency ω).

The function $\varphi(\omega)$ (called the phase–frequency characteristic) determines the phase shift. For instance, if the input and output signals are $f_{in}(t) = I_0\sin(\omega t)$ and $f_{out}(t) = I_0\sin(\omega t + \varphi) = I_0\sin[\omega(t–a)]$, respectively, then the function $\varphi(\omega)$ can be found with high accuracy by measuring the delay $a(\omega)$ between the input and output signals as a function of ω. (The methods for time interval measurements will be discussed in the next chapters.) It is clear that for steady sinusoidal input signals, the value of φ is calculated as $\varphi(\omega) = \omega a(\omega)$, where the delay $a(\omega)$ is obtained from experimental measurements. Obviously, the value of φ can be found with an accuracy of $2\pi n$. However, this circumstance is not important, as we are interested in $K(\omega) = A(\omega)\exp[–i\varphi(\omega)]$, since $\exp(–i2n) \equiv 1$. Finally, by conducting the quite simple measurements described above, one can obtain the transfer ratio function as $K(\omega) = A(\omega)\exp[–i\varphi(\omega)] = A\omega/A_0\exp[–i\omega a(\omega)]$.

In optics, the problem of measuring $\varphi(\omega)$ is not a simple problem due to the low positioning accuracy of the comparison input and output images, but it is necessary to do if transfer function is not even as it is mostly in optics when $\varphi(\omega) \equiv 1$. In those cases, the only way out is to measure the transfer function $g(x)$ and then to calculate $K(\omega)$ by taking the Fourier transform of the transfer function.

REFERENCES

1. Shannon, C.E. 1948. A mathematical theory of communication. *The Bell System Technical Journal* 27: 379–423, 623–656.
2. Pergament, M.I. 1998. Real features of the image registration systems and the reconstruction problems in light of information theory. *SPIE* 3516: 465–473.

CHAPTER 3

Studying Pulse Processes

IN THIS CHAPTER, WE are going to explore the principles for measuring time, currents, and voltages, relative to the study of high-speed pulse processes. Let us note here that the time measurement of the current and voltage represents a problem of self-contained interest every now and then. However, more often these measurements are valuable because they contain the information about the object under investigation, obtained in various ways, as in the output of the majority of modern measuring devices, one deals with analog or digital electrical signals. Let us also note that the widespread cliché "high-speed processes" belong to the literary, rather than physical realm. Hence, let us stress that the scope of the further discussion is restricted to nonstationary physical processes, where the characteristic time of the object parameter change is given by some quantity Δt. Naturally, it is not just the hardware used, but essentially the whole principle underlying the measurement of the above-mentioned observables, which should depend cardinally on the characteristic time Δt. Thus, let us specify from the very outset that $10^{-2} > \Delta t > 10^{-15}$ s, which is the range of Δt that is of key interest for modern experimental physics. Even though this time range ignores centuries, years, hours, minutes, and even seconds, it is certainly extensive enough and certainly cannot be covered by a single family of devices. Hence, in the analysis of the measurement strategies and devices used, we shall progress step by step in our understanding from large time intervals (10^{-2} s) to smaller and smaller ones. As far as the hardware per se is concerned, we are primarily interested in its underlying operation principle, rather than engineering nuances. The former principle is of primary importance for experimental planning and design. This approach will also confine this chapter within its present limits, rather than becoming a voluminous technical encyclopedia.

3.1 TIME INTERVAL MEASUREMENT AND SOME ELEMENTS OF COUNTING SCHEMES

In this section, we consider the means for measuring the time intervals between events, occurring at the instants, say ti and tj. In addition, we will discuss the corresponding elements of the counting schemes. The methods for measuring the time dependencies of the current $I(t)$ and the voltage $U(t)$ will be discussed in Section 3.2. In modern practice, reasonably long time intervals are measured by means of discrete digital technology. As a rule, these devices in question use some time standard provided by signals from an alternating current (AC) generator, which is stabilized by a quartz crystal. The idea is that on the one hand quartz is a piezoelectric and can be easily integrated into positive feedback generator circuits; on the other hand, it possesses an assortment of remarkable qualities (primarily an extremely low linear expansion constant), which ensure an extraordinary stability of the natural frequency of the crystal's mechanical oscillations. The piezoelectric effect then creates an alternating electric potential difference between the two sides of the crystal, which is synchronized with its mechanical oscillations, thus yielding the radio-electronic generator frequency. If the frequency is stable, then to measure the duration of a time interval, one should simply calculate the number of periods of the oscillation, covered by this time interval. Quantitatively, the dependability of such a device can be easily determined by anyone who has dealt with the so-called quartz watch. This term has clearly originated from the principle of time measurement that is discussed here. If you check your quartz watch regularly every month, you can find that itself running or lagging is behind by one or two seconds. In other words, the accuracy of a one-month time interval measurement turns out to be some 5×10^{-7}, which is higher than one-tenth of a thousandth percent! Why, it seems, would modern consumer electronics need such accuracy? The answer is very simple: One just cannot do worse with such a measuring principle. One should notice at this point that among other things, this high accuracy is ensured by the very substantial sample duration.* For shorter intervals, it is not so easy to provide such high accuracy, even two orders of magnitude

* The accuracy of the watch over a long period is an example of a systematic error arising from the accuracy of the quartz crystal.

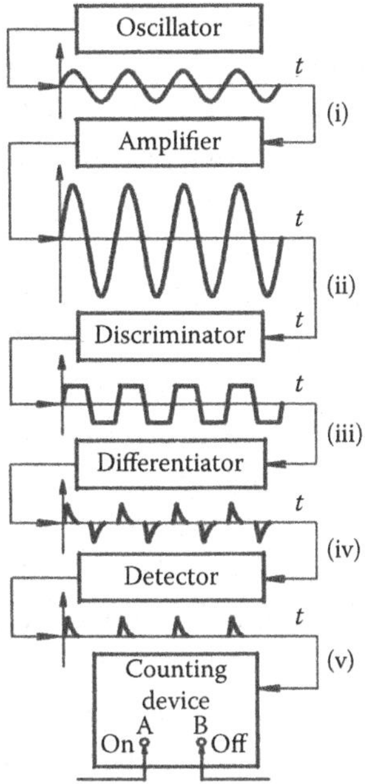

FIGURE 3.1 Operational principle of an electronic stopwatch.

lower. However, difficult is not impossible, as it will become clear from the following discussion.

Figure 3.1 explains the operation principle of an electronic stopwatch based on the above concept. The standard oscillator signals (i) are amplified (ii) and further discriminated so that the amplitude is restricted to obtain a series of rectangular pulses (iii). These pulses are differentiated (iv) and then detected, and the resulting unipolar pulses (v) are sent to a counting device, which is coupled with a control unit (CU). The counting device is required to turn on the pulse counter when a signal is applied to the input A, and then to turn it off after a signal is applied to the input B. Clearly, the accuracy of measuring the duration T of an interval is t/T, where t is the oscillation period of the standard generator. This "counting" branch of electronics is developing rapidly as it is directly related to the speed of modern computers. Today the frequency $1/t$ has exceeded 2 GHz, and the value $t = 1$ ps (10^{-12} s) is far from being a limit. Further extrapolation is simple: A microsecond time interval can be measured with a 10^{-3} accuracy, which falls down to 10^{-1} for a time interval with duration of 10 ps. There exist devices with built-in mechanisms, which measure the time intervals between the starting signal and the first countable pulse, as well as the last counted pulse and the count termination signal. If the duration, or at least the leading edges of all these signal pulses, is considerably shorter than t, then the accuracy under discussion can be improved. However, one way or another, one has to conclude that instruments of the type discussed here would fail to measure the time intervals whose duration lies in picosecond range. Understandably enough based on such a stopwatch, one can also build various experiment control units. For instance, Stanford Research Systems, Inc, Sunnyvale, CA, manufactures a digital delay pulse generator DG535, which produces a series of four pulses, designed to trigger pieces of experimental equipment at strictly prescribed times. The delay time between each pair of consecutive pulses can be adjusted in the range from 0 to 1000 s. The absolute error of marking a time interval T (in seconds) between any two pulses, relative to a fixed value, does not exceed the quantity $\Delta T = 1.5 \times 10^{-9} + (\Delta t/t)T$, where Δt is the discrepancy between the nominal period of the frequency generator and its real value t. The device exists commercially in two versions: the standard model with $\Delta t/t = 2.5 \times 10^{-5}$ and the precision model with $\Delta t/t = 1 \times 10^{-6}$. Therefore, for a time interval T of 1 ms, the absolute error will not exceed 26.5 ps for the standard model and 2.5 ps for the precision model.

Before progressing any further, let us discuss how the amplitude restriction, or discrimination, differentiation, and integration of signals in electric circuits is fulfilled. The discriminator operation principle should be clear from Figure 3.2a. As one can see in the figure, the discriminator consists of a resistance R_1, a diode (which has resistance R_2 when in the conducting state, with $R_2 << R_1$), and a reference voltage generator,* producing a potential difference of U_0. As long as the input voltage does not exceed U_0, no signal amplitude restriction occurs; as the diode is not conducting, there is no current and hence no voltage drop across the resistance R_1, and the output

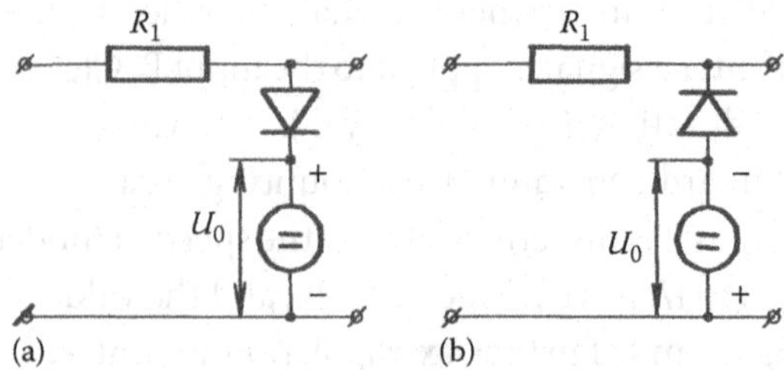

FIGURE 3.2 Principles of discrimination for (a) positive and (b) negative pulses.

* The concepts *current generator* and *voltage generator* imply by definition that the former has an internal resistance equal to ∞, whereas for the latter it is 0.

voltage equals the input voltage, that is, $U_{in} = U_{out}$. However, as soon as U_{in} exceeds U_0, the situation changes: The current will flow and the potential drop across the resistance R_2 will be added to the base voltage. It is not difficult to calculate that in this case $U_{out}(t) = U_0 + [R_2/(R_1 + R_2)][U_{in}(t) - U_0]$, but as long as the ratio $\left[R_2/(R_1 + R_2)\right]$ is extremely high, the output signal will be effectively restricted by the quantity U_0. In other words, over the time interval when $U_{in} > U_0$, one has $U_{out} \approx U_0$. Clearly, the above discriminator restricts only the positive component of the pulses. To restrict the negative component, one should invert the diode and the base voltage generator polarities, as shown in Figure 3.2b. Consecutive use of the two schemes results in the restriction of both the positive and negative voltage components; see, for instance, step iii in Figure 3.1, the cutoffs being determined by the voltages of the base generators.

The simplest devices that fulfill the requirements of electrical signal integration and differentiation are shown in Figures 3.3 and 3.4, respectively. Let us write down the Kirchhoff equations for the input contours of these circuits.

$$R\dot{q}(t) + \frac{q(t)}{C} = U_{in}(t) \tag{3.1}$$

where:

- $q(t)$ is the charge
- $\dot{q}(t)$ is the input current
- $R\dot{q}(t) = RI(t) = U_R(t)$ (the first term on the left-hand side of the equation) is the voltage drop across the resistance
- $q(t)/C = U_C(t)$ (the second term on the left-hand side of the equation) is the potential difference across the capacitor

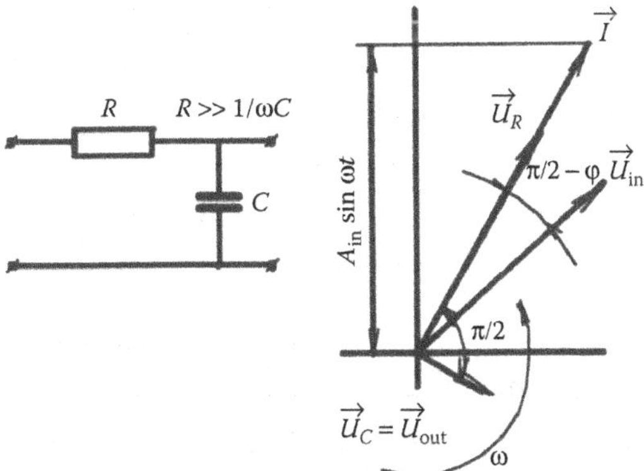

FIGURE 3.3 Vector diagram of the current and voltage when $U_R >> U_C$.

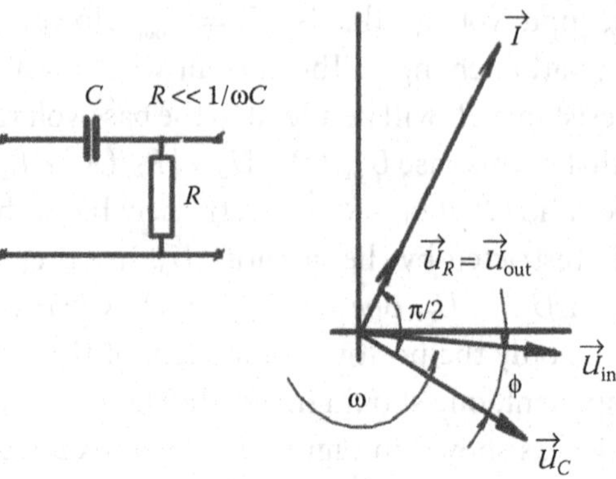

FIGURE 3.4 Vector diagram of the current and voltage when $U_R << U_C$.

Consider the following two limiting cases: $U_R >> U_C$ and $U_R << U_C$. Clearly, one can disregard the second and first terms in Equation 3.1, in the former and latter cases, respectively. (It will be shown below that the resulting error is negligibly small.) In other words, in both of the above limiting cases, one can write Equation 3.1 either as $R\dot{q}(t) \approx U_{\text{in}}(t)$. or as $[q(t)/C] \approx U_{\text{in}}(t)$ Clearly, the condition $U_R >> U_C$ is equivalent to the requirement that $R >> 1/\omega C$, as $U_R = I(t)R$, while $U_C = I(t)/\omega C$. Of course, the requirement $R >> 1/\omega C$ should hold for each spectral component of the input signal. It is sufficient that this requirement be fulfilled up to the value $\omega = 2\pi/T$, where T is the full duration of the input signal, that is, the domain of the function $U_{\text{in}}(t)$. From the equation $R\dot{q}(t) \approx U_{\text{in}}(t)$, it follows that the current in the circuit is $I(t) \approx U_{\text{in}}(t)/R$. Hence, the charge $q(t)$ that is accumulated in the capacitor C over the time interval from zero (the beginning of the pulse) to t is $q(t) = \int_0^t I(\tau)d\tau = (1/R)\int_0^t U_{\text{in}}(\tau)d\tau$. At the same time, for any capacity, the voltage $U = q/C$. Therefore, in the scheme in Figure 3.3, one has $U_{\text{out}}(t) = q(t)/C$ for the output voltage. Substitution of the above expression for $q(t)$ yields

$$U_{\text{out}}(t) \approx \left(\frac{1}{RC}\right)\int_0^t U_{\text{in}}(\tau)d\tau \tag{3.2}$$

Thus, the output signal of the circuit under consideration equals the time integral of the input signal divided by the constant factor $1/RC$.

Conversely, consider the situation when $U_R << U_C$, or $R << 1/\omega_c C$, where $\omega_c = 2\pi\nu_c$ is the critical frequency determining the upper limit on the spectrum of the input signal. In this case, following on from what has been discussed above, $[q(t)/C] \approx U_{\text{in}}(t)$, that is, $q(t) \approx CU_{\text{in}}(t)$. Differentiating both sides of the latter equation, one gets $\dot{q}(t) \approx C\dot{U}_{\text{in}}(t)$. But $\dot{q}(t)$ is the current in the circuit, and therefore, the voltage drop across the output resistance of the circuit in Figure 3.4 is $RI(t) = R\dot{q}(t)$, that is, the output signal is

$$U_{\text{out}}(t) \approx RC\dot{U}_{\text{in}}(t) \tag{3.3}$$

In other words, in this case, the output signal is the time derivative of the input signal multiplied by the factor *RC*.

Let us evaluate the measurement error appearing in circuits of this kind and in doing so find out what exactly has been implied by saying that *R* be "sufficiently" greater or smaller than $1/\omega C$. What is sufficient? Consider the vector diagram in Figure 3.4. The angle between the vectors $\boldsymbol{U}_{\text{in}}$ and $\boldsymbol{U}_C$, that is, the angle $\varphi = U_R/U_{\text{in}}$; thus, $U_R = \varphi U_{\text{in}}$. Note by the way that U_R has been neglected in order to arrive at Equation 3.3. Therefore, the error acquired is $\Delta U = U_R = U_{\text{in}} - U_C$. However, $U_C^2 = U_{\text{in}}^2 - U_R^2 = U_{\text{in}}^2(1-\varphi^2)$ and hence $\Delta U = U_{\text{in}} - U_C = U_{\text{in}} - U_{\text{in}}\sqrt{1-\varphi^2} \approx U_{\text{in}}\varphi^2/2$. Then the relative error is $\Delta U/U = \varphi^2/2$. Using this equation, it is not difficult to estimate that already for values of *R* that are three times smaller than $1/\omega C$, one has $\Delta U/U < 5\%$. If the above ratio is equal to 10, the relative error $\Delta U/U$ turns out to be less than one-half of 1%.

Therefore, the accuracy of the measurement does not represent such a great problem *per se*. The situation gets worse, however, if one looks at the input/output signal ratio, as the circuit in consideration is nothing but a voltage divider. This ratio can be found from the amplitude–frequency characteristic of the circuit in consideration. By definition, $A(\omega) = |U_{\text{out}}|/|U_{\text{in}}| = IR/I\sqrt{R^2 + R_C^2} = R/\sqrt{R^2 + R_C^2}$. Since $(|U_{\text{out}}|/|U_{\text{in}}|)^2 = R^2/[R^2 + 1/(\omega C)^2] = (RC\omega)^2/(RC\omega)^2 + 1$, one has $|U_{\text{out}}|/|U_{\text{in}}| = RC\omega/\sqrt{(RC\omega)^2 + 1} \approx RC\omega$ because the condition $R << 1/\omega C$ implies simultaneously that $RC\omega << 1$. Suppose $1/\omega_c C$ is *M* times greater than *R*, that is, $MR = 1/\omega_c C$. This implies that $RC\omega_c = 1/M$ and $|U_{\text{out}}|/|U_{\text{in}}| \approx RC\omega = 1/M$. In other words, the output signal amplitude will be at least *M* times smaller than the input signal amplitude. If you want to reach, say, a 0.5% measurement accuracy ($M = 10$), one has to accept a tenfold drop in the amplitude of the output signal: Accuracy always comes at a price. Let us point out that the estimates above are valid only if $R << 1/\omega_c C$.

Similar formal equations can be written for the integrating circuit. Indeed, as in this case the amplitude–frequency characteristic is $A(\omega) = |U_{out}|/|U_{out}| = R_C/\sqrt{R^2 + R_C^2}$, $(|U_{out}|/|U_{in}|)^2 = 1/(\omega C)^2[R^2 + 1/(\omega C)^2] = 1/(RC\omega)^2 + 1$, and thus $|U_{out}|/|U_{in}| = 1/\sqrt{(RC\omega)^2 + 1} \approx 1/RC\omega$, as in this case $(RC\omega)^2 >> 1$, for $R >> 1/\omega C$. Since we are dealing with the situation $R >> 1/\omega C$, $R = M/\omega C$ then $RC\omega = M$, that is, $|U_{out}|/|U_{in}| \approx 1/RC\omega = 1/M$. Therefore, in this case also, the output signal amplitude will be M times smaller than the input signal amplitude. This state of affairs is painfully familiar to any experimental physicist. In some cases, when the input signals are large enough, it does not cause a problem. However, what is one supposed to do when they are weak? Well, modern electronics has the equipment that is necessary to deal with this problem. Let us see how this is put into effect by means of the example of an integrating scheme that is shown in Figure 3.5.

The illustrated device is not very complicated, consisting of an operational amplifier with a high gain and strong negative feedback, that is, an amplifier in which the phase of the output signal is shifted by π with respect to the input. If the negative feedback were 100%, the overall amplification coefficient of the whole system would be close to one, no matter how large would be the gain of the operational amplifier, simply because of the cancellation between the input and output signals, due to the above phase shift. However, as the feedback circuit, comprising the capacitor C and the resistance R, represents a *bona fide* high-frequency filter,* as a result only the high-frequency components with $\omega > 1/RC$ are actively suppressed. In other words, the latter device operates in the same way as the simple integration scheme, depicted earlier in Figure 3.3, which also suppresses the high-frequency signal components. However, the device in Figure 3.5 lets

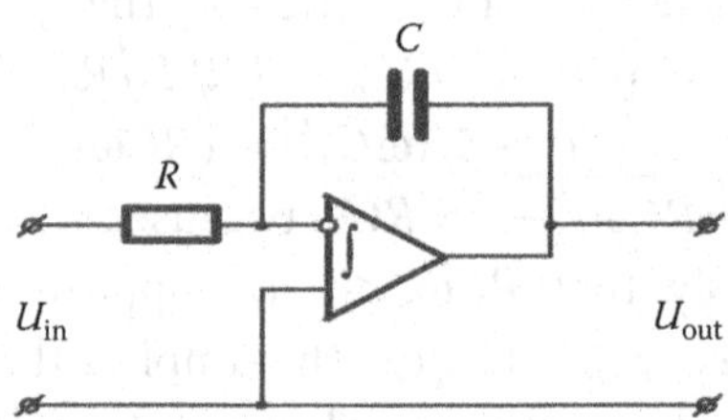

FIGURE 3.5 Schematic of an integrator.

* A *high-frequency filter* is a device that transmits the high-frequency signal components, cutting off the lower part of the spectrum. It is this type of filter that is used in feedback circuits. At the same time, the system as a whole is a low-frequency filter, namely, it transmits only the low frequencies.

the lowest harmonics pass through virtually unaffected. There is another, simpler explanation of the principle of the operation of the latter integration scheme. The point is that the operational amplifier represents a proper *current generator* (in the sense of the footnote on page 66) charging the capacity C. In point of fact, the current $I(t)$, generated by the operational amplifier, is proportional to the input voltage $U_{in}(t)$ at any instant of time. What happens further can be easily understood if one goes back to the discussion accompanying the derivation of Equation 3.2. The only difference is that, contrary to a simple integrating circuit, such an electronic scheme allows a high-precision integration, without attenuating the lowest signal component by a factor of $M = RC\omega$.

Finally, let us try to understand, at least in general terms, the operation principle of a comparator, that is, a device whose function is to compare voltages. A comparator consists of an operational amplifier with a very high gain, that is, with a very steep volt–ampere characteristic, and a logical circuit, which returns a binary value (0 or 1) as a result of the comparison between the measured and the reference voltage. The reference voltage and the voltage that is to be compared with the reference voltage are applied to a pair of input contacts of the comparator. Accordingly, the difference between the two voltages is the input signal of the amplifier. The volt–ampere characteristic of the amplifier is so steep that an input of several millivolts is sufficient to change its output voltage from the minimum to the maximum value. In effect, such an amplifier has only two discrete states: $U = U_{min}$ and $U = U_{max}$. These are the states that are detected by the logical circuit. This means that as soon as the measured voltage exceeds the reference voltage by several millivolts, the amplifier will change from one state to the other, and the logical circuit will return the value 1 instead of 0. Naturally, if the voltages measured are in the range of several volts, the accuracy of the comparison is of the order of one-tenth of 1%.

3.2 MEASURING TIME DEPENDENCE

In this section, we consider the strategies for measuring time-variable currents $I(t)$ and potential differences $U(t)$, as well as other time-dependent physical observables, which are eventually transformed into currents and voltages by their measuring devices. In modern experimental practice, these measurements are carried out by means of discrete digital instruments. By the Kotelnikov–Shannon theorem (Chapter 2), in order to measure signals whose spectrum is situated within the interval $[0, \nu c]$, it suffices to take successive measurements of the quantities $I(t)$ or $U(t)$, separated by time intervals $\Delta t \leq 1/2\nu_c$, which means that the nominal frequency F (sample rate) of the corresponding electronic device should exceed 2νc, i.e. $F = 1/\Delta t \geq 2\nu_c$. A reader familiar with the technical characteristics of modern personal computers will know that machines with the sample rate $F = 2$–4 GHz are available commercially. The state of the art in discrete digital instruments is the single-shot oscilloscope DSAX96204Q with a sample rate F = 160 GHz and a spectrum bandwidth up to 63 GHz produced by the company Agilent Technologies, Santa Clara, CA. Unfortunately, the bandwidth limit for a traditional oscilloscope is no more than 5 GHz, and the sad reality is that we may be seeing the end of the history of the electronic oscilloscope, as one of the most distinguished veterans of physical laboratories. In addition, in the same way that it is often impossible to distinguish a quartz watch from a mechanical one by a mere appearance (despite the existence of the fundamental difference separating not only the interior design of the two watches but also the two chronometry principles involved), it is not quite easy to distinguish a digital oscilloscope from an analog one. The former is just much more convenient to work with, and most importantly, it displays the data as a numerical array, which is easily stored by a computer and ready for further processing.

In Section 2.5, we have discussed the discretization (digitization) principles, namely, the ways to represent the values of continuous variables and functions by discrete arrays of numbers. Now we are going to consider the technical means of discretization, concentrating mainly on the key ideas rather than the technical gimmicks particular to one instrument or another. Continuous current or voltage values are transformed into a series of discrete values, using the so-called *analog–digital converters* (*ADCs*). Let us remark here that electronics always provides and has provided a boundless realm for one's fantasy and creation. To this effect, flipping through the pages of a mainstream professional reference book, one easily encounters up to 100 ADCs of various types. However, a closer look reveals that there are only few fundamental principles, while the whole wide range of ADCs stems from various combinations of these basic ideas. Only these fundamental principles are discussed here:

1. Let us start out with the *parallel ADC*. The devices of this type convert the signal by means of a set of *parallel* comparators, in the sense that the same voltage that is to be measured is applied to a series of many identical inputs, whereas the other inputs are connected to a

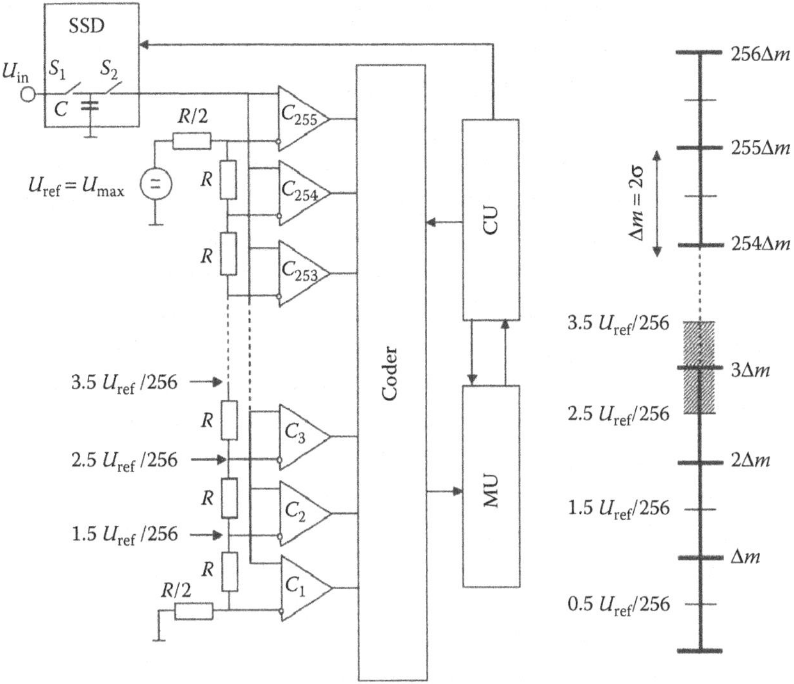

FIGURE 3.6 Principle of the operation of an 8-bit ADC.

reference voltage divider. Figure 3.6 illustrates the principle of an 8-bit ADC (this means that the maximum discrete number of measured value is $2^8 = 256$). This device with $N = 8$ consists of $2N - 1 = 255$ comparators, a reference voltage source, $2N = 256$ autocorrelated voltage divider resistances, a sampling/storage device (SSD), a coding device, a control units (CU), and a memory unit (MU). The simplest SSD would be simply a capacitor with two switches (Figure 3.6). It is intended to temporarily store the measured voltage value $U(t)$ at the instant ti, which is determined by the CU. Upon a signal from the latter, first the switch S_2 cuts off the capacitor C from the comparators. Then the switch S_1 is turned on, causing the capacitor C to charge, until the capacitor voltage matches the voltage to be measured, whereupon S_1 is switched off, while the switch S_2 connects the capacitor to the comparators, representing the measuring scheme per se. (Some integrated circuit chips for parallel ADCs, such as the model MAX100, are equipped with super-high-speed SSDs, with sampling time less than 0.1 ns.) As the scheme shows, following the increase of the input signal, the comparators change their state from 0 to 1, ascending order one after the other. For instance, if the applied input voltage is in the range between 2.5 Δm and 3.5 Δm (the shaded section of the scale in Figure 3.6), where $\Delta m = \Delta U_{ADC}$ is an input voltage of value unity, corresponding to the lowest ADC bit; then the first through the third comparators are set into the state 1, whereas the comparators numbered 4 through 255 remain in the state 0. The logical coding device interprets this code group as an 8-digit binary number, thus translating it into a form that can be manipulated by a computer, and stores the measurement result $U(t)$ at the time instant *ti* in the MU, where it remains until required. The procedure then repeats itself: the CU cuts off the capacitor C from the comparators via the switch S_2, then the switch S_1 is turned on, and so on. Owing to the parallel function of the comparators, such a parallel ADC is the fastest type of ADC. For example, an 8-bit commercially available model such as the MAX104 enables one to achieve some 10^9 counts per second, with the signal's delay not exceeding 1.2 ns. The flaw of this scheme consists in its high complexity. Indeed, as we have seen, an 8-bit parallel ADC contains 255 comparators and 256 autocorrelated resistors, which results in high costs and a considerable power consumed—typically, this is 4 W for the model MAX104 that is mentioned above.

Consecutive parallel ADC: This represents a compromise between the desire to have a high speed and the constraints of cost. Typical examples are multistep ADCs, one of which (a two-step 8-bit ADC) is shown schematically in Figure 3.7. The upper part (ADC-1) consists of a parallel ADC, which affects a "coarse" conversion of the input signal into four higher binary digits of the output signal, affecting a measurement with the step $\Delta U_{ADC-1} = U_{max}/2^4 = U_{max}/16$. The digital output from ADC-1 is passed to an output register and simultaneously to the input of a high-speed 4-bit digital–analog converter (DAC), which transforms the binary values of ADC-1 output into a corresponding analog voltage, which goes to the integrator (Σ). The difference between the original input voltage and the DAC output voltage serves as the input signal for the ADC-2, in which the steps of reference voltage are 16 times smaller than that of ADC-1, that is, $\Delta U_{ADC-2} = U_{max}/256$. This difference, converted by the ADC-2 unit into a digital form, represents the four lowest binary digits of the output code. Forwarding the latter 4 bytes to the corresponding positions of the 8-byte output register thus completes the formation of an 8-bit number. First of all,

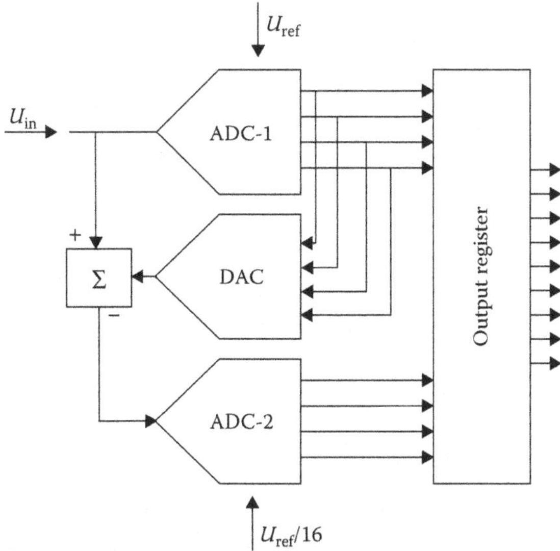

FIGURE 3.7 Consecutive parallel two-step 8-bit ADC.

the difference between ADC-1 and ADC-2 is determined by the accuracy requirements: ADC-1 should have the precision of an 8-bit converter, whereas a 4-bit converter precision is sufficient for ADC-2. Clearly, if the requirements are the same, a consecutive parallel converter turns out to be slower than a parallel one. Indeed, the units ADC-1, DAC, and finally ADC-2 switch on one after the other, the operation time of the DAC being at least that of an ADC. In addition, the digital control chips are characterized by some extra operation time and, all things considered, the sampling time for a consecutive parallel converter turns out to be approximately one order of magnitude longer than that for a parallel converter. Nevertheless, it is simpler and cheaper. Instead of 255 comparators and 256 autocorrelated resistors, there are 30 comparators and 32 autocorrelated resistors, respectively. In addition, ADC-2 has more relaxed accuracy requirements and hence may be cheaper. In conclusion, let us add that the coarsely and finely measured quantities should naturally correspond to the same input voltage $U(t_i)$. This implies that during the whole time it takes the signal to travel through the sequence ADC-1, DAC, ADC-2, output register, and the MU, one should maintain a constant input voltage by means of an SSD.

Consecutive count ADC: This is another member of the family, whose typical representative is shown in Figure 3.8. The device

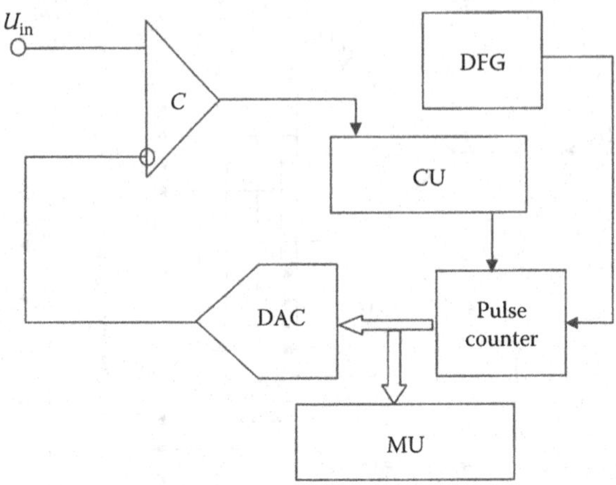

FIGURE 3.8 Structural schematic of the consecutive count ADC.

consists of a comparator (C), a digital frequency generator (DFG), a CU, a counter, a DAC, and an MU. The input signal is applied to one of the inputs of the comparator. The other inputs receive the feedback signal from the DAC. The operation of the converter starts when a signal from the CU at some instant *ti* turns on the counter, whose purpose is to add up the number of pulses coming from the DFG. The counter's output code is sent to the DAC, which transforms it into the feedback voltage U_f. The pulse counter will operate until the feedback voltage matches the input voltage. At that moment, the comparator changes from its state 0 to state 1; the corresponding nonzero output signal causes the counter to stop. Clearly, the number of pulses counted, which is to be recorded in the MU as a binary code, is proportional to the input voltage $U(ti)$ at the time instant *ti*. After a predetermined time interval Δt, the CU will reset the counter, and the cycle will repeat itself and carry out the measurement of the input voltage $U(tj)$ at the instant $tj = ti + \Delta t$. A single cycle/measurement duration for this ADC is a function of the ADC bit number N and the sample rate F. For an N-bit ADC, one has $t_{max} = (2^N - 1)/F$. For example with $N = 10$ and $F = 1$ GHz, one has $t_{max} = 1023$ ps, which fixes the maximum sample frequency at about 1 MHz. Note that in an ADC of this type, we are essentially measuring the time it takes for the feedback voltage to match the measured one, and then we interpret this time to determine a value for the measured voltage. Hence, the accuracy of this measurement depends on that of the time measurement, which is reflected primarily by the stability of the

DFG. As a result, if one is pursuing an accurate voltage measurement, one should not economize and save on the DFG but go perhaps for a more expensive but highly stable unit.

Two-cycle integration ADC: Each operation cycle of such an ADC (Figure 3.9) splits into two stages: the integration of the measured voltage and its measurement. The input signal's integration interval Δt_1 is constant. As a timer, one uses a pulse counter. The key element of the device is the electronic integrator (*I*) that has been described above. During each ADC operation cycle, the integrator is connected with the input of the ADC via a switch S_1. Simultaneously with turning on the switch, a CU activates a pulse counter, which comes to a stop after a fixed number of pulses *n*, corresponding to the time interval $\Delta t_1 = n/F$. Thereupon, the counter sends a response signal to the CU, thus finishing the input signal integration stage. (Let us emphasize that the interval Δt_1, that is, the time it takes the electronic integrator capacitor *C* to be charged from the input voltage via the resistor *R*, is determined by the fixed number $n = \Delta t_1 F$ of the counter's pulses, where *F* has been specified earlier as the sample rate.) Upon receiving a response from the pulse counter, the CU turns off the switch S_1 and opens the switch S_2, thence connecting the base voltage source to the integrator. Simultaneously, a command is sent to the pulse counter to start a new pulse sequence of total number *m*. As one can see from Figure 3.9, the polarity of the reference voltage U_{ref} is the opposite of the polarity of the input voltage. Thus, the capacitor will be discharging at the measurement stage, namely, the integrator's output voltage will be decreasing linearly in absolute value. As soon as the output voltage of the integrator reaches zero, the comparator C will change from one state to the other and

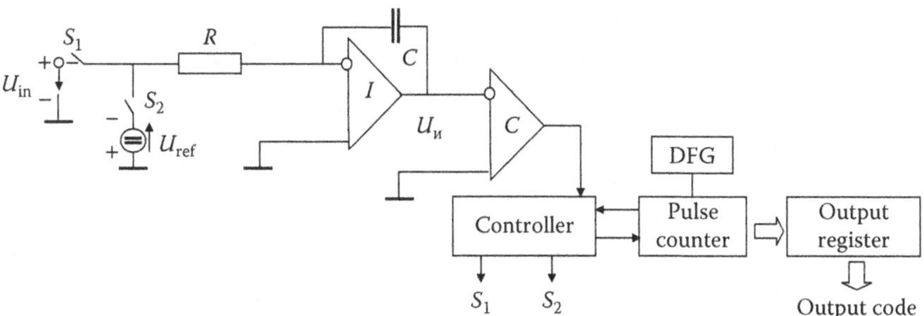

FIGURE 3.9 Simplified schematic of a two-cycle integration ADC.

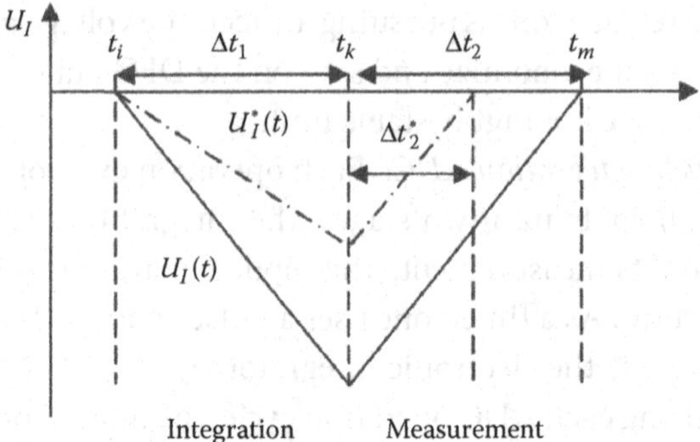

FIGURE 3.10 Time dependence of the integrator's output voltage U_I.

signal the CU to stop counting and to pass the resulting number of pulses m to the output register. It should be clear that during the time $\Delta t_2 = t_m - t_k$, namely, the time it takes the capacitor to discharge completely, the number of pulses counted will be $m = \Delta t_2 F$.

The reader is now referred to Figure 3.10, which shows the time dependence of the integrator's output voltage U_I. It is clear that by the instant t_k, the output voltage of the integrator will reach the value:

$$U_I(t) = -\frac{\bar{U}_{\text{in}}(t)\Delta t_1}{RC} = -\frac{\bar{U}_{\text{in}}(t)n}{RCF} \tag{3.4}$$

where:

$\bar{U}_{\text{in}}(t) \approx [U_{\text{in}}(t_i) + U_{\text{in}}(t_k)]/2$ is the input voltage averaged over the time interval $\Delta t_1 = t_k - t_i$*

From the above equation, one can see that if the quantity $\bar{U}_{\text{in}}$ is large, then so will be the output voltage U_I of the integrator. This is also illustrated in Figure 3.10 (the dependence $U_I^*(t)$ corresponds to a lower input voltage). But if $|U_I|$ is large, then it will take more time for it to go down to zero, as one can see from Figure 3.10. During the measurement stage, to which there corresponds the time interval $\Delta t_2 = t_m - t_k$, the integrator output voltage will decrease in value linearly with time, that is, in direct proportion to the number m:

$$U_I(t) = \frac{U_{\text{ref}} m}{RCF} - \frac{\bar{U}_{\text{in}}(t)n}{RCF} \tag{3.5}$$

* The exact value is $\bar{U}_{\text{in}}(t) = (1/\Delta t_1)\int_{t_i}^{t_k} U_{\text{in}}(t)\text{d}t$, where $\Delta t_1 = t_k - t_i$.

The voltage at the integrator's output will be zero at the instant t_m, which implies that the two terms on the right-hand side of the above equation will become equal to each other, and thus,

$$U_{\text{in}} = \frac{U_{\text{ref}} m}{n} \tag{3.6}$$

From the above equation, it follows that the result of such a measurement is independent of the sample rate F, as well as the integration constant RC. What is necessary is that the generator's frequency should remain constant during the time interval $\Delta t_1 + \Delta t_2$. This can be ensured even if one is dealing with a simple and therefore cheap generator because significant temporal or temperature frequency drifts would be characterized by orders of magnitudes many times greater than $\Delta t_1 + \Delta t_2$. The second considerable advantage of the measurement approach in question is averaging out the noise component, or noise filtering, which takes place during the integration stage. Indeed, these are not the instantaneous values of the measured voltage which are being recorded, but rather the average value over the time Δt_1. As a rule, ADCs of this type have high precision, but their sample frequency rarely exceeds 1 kHz.

2. In the process of studying ADCs, we have also encountered DACs. The latter widespread device is in a way the inverse of the former. Whereas an ADC makes the conversion of a sequence of analog voltage values into a corresponding string of N-bit binary numbers; the function of a DAC is exactly the opposite and it translates an input in binary number code into its equivalent analog voltage. Without noticing the fact, we constantly make use of these devices, for example, while listening to a compact disk (CD), staring at a graph on a computer screen, or surfing the Internet. What is more, the vast majority of computer-controlled apparatus and equally the working parts and functional components of much laboratory equipment are not capable of directly "understanding" the digital codes and require the input data to be in the analog form. For this reason, the DAC is not at all a rare species, and may be even more common than the ADC.

   DACs have a simpler design than ADCs, and hence the number of ways to build them is not so diverse. As with ADCs, the DACs can be classified into consecutive (relatively slow) and parallel (considerably faster) types. The parallel devices are in turn categorized into two categories—DACs with current or DACs with voltage summation. Let us first consider a DAC with voltage summation.

A block diagram of a *parallel 8-bit DAC* with voltage summation is shown in Figure 3.11. The integrated chip consists of the 256 consecutively connected resistors of the reference voltage divider. A decoder serves to transform an input binary number into a signal, controlling the switches. One can see from the schematic that the output voltage on the contacts *A* and *B* of the device is determined by the address of each particular switch, which will switch on, following the instruction from the decoder. Naturally, such a DAC design requires the placing of 2^N switches and 2^N correlated resistors on a single integrated chip. However, today there are 8-, 10-, and 12-bit DACs of this design commercially available.

A key element of the majority of the DACs with weighted current summation is the so-called *constant impedance resistive matrix* (Figure 3.12). Such a matrix possesses a variety of remarkable properties. First, its input resistance, that is, the resistance measured across the input contacts of the matrix, remains constant and equal to the same value *R*, no matter how many cells are added. It is easy to convince oneself of this fact by comparing the schematics in Figure 3.12a, b, and c. Hence, if the matrix is powered via a *current generator*, the input voltage

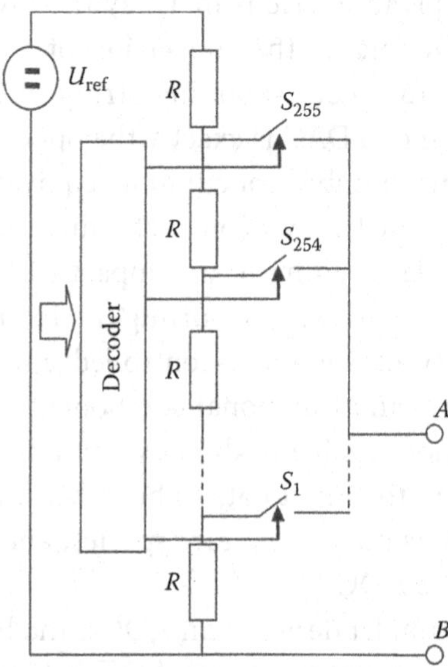

FIGURE 3.11 Block diagram of a parallel 8-bit DAC with voltage summation.

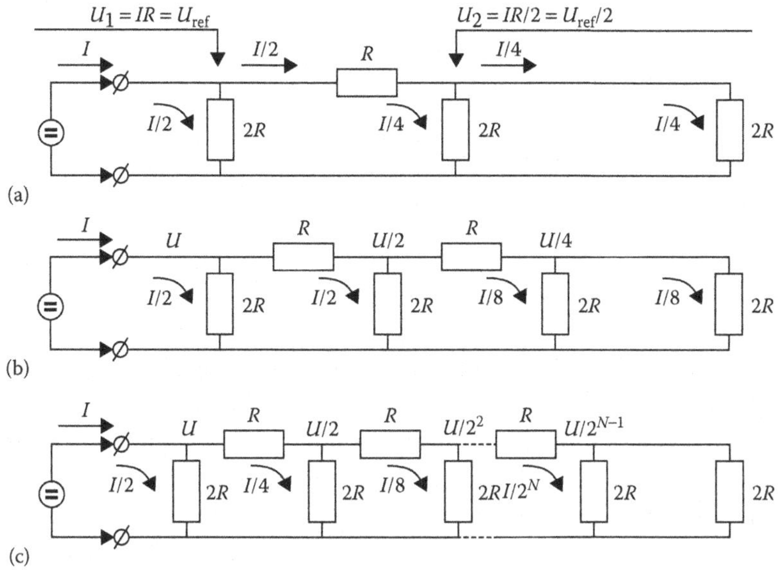

FIGURE 3.12 Constant impedance resistive matrix a, two cells; b, three cells; c, four cells.

of the matrix will always be $IR = U_{ref}$. Second, the current in each node of the matrix is split exactly in halves, and therefore, the voltage in each consecutive node will be two times smaller than that in the preceding one. Consequently, in the *N*th node, it is $U_N = U_{ref}/2^{N-1}$, where $N = 1$, 2, 3,…, and the net current flowing out of some node of the scheme (before being divided in halves) is $I_N = I/2^N$. Third, if one uses the notation $\Delta m = I_{min}$ for the minimum current flowing in the system, then the currents passing through the resistances of the magnitude $2R$ will form a sequence Δm, $2\Delta m$, $4\Delta m$, $8\Delta m$,…. Thus, each time the currents will get doubled, which is just one order of magnitude, as far as the binary numerology is concerned. In other words, the normalized (divided by Δm) "weight" of each current will match the corresponding position of the binary number.

The schematic of a DAC with weighted current summation, using the constant impedance matrix, is shown in Figure 3.13. It can be seen from the schematic that the output current I_{out} is in one-to-one correspondence with the states of the sequence of switches S_0, S_1, S_2,…, SN_{-1}. One can also see that every switch can be only in either of two different positions—left or right. The "left" position is equivalent to the "turned on" position and the current of the corresponding

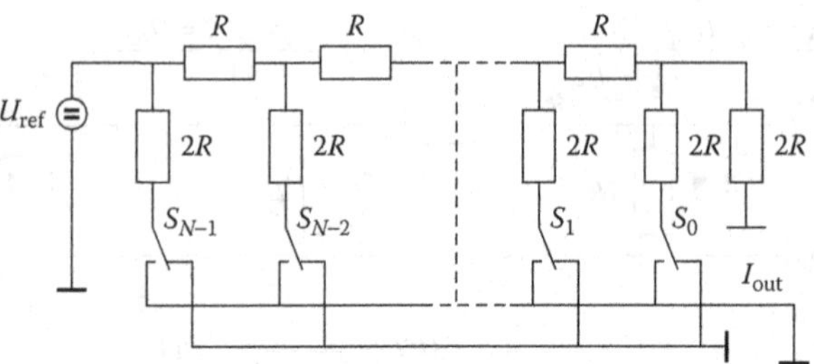

FIGURE 3.13 Schematic of a DAC with weighted current summation.

circuit will be directed into the output in this case. The switches are set in the on/off positions, depending on whether there is 1 or 0 at the corresponding bit site (orders of binary code) in the transformed binary number. For instance, if this number is equal to 1, that is, the lowest bit is the only one with the value one, then the only switch turned on is S_0 and the output current $I_{out} = \Delta m$, which corresponds to its minimum (nonzero) value. If only the two lowest numbers are equal to 1, then only the switches S_0 and S_1 are turned on and the output current $I_{out} = \Delta m + 2\Delta m = 3\Delta m$. Let us assume that we are dealing with an 8-bit DAC ($N = 8$). If none of the switches are turned on, then clearly $I_{out} = 0$. Conversely, if all of them are turned on, then $I_{out} = I_{max} = 255\Delta m$, which constitutes 256 different discrete current values (zero included). Choose a number, say 147 (in the decimal representation). Its binary representation is 10010011. If this number comes as a DAC input value, it will turn on the switches S_0, S_1, S_4, and S_7; thus, $I_{out} = \Delta m + 2\Delta m + 16\Delta m + 128\Delta m = 147\Delta m$. Contrary to the 8-bit DAC with voltage summation considered earlier, this device contains 16, rather than 256 resistors, and only 8, rather than 256 switches.

Concluding our rather cursory journey into ADC–DAC territory, we would like to refer the reader who is interested in more detail to the Internet, where one can find a plethora of useful (and sometimes not so useful) information.

3. We have considered various ways, in which one can transform input signals, representing analog voltages that are continuous in time, into a sequence of their discrete values, store these sequences as a digital value, and if necessary restore their original analog form using a DAC. In our case, the ADC input data is the output of the data reading sensors, which indirectly represents the state of the underlying

physical processes. Thus, the sensor signals have to be matched to the ADC capabilities, and because the sensors and the ADC are not necessarily located close to each other, they have to be connected to the ADC input. In a real-life experiment, the sensors' signals, apart from having to obey the laws of physics, are often in full compliance with Murphy's law, being (among other things) many times larger or smaller than the range of several volts that is required to gain the desired measurement accuracy using an ADC. If the sensor signals are too weak, then there is no other way around but to amplify them. This is often not so easy. It may appear that as far as excessively large signals are concerned, their amplitude can easily be reduced, especially as many ADCs are equipped with input voltage dividers. Unfortunately, the capabilities of these dividers are limited and, unless you have a neo-Luddite mission to destroy the experimental equipment, even several kilovolts are too much for the ADC input. The task of designing and constructing a voltage divider to handle and measure the shape of pulses with a characteristic voltage of, say, 100 kV, is not an easy one. But in the long run all these difficulties are surmountable, and let us assume that our only task is to transmit these sensor output signals to the ADC input. At the first sight, it cannot possibly pose a serious problem, but unfortunately it does.

In experimental physics, data are transmitted from sensors to recording devices via either coaxial radio cables or fiber optic channels. A coaxial cable contains a central solid conductor. The conductor is surrounded by a cylindrical insulation layer, most often made of polyethylene, then an outer shield of braided wire in the shape of a knitted stocking that is woven out of thin copper wires, and finally, this braided layer is covered by a protective insulating coating. The cable is characterized by its wave resistance $\rho = \sqrt{L/C}$, where L is the inductivity and C is the capacity per unit length, as well as by the so-called attenuation coefficient, which is determined by the losses of the transmitted energy in the conducting and dielectric materials. When designing and analyzing cable assemblies for use in experimental physics, it is more expedient to use the transfer ratio $K(\omega)$ instead of the attenuation coefficient (a parameter that was introduced by telegraphists). In any case, it is not only the terminology. It is more convenient to use the transfer ratio $K(\omega)$ instead of the attenuation coefficient. But the fact is that the attenuation increases and the transfer ratio decreases as the signal frequency is increased.

This dependence is such that the shape of a pulse of duration, say 10^{-9} s, will be considerably distorted when the length of a conventional coaxial radio cable exceeds 1 m. Of course, there exist special cables with smaller high-frequency attenuation, but they are very expensive. Their conductors have silver coatings and spaced ceramic rings are used to insulate the central conductor. Their appearance is more reminiscent of a vacuum cleaner hose rather than a conventional TV antenna cable. In the past years, they were used as the last resort in the physics laboratory until the situation was fundamentally improved due to the advent of fiber optical communications.

In order to be used to investigate high-speed physical phenomena, all the cables should be matched. This means that they should be terminated with a resistance $R = \rho$, whenever the characteristic time of the signal $\Delta t \approx 1/\nu_c$ (νc as usual is the upper critical frequency in the signal's spectrum) is less than twice the signal propagation time in the cable $2L/c\sqrt{\varepsilon}$ (where L is the length of the cable, c is the speed of light in vacuum, and ε is the dielectric constant of the material insulating the central conductor of the cable from the outer sheath). Otherwise the signal will be reflected from both ends of the cable, which act like mirrors, and any imperfection or voltage fluctuation will repeatedly return to the ADC input, distorting the signal. A similar effect is often observable with a home TV antenna, when the signal is bounced off buildings at various distances creating multiple images on the TV set. Thus, the cables have to be matched with the corresponding resistances. This is not an entirely harmless undertaking. First of all, the cable matching will usually cause the amplitude of the measured signals to decrease. Second, in cases when the sensors have a high output resistance, more electronics becomes necessary to match the high output resistance of the sensor with the low input resistance of the cable. Unfortunately, this is not the only headache with cables. The hundreds of meters or even kilometers of cable that are required for large-scale experimental facilities form a sort of a cobweb, with numerous closed electric loops, in which the so-called roaming currents are excited—these cause distortions that are difficult to suppress. The standard way of dealing with this phenomenon is well known: All the cables should be "grounded," that is, connected to the ground, at a single point. This is not difficult to fulfill when dealing with a small tabletop facility, but it is nearly impossible to do for a large experimental complex.

Practically, all the difficulties listed above disappear if fiber optics is the chosen medium for transmitting the signals from the sensor to the recording device. Naturally, the fiber optical channels are considerably more complex than simple and relatively cheap radio-electronic cables. A fiber optical line contains a transmitter, a fiber optical cable, and a photoreceiver. The transmitter includes a semiconductor laser and a modulator (or another coding device), for converting the electric signals into time-modulated visible or infrared light, which is transmitted to the receiver via the optical cable. The fiber optical cable is a thin (~100 μm) glass filament enclosed in a protective coating. The filament surrounds an even thinner (~10 μm) core. The refraction coefficients of the material forming the core and the filament are substantially different. The complete internal reflection of light off the boundary separating the two types of glass allows the light to propagate within the optical fiber. Hence, the only way for the light to leave the fiber is to reach one of its ends where the photoreceiver is placed. Today, the number of kilometers of fiber optical cables used in telecommunications has already exceeded the lines of communication based on traditional radio-electronics cables. Fiber optic technology is in a rapid stage of development. Fiber optical cables with bandwidth of several 10 GHz are routine and there are samples of experimental fiber optical cables with bandwidth of up to 400 GHz. This implies that if ADCs with sufficient speed were available, we would be able to investigate the physical processes of parameters with characteristic measurement times of the order of $\Delta t \approx 1/\nu_c = 10^{-10}\,\text{s}$. In other words, today it is the ADC design rather than the communication channel that imposes the limit on the speed of recording electrical signals. However, the most significant advantage of the fiber optical cables is not simply the bandwidth mentioned above, but rather their high noise-reducing properties. This is derived from three factors: First, the sensor and recording devices in such measurement schemes are electrically decoupled, that is, safely isolated from each another. Second, due to their small size as well as their low appetite for electrical power, the sensors and photo receivers can be powered by small local batteries, and they can be easily and effectively shielded from external electromagnetic fields. Third, because they are made of dielectric materials, fiber optical cables are immune to all sorts of inductive currents, as well as to parasitic closed circuit loops.

4. As an example, let us consider the two types of sensor, namely Rogowski coils and voltage dividers, which are used for measuring high currents and voltages, respectively, in the high-voltage circuit facilities of a physics laboratory, it is so-called *Rogowski belt* and a voltage divider for measurements in circuits with an ultra-high-frequency (UHF) components. To measure currents, the Rogowski belt is commonly used. Its design, as well as the principle of measuring the current, is shown in Figure 3.14. The principle of the schematic (Figure 3.14a) makes it clear that one is dealing with a current transformer. However, here is a brief, but dogmatically important remark. This device operates as a current transformer only if the following condition is satisfied: $\omega L >> R$, where L is the inductance of the secondary winding. Indeed, the induced electromotive force (EMF) in the secondary winding is proportional to the rate of change of the magnetic flux piercing it ($E = -\mathrm{d}H/\mathrm{d}t$), that is,

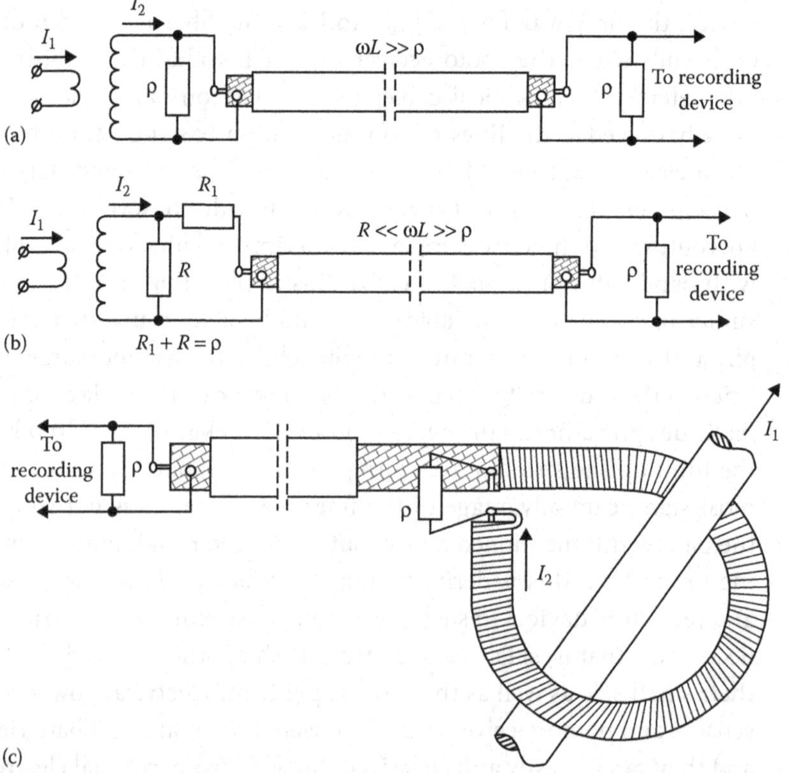

FIGURE 3.14 Rogowski belt. Design (c) and the electric circuits for measuring the current in the two cases when (a) $\omega L >> \rho$ or when (b) $\omega L << \rho$, where L is the inductance of the secondary winding and ρ is cable wave resistance.

the rate of change of the measured current. Therefore, if we were measuring the voltage at the endpoints of the winding, we would obtain the time derivative of the current in the primary winding rather than the current itself. Clearly, if the objective is to measure the current rather than its derivative, one should integrate the derivative. This is done by the *R–L* circuit, when $\omega L >> R$. If this condition is satisfied, the current in the transformer's secondary winding is simply N_2/N_1 times smaller than the current that is being measured, where N_1 and N_2 denote the number of loops in the windings, respectively.

Usually, the Rogowski belt is constructed as follows (Figure 3.14c): The outer insulation and the braiding are removed from part of a suitable length of coaxial cable leaving the central core conductor and its surrounding insulation. A wire is coiled tightly and helically around the insulator: One end of it is soldered to the remaining section of braiding and the other end is soldered to the central conductor. Then this piece of cable is wrapped around the conductor that carries the current that is to be measured, and it is shunted by a resistance ρ. This arrangement is nothing more than a current transformer, in which the role of the primary winding is played by the conductor carrying the current that is to be measured, while the secondary winding is nothing more than the "coil" that has been fashioned out of the piece of coaxial cable. Finally, let us say a word about matching our sensor with the input of the cable line. As we have already mentioned, the cable should be terminated into a matched load. Therefore, the ways to make the connection differ in the two cases when $\omega L >> \rho$ or when $\omega L << \rho$ but $\omega L >> R$, as illustrated in Figure 3.14a and b.

The cable line matching requirements are extremely important as far as measuring a voltage is concerned. This is not surprising, as integration, in contrast to a current measurement, essentially boils down to averaging over some time interval and always results in some sort of filtering, generally suppressing random fluctuations with zero mean.* Hence, measuring voltages is trickier than measuring currents. This fact is reflected in Figure 3.15. The figure shows the schematic construction of a voltage divider for measurements in circuits with an ultra-high-frequency (UHF) component. We would like to draw the reader's attention to the fact that both matching resistances ρ_1 and ρ_2 are made of hollow ceramic cylinders, placed around the central wires of the coaxial cables

* Random variable functions with zero mean are such that $\int_{-\infty}^{\infty} N(t)\mathrm{d}t = 0$.

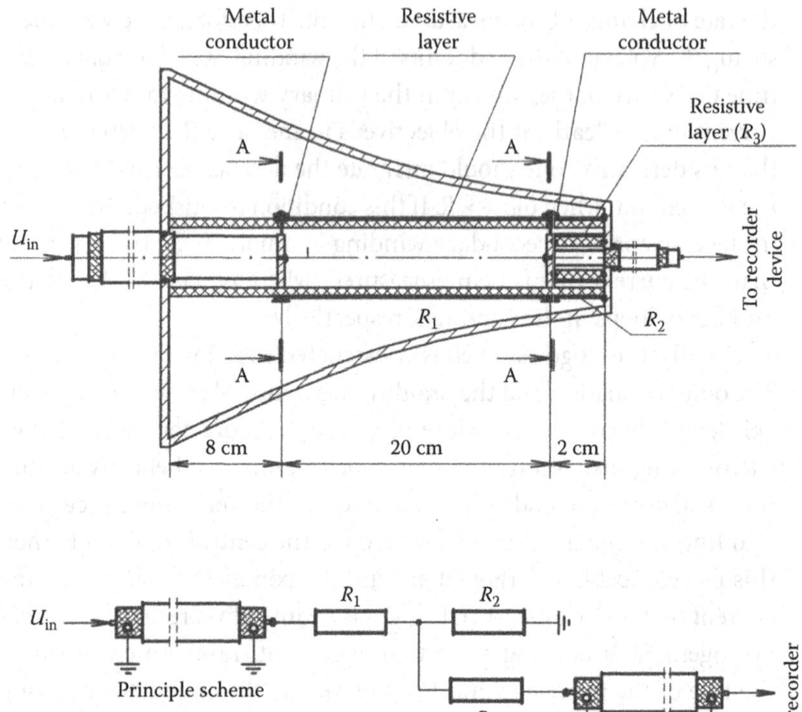

FIGURE 3.15 Schematic construction of the voltage divider for measurements in circuits with UHF components and divider electrical schema.

that have been stripped of their braiding, but have retained the central insulator. The upper surface of each ceramic cylinder is covered uniformly with a resistive layer. In addition, two thin wires are wrapped around the surface of the larger cylinder. The first one connects to the core of the high voltage cable, whereas the second one separates the resistance ρ_1 into two parts, R_1 and R_2, such that $R_1 + R_2 = \rho_1$. (In the same fashion, $R_3 + R_2 = \rho_2$.) R_2 is much smaller than R_1 and R_3, and so with sufficient accuracy, the output voltage of such a voltage divider $U_{out} = U_{in} \times R_2/R_1$. The outer metal shell ensures the continuity of the wave resistance of the cable along the section where the resistances are located. Overall, looking at this picture is rather edifying, and a close look at it reveals a number of remarkable details. We invite the reader not to hesitate to ponder a while over this picture.

So far we have considered the ways to measure time-dependent physical observables, whose values are transformed by sensors into currents $I(t)$ and voltages $U(t)$. We have discovered that one can use modern computer facilities to make measurements that cover a wide time range, including the nanosecond regime (10^{-6}–10^{-9} s) and partly subnanosecond regime. In Chapter 4, we will show that electron optical devices enable us to operate in the picosecond (10^{-9}–10^{-12} s) time range as well, at least as far as light pulses rather than electrical signals are concerned. Can one progress even further and enter the femtosecond (10^{-13}–10^{-15} s) territory? The methods for doing this are currently being developed. However, to understand some of them, one has to use rather sophisticated mathematical methods. In Section 3.3, we will expose the reader to several cases when the crux of the matter hopefully can be understood without too much mathematical complexity.

3.3 MEASURING LIGHT IMPULSES OF FEMTOSECOND DURATION

The conventional methods in this time range would work only if there existed instruments with the transfer function full width at half maximum (FWHM) β of the order of femtoseconds or even shorter. Unfortunately, such devices do not exist and their advent is not anticipated.* From the point of view of any real-life device, a pulse of such an extremely short duration would be a delta-function. Therefore, no matter whether we feed the device with a pulse of duration Δt equal to 10 or 100 fs, the recorded output will be the same, always reproducing the device transfer function $g(t)$. Therefore, the only information provided by such a measurement is the upper estimate $\Delta t < \beta$, assuring one that the signal's duration is at least shorter than the transfer function FWHM β. Using spectral measurements, which are not too difficult if one deals with relatively short light pulses, whose spectral width can be several tens if not hundreds of angstroms, one can also get a lower limit. Indeed, as mentioned already in Chapter 2, one of the main results of the spectrum theory is the *uncertainty relation* $\Delta\nu\ \Delta t \approx 1$. This implies that if the temporal duration of the measured pulse is Δt, the order of magnitude of the corresponding spectral width measured by a spectrometer will not be smaller than the reciprocal quantity $\Delta\nu \approx 1/\Delta t$.† Unfortunately, this is only a necessary, but not a

* Further in Chapter 4 it will be demonstrated that as far as electron optical devices are concerned, the narrowest achievable transfer function FWHM has a lower limit of $\beta > 0.3$ ps.

† Spectrometers are specified in terms of wavelength rather than frequency. Therefore, the corresponding frequency bandwidth is to be found as $\Delta\nu = c\Delta\lambda/\lambda^2$, where c is the speed of light, $\Delta\lambda$ is the corresponding wavelength bandwidth, and λ is the mean wavelength or the wavelength corresponding to the center of the spectral distribution.

sufficient, condition in the sense that the frequency components of the pulse should also satisfy a certain phase correlation. In other words, the fact that the frequency bandwidth of the light pulse, obtained through the spectral measurements is $\Delta\nu$, generally speaking does not imply that the pulse's duration $\Delta t \approx 1/\Delta\nu$. However, the lower limit does hold: If the frequency bandwidth of the pulse is $\Delta\nu$, its duration cannot Δt be shorter than $\sim 1/\Delta\nu$. Incorporating this upper estimate, one gets $1/\Delta\nu < \Delta t < \beta$. However, this interval can often incorporate a whole order of magnitude or even more. Therefore, the above simple inequality gives only a very coarse estimate. There are other ways to determine the duration of light pulses in the femtosecond time range with more accuracy.

The principle of one such device is shown schematically in Figure 3.16. The device includes a system intended to split the measured pulse into two pulses with control over the delay of one pulse relative to the other, as well as a nonlinear crystal and a *photorecorder*. The propagation path of the light ray in this device is clear from the figure. The measured pulse enters the device through a 50% semitransparent mirror M_1. After the pulses have been separated, one of the pulses, after being reflected from the mirror M_2, falls onto the nonlinear crystal at an angle φ with respect to the z-axis, whereas the other pulse passes into the delay unit, consisting of the mirrors M_3–M_5. This latter pulse, after leaving the delay unit, also falls onto the crystal, symmetrically to the first pulse with respect to the z-axis, that is, at the angle $-\varphi$. The mirrors M_4 and M_5 of the delay unit are fixed within the common frame, which is free to move along the direction of the propagation of the second light pulse. This enables one to change the optical path of the latter, without changing the angle of incidence with the crystal. The delay unit is necessary to match the two optical

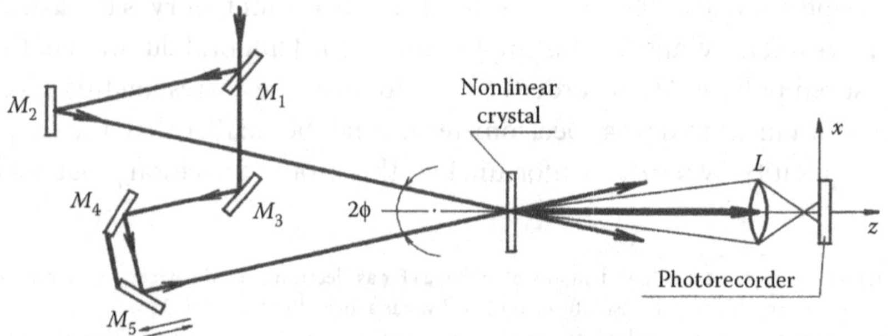

FIGURE 3.16 Principle of the scheme for measurements of light pulses with femtosecond duration.

path lengths so that both pulses reach the crystal simultaneously. The two pulses, that is, two electromagnetic waves with some frequency ω_0 intersecting within the nonlinear crystal, may, under certain conditions, cause the luminescence of the crystal with the doubled (or second harmonic) frequency $2\omega_0$. This secondary light signal is recorded by a photoreceiver with spatial resolution. What are the conditions that cause the frequency of the electromagnetic wave to double after passing through the nonlinear crystal? Well, there is more than one: The crystal plate targeted by the radiation should be carved out of a single crystal in such a way that the plate's surfaces are specially aligned with respect to the crystal axes; the radiation causing the crystal's luminescence should be incident almost at a right angle with respect to the crystal axis*; the incident radiation has to be properly polarized with respect to the crystal axes; and so on. We do not wish to go into a discussion of the complicated wave interaction physics inside the crystal, as this would lead us straight into the middle of the dense forest of modern nonlinear optics. Instead, let us pay attention to the ostensible simplicity and elegance of the phenomenon. On the path of a laser beam, perhaps on the infrared scale that is invisible to the eyes, we place a thin transparent plate, and out of the crystal, there comes a (weakly divergent) bright lettuce-green beam, as if born inside of it: lo and behold! For the practical application of this beauty, which we are concerned with, another thing is important. As the further analysis will show, the spatial dimension of the luminescent region inside the crystal depends only on the measured pulse duration and the crossing angle φ. As the latter is known, it is the dimension of the luminescent region that enables one to estimate the duration of the light pulse under investigation.

Let us discuss Figure 3.17. Note that for the benefit of clarity the proportions in the figure have been exaggerated. Before facing the crux of the matter, let us evaluate the spatial volumes, occupied by the bunches of energy, alias our pulses, traveling within the experimental arrangement. The transverse dimension of the ray is determined by the measuring device's aperture and usually amounts to between 3 and 5 mm. The longitudinal "size" of our pulse in the direction of its propagation can be estimated as $l = c\tau$, where c is the speed of light and τ is the duration of the pulse. The duration is between 10^{-14} and 5×10^{-13} s corresponding to $l = 3$–150 μm. Thus, the spatial domain, supporting the energy of the

* The second harmonic emission is possible only if the angle between the wave vector $\boldsymbol{K}$ and normal to the surface of the crystal does not exceed approximately one angular minute.

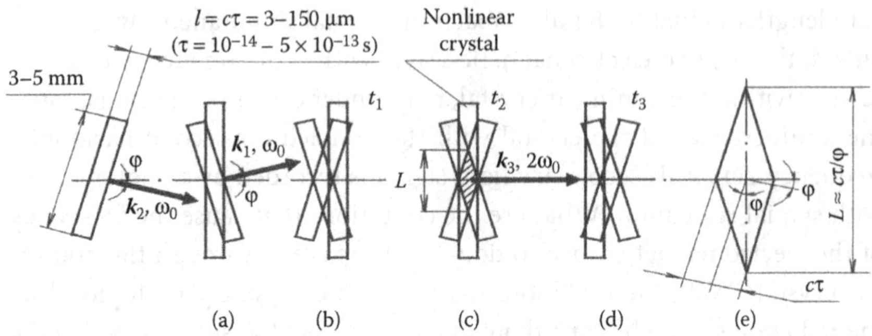

FIGURE 3.17 Measurement principles of light pulses with femtosecond duration; (a) The optical path lengths of the two separated rays do not match; (b–d) different positions when both pulses are intersecting the nonlinear crystal simultaneously; (e) the luminescent region of the nonlinear crystal.

femtosecond pulse, has a special feature that its longitudinal dimension is at least 20 times smaller than the characteristic width. The reader should keep this in mind while viewing Figure 3.17. Figure 3.17a illustrates the situation when the optical path lengths of the two separated rays do not match. One can see that as the first pulse is intersecting the crystal, the second one has not yet reached its surface. As the incidence angle is considerably large, the first pulse will pass through the transparent crystal without being in any way affected. The same thing will happen to the latter pulse as the first one, which will be intersecting the crystal all alone. If the crystal does not radiate, that is, the second harmonic is not generated, we can match the lengths of the optical paths by means of the delay unit. Now both pulses will be intersecting the crystal simultaneously (Figure 3.17b), and the resulting net wave vector (the vector sum of the two waves corresponding to each ray) becomes normal to the surface of the crystal, as the two incident components are symmetric with respect to the surface normal. (Accordingly, the net electric field of two electromagnetic waves gains the necessary alignment with respect to the main crystal axes.) However, the luminescence will occur only in that region inside the crystal, where these two waves intersect (the shaded region in Figure 3.17b). The largest size of the luminescent region is achieved at the moment t_2, when the center of the intersection of the two waves coincides with the center of the crystal (Figure 3.17c). As has been mentioned earlier, the size of the luminescent region is determined only by the duration of the measured pulse and the angle between the wave vectors of the two split waves. This can be clearly seen from Figure 3.17e. If the measured pulse was rectangular [the function $I(t)$ describing the time dependence of the pulse's intensity would have a rectangular shape as shown in Figure 3.17e], it would be possible to determine the pulse's duration from the dimension L of the luminescent region as follows: $\tau = L\tan\varphi/c \approx L\varphi/c$, since the angle φ is sufficiently small. The dependence of the intensity of the second harmonic radiation $I_{2\omega}(x)$ with respect to the coordinate x (the x-axis coincides with the larger diagonal of the rhombus) will be recorded by the photorecorder, where the luminescent region is projected in such a way that the planes of the crystal and the photodetector are conjugate to one another. Then, in the case of a rectangular pulse, the magnification of the projecting system $M = 1$, while the graph of the function $I_{2\omega}(x)$ represents an isosceles triangle with the base L and the FWHM $L_{1/2} = L/2$. It easily follows that the pulse's duration $\tau = 2L_{1/2}\varphi/c$.* Unfortunately, things are not so perfect for real-life pulses. The FWHM of the pulses obtained in this way from the recorded FWHM of the luminescence turns out to be about 1.5–2 times greater than the real ones. Moreover, it is impossible in principle to determine the fact whether it is 1.5 or 2 times greater if the actual shape of the function $E(t)$ [or at least the function $I(t)$] is unknown. Of course, it would be desirable to end up actually having the unknown function $E(t)$ itself, rather than only its FWHM, and with an error of the order of 30%. How can one possibly get there?

First of all, let us make clear what we should know in order to unambiguously determine the function $E(t)$ in question, namely,

$$|E(t)| = A(t)\exp\{i[\omega_0 t - \varphi(t)]\} \tag{3.7}$$

where:

ω_0 is the frequency
$A(t) = \sqrt{I(t)}$ is the amplitude
$\varphi(t)$ is the phase of the electromagnetic wave at time t

As the quantity ω_0 is usually known, we are left to find "only" the quantities $A(t)$ and $\varphi(t)$, that is, our objective is to record the time dependence of the amplitude and the phase of our ultrashort light signal. To understand the methods used for that purpose, let us discuss in more depth the situation that occurs when a very thin nonlinear crystal is intersected by two

* Note that in data processing, the FWHM of the luminescent figure usually is used to obtain an acceptable accuracy instead of the quantity L itself because the light intensity and thus the signal-to-noise ratio are low in the vicinity of the image's boundary.

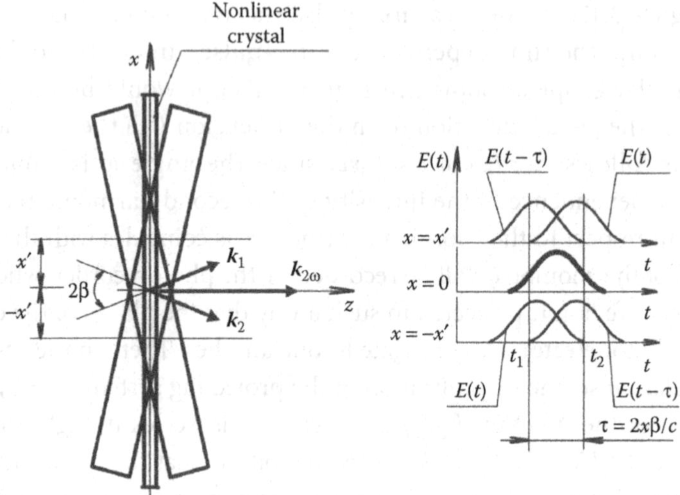

FIGURE 3.18 Intersection of a very thin nonlinear crystal by two otherwise identical light pulses.

otherwise identical light pulses with the wave vectors $\boldsymbol{k}_1$ and $\boldsymbol{k}_2$, traveling at the angles $\pm\beta$ with respect to the z-axis (Figure 3.18). It follows from the figure that both pulses intersect the crystal simultaneously only on the system's axis, namely, where $x = 0$. In the region where x is positive, the first pulse is delayed with respect to the second one, and for negative values of x, it is the other way around (from symmetry). Given a fixed value x', it follows from the figure that the delay is $\tau(x) = 2\beta x'/c$. Recall that both pulses originated after the input pulse had been split in halves after having entered the measuring device. Thus, for a fixed x', the net electric field of the pair of electromagnetic waves in consideration can be represented as $E_\Sigma(t) = E(t) + E(t - \tau)$. In this case, which is referred to as *noncollinear interaction*, the electric field of the second harmonic emission, generated inside the crystal, is

$$E_{2\omega}(t,\tau) \propto E(t)E(t-\tau) \tag{3.8}$$

and the intensity

$$I_{2\omega}(t,\tau) \propto I(t)I(t-\tau) \tag{3.9}$$

In Equations 3.8 and 3.9, a number of coefficients, describing the efficiency of the energy transfer into the second harmonics, the optical tract energy losses, the photorecorder's sensitivity, and so on, have been omitted, and this is why instead of the equality sign, the proportionality sign has been used. These coefficients are not so important in order to expose the crux of the matter. Note that modern photoreceivers,* rather than being sensitive to the electric field E of the light wave, are capable of measuring only the intensity I, or more precisely the energy $\mathcal{E} = I\Delta t$,which has arrived at the photoreceiver during the exposure time Δt. What is more, although in the femtosecond time range, the receivers may, if necessary, possess spatial resolution, temporal resolution is out of the question, that is, it is not possible to record either the time dependence of $E(t)$ or $I(t)$. In other words, in the measuring scheme discussed here, it is possible to measure the net energy of the second harmonic only over the whole time interval during which it has been generated, namely, when the two intersecting waves have been present simultaneously inside the crystal, causing the radiation. In Figure 3.18, the endpoints of this time interval are denoted as t_1 and t_2. Thus, for the second harmonic emission energy, relevant to a point x of the nonlinear crystal, with $x = \tau c/2\beta$, one gets

$$\mathcal{E}_{2\omega}(\tau) \propto \int_{t_1}^{t_2} I(t)I(t-\tau)\mathrm{d}t = \int_{-\infty}^{\infty} I(t)I(t-\tau)\mathrm{d}t \tag{3.10}$$

The integrals in the above equation are equal to one another, simply because outside the interval $[t_1, t_2]$ one has $I(t)I(t - \tau) = 0$, and the second integral being simply the *autocorrelation function* of $I(t)$. Because we know $\tau(x) = 2\beta x/c$, a simple change of variable yields the energy density distribution over the photoreceiver's input aperture. Thus, if the magnification coefficient of the projecting system $M = 1$, then

$$\mathcal{E}_{2\omega}(x) \propto \int_{-\infty}^{\infty} I(t)I(t-2\beta x/c)\mathrm{d}t \tag{3.11}$$

Note that Equations 3.10 and 3.11 do not contain any information about of the phase $\varphi(t)$. Nevertheless, if the mode of function $I(t)$ or $\mathbf{E}(t)$ is known, it is easy to derive $I(t)$ from the FWHM of the function $\mathcal{E}_{2\omega}(x)$ or $\mathcal{E}_{2\omega}(\tau)$. Unfortunately, if one does not possess enough information about the former functions, this cannot be done, as the relation (3.11) enables one to

* See Chapter 4.

determine only the spectral density of the function $I(t)$. In principle, one can try to obtain $I(t)$ from the integral Equation 3.10 or 3.11, say, by minimizing the functional:

$$Z \equiv \int_{-\infty}^{\infty} \left| \mathcal{E}_{\text{sig}}(\tau) - \int_{-\infty}^{\infty} I(t)I(t-\tau)\text{d}t \right|^2 \text{d}\tau \tag{3.12}$$

where:

$\mathcal{E}_{\text{sig}}(\tau) = k\mathcal{E}_{2\omega}(\tau) = k\int_{-\infty}^{\infty} I(t)I(t-\tau)\text{d}t$ is the result of the experiment

k is a coefficient that takes into account the efficiency of the energy transformation into the second harmonic, the optical tract energy losses, the photoreceiver's sensitivity, and so on

However, the above equation represents an archetypal ill-posed problem with all the consequences. This implies in particular the nonuniqueness of the minimization of Z, that is, there exists some family of functions $Ik(t)$, which equally well minimize Equation 3.12, but in a sense are arbitrarily different from one another. A working strategy to overcome these difficulties is a radical increase in the amount of the available experimental information,* which is achievable by means of moving from the one-dimensional $[\mathcal{E}_{2\omega}(x)]$ to the two-dimensional (for instance, spectral) measurement techniques. To do so, one should place a spectrograph in front of the photoreceiver in such a way that the spectrograph's slit is aligned along the x-axis (accordingly, the dispersion direction will be perpendicular along the y-axis) and projects the second harmonic generation (SHG) region onto the slit. This will enable us to obtain the set of spectra of the second harmonic, emitted from each point of the crystal, with the coordinate $x = \tau c/2\beta$†:

$$S(\omega,\tau) \propto \left| \int_{-\infty}^{\infty} E(t)E(t-\tau)\exp(-i\omega t)\text{d}t \right|^2 \tag{3.13}$$

* It has been stressed more than once that this is the lack of information that prevents one from choosing the one physically "correct" solution from the whole family of solutions of a mathematically ill-posed problem.

† This method of measurements of ultrashort pulses, which is called frequency-resolved optical gating (FROG), was proposed by Rick Trebino [3].

Clearly, if the system's magnification coefficient $M = 1$, the energy density distribution in the recording plane will be

$$S(\omega, x) \propto \left| \int_{-\infty}^{\infty} E(t)E\left(t - \frac{2\beta x}{c}\right)\exp(-i\omega t)\mathrm{d}t \right|^2 \tag{3.14}$$

Equations 3.13 and 3.14 make it possible to find $E(t)$ by iteration [3]. Indeed, this means that in this process we calculate both $A(t)$ and $\varphi(t)$. At each step of iteration after solving the inverse problem and finding $Ai(t)$ and $\varphi i(t)$, the deviation $Z \equiv \int_{-\infty}^{\infty} |\mathcal{E}_{\text{sig}}(\omega,\tau) - \int_{-\infty}^{\infty} E(t)E(t-\tau)\exp(-i\omega t)\mathrm{d}t|^2 \, \mathrm{d}\tau$ in Equation 3.13 should be calculated. In Figure 3.19 examples of experimental and reconstructed spectrograms are compared taken from reference [3].

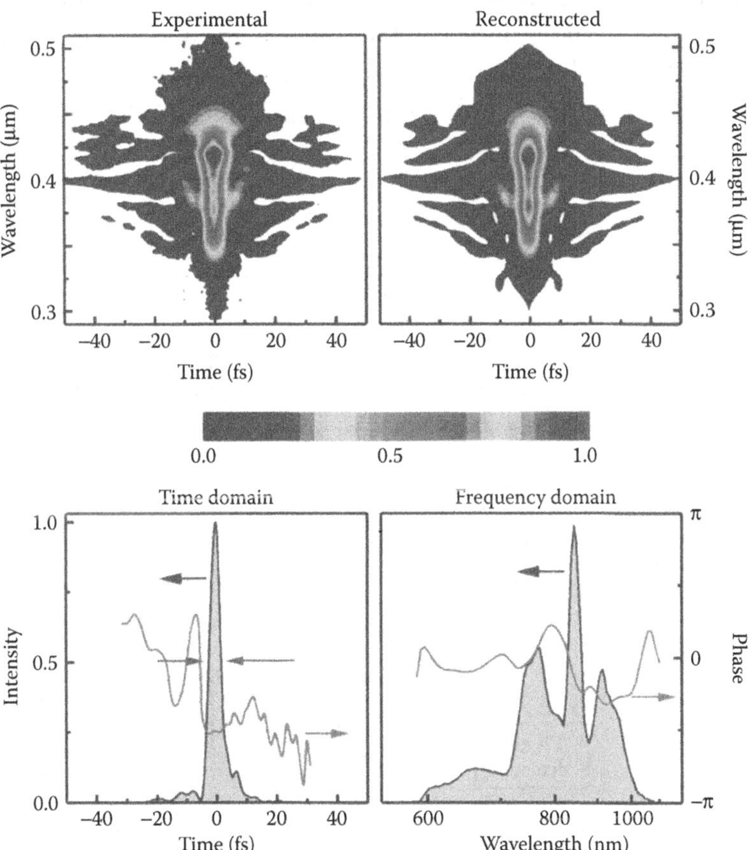

FIGURE 3.19 Comparison of the experimental and reconstructed spectrograms [3].

The above discussion refers to the methods and devices for measurements with single pulses (so-called single-shot correlators). Other possibilities appear when there is a train of femtosecond pulses. (Usually, femtosecond master oscillators generate a continuous train of identical femtosecond pulses following one by one with the time interval between the pulses of 10–20 ns.) Figure 3.20 shows the arrangement used for measurements with a train of pulses. As one can see, the basis of this scheme is a Michelson interferometer in which one of the mirrors (M_3) is movable lengthwise along the optical axis. (The optical paths are clear from the picture.) This scheme uses a photodetector whose output is displayed on a relatively slow oscilloscope. The oscilloscope is synchronized with the motion of the mirror (M_3 in Figure 3.20) in the delay arm of the interferometer in such a manner that each point of the oscillogram corresponds to a specific position of the mirror. As a result, each point of the oscillogram corresponds to the specific time delay between the pulses arriving to the SHG crystal. Note that in the case when the duration of the oscilloscope trace is, for instance, 10 ms and a train of pulses with period of 10 ns is measured, as many as 10^6 pulses gradually delayed with respect to each other by some time τ will

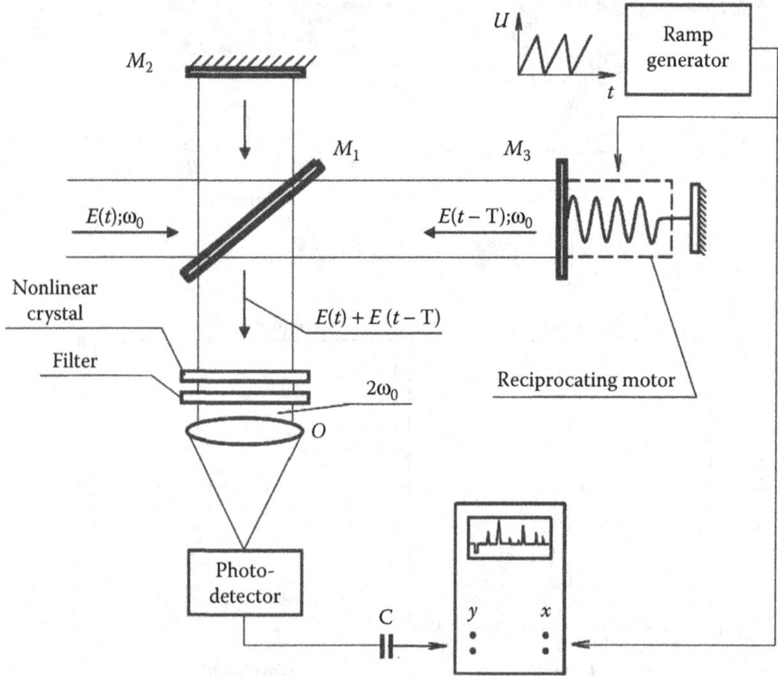

FIGURE 3.20 A typical autocorrelator.

be recorded by the photodetector and then will construct the oscillogram. Obviously, the curve on the screen will be observed with the difference that the timescale will run along the x-axis of the oscillogram. It is not a *real* timescale. The true scale can be found by the following considerations. For providing a delay τ between pulses in autocorrelation measurements, one should shift the delay mirror M_3 by a length $l = c\tau$, where c is the speed of light. The maximum delay time τ_{max} is selected so that it would exceed the expected FWHM of the measured pulse by a factor of 2–3 (note that $l_{max} = c\tau_{max}$). As the delay mirror is shifted from a zero position by the length l, the oscilloscope trace will move along the screen by the distance $L = vt$, where v is the trace speed (speed of the electron beam motion along the x-axis of oscillogram) and t is the time elapsed from the beginning of the measurement. For the measured function to fit onto the screen, one should provide that the whole length of the trace record would equal or exceed the value $L = vt_{max}$. For calculating L, we need to find the relationship between τ_{max} and t_{max} or, equivalently, between τ and t. For doing that, let us use the scaling coefficient $M = l/L$, which is actually the ratio of electron beam shift along the oscillograph screen to the corresponding displacement of the delay mirror. Because $l = c\tau$ and $L = vt$ $(M = l/L = c\tau/vt)$, one can find the true delay time:

$$\tau = M\frac{v}{c}t \tag{3.15}$$

where:

t is the time measured over the oscilloscope screen

Rewriting the above equation in the form $c\tau_{max} = vt_{max}M$, one can readily see that $vt_{max} = \text{const}$ for any given τ_{max}.

A typical autocorrelation function obtained with this technique is shown in Figure 3.21. In this figure, the timescale is plotted in accordance with Equation 3.15. Indeed, the maximum of the autocorrelation function corresponds to the situation when both pulses arrive at the crystal simultaneously. The oscillations in the behavior of the function can be accounted for by the periodical variation of the mutual phase of the pulses. Let us write the electric fields of the both waves; they are as $U_1 = Ae^{i\omega(t+\varphi)}$ and $U_2 = Ae^{i\omega(t+\varphi+\tau)}$ (delayed pulse). The total electric field and intensity can be written as $U_\Sigma = U_1 + U_2$ and $I_\Sigma = U_\Sigma U_\Sigma^* = 2A^2 + A^2e^{i\omega\tau} + A^2e^{-i\omega\tau} = 2A^2 + 2A^2\cos\omega\tau$, respectively. It should be pointed out that this equation does not describe the

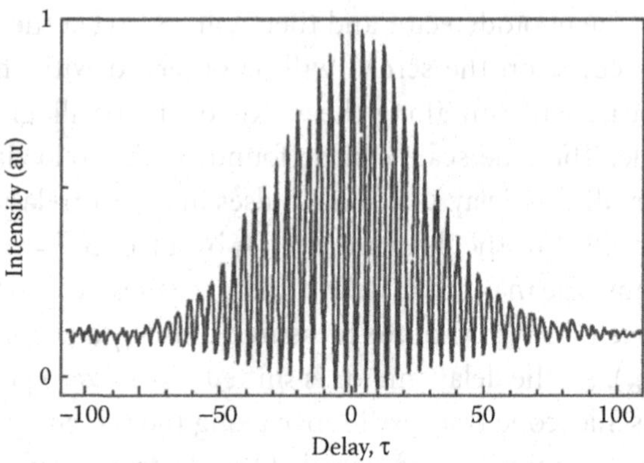

FIGURE 3.21 Typical autocorrelation function (horizontal scale in femtoseconds).

intensity of the second harmonic emission (Figure 3.21) because its field is proportional to the product of the incoming waves $U_{2\omega} \propto U_1 U_2$ but not to their sum. However, at any delay when the product $\omega\tau$ satisfies the relationship $\omega\tau = n\pi$, $n = 1, 3, 5,\ldots$, the intensity of the field in the crystal will be equal to zero. Indeed, at these moments, the conversion efficiency will reduce as well. Finally, it is worth mentioning that by measuring the oscillation period one can determine the frequency and thus the wavelength of the laser under investigation.

The method described above is a good, rather simple and reliable way for estimating the duration of an ultrashort pulse in a train of pulses. However, what we obtain by using this method is just an *estimate* of the pulse duration. For instance, for "rectangular" pulses, the FWHM of the autocorrelation function is equal to the FWHM of the original pulse. In other cases, the FWHM of this function can exceed the true FWHM of the laser pulse by a factor of between 1.4 and 2.0.

Nevertheless, it is possible to obtain the Fourier image of the pulsed electric field $E(t)$, that is, $E(\omega) = A(\omega)\exp[-i\varphi(\omega)]$, directly from interferometric measurements. This method [1,2], which its inventors call *spectral phase interferometry for direct electric-field reconstruction* (SPIDER), allows the reconstruction of both $A(\omega)$ [or $A(t)$] and $\varphi(\omega)$ without the need to use an iterative procedure. The principle of this technique is shown schematically in Figure 3.22. First, the input pulse is divided into three replicas, which then are directed onto a nonlinear β-barium borate (BBO) crystal. Before they reach the crystal, the first two replica pulses are delayed with respect to each other by a fixed delay τ and the third replica is stretched in time and its polarization is rotated

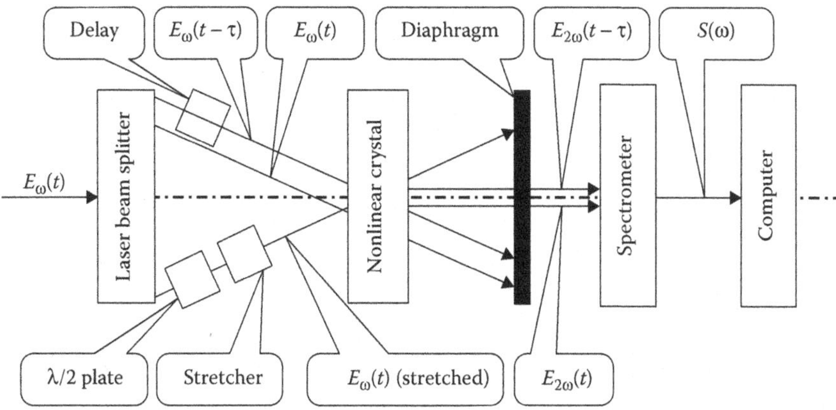

FIGURE 3.22 Schematic of the SPIDER technique.

so that it is orthogonal to the polarization of the two other pulses as required for type II SHG. The resultant SHG pulses $E_{2\omega}(t)$ and $E_{2\omega}(t-\tau)$ are frequency shifted with respect to each other by a spectral shear $\delta\omega$ due to the delay τ, because they are mixed with different slices of the stretched replica. The interferogram $S(\omega)$ will contain two collinear frequency components $E_{2\omega}(\omega)$ and $E_{2\omega}(\omega+\delta\omega)$, which interfere with each other, and therefore

$$\begin{aligned} S(\omega) &= \left|E_{2\omega}(\omega)\right|^2 + \left|E_{2\omega}(\omega+\delta\omega)\right|^2 \\ &\quad + 2\left|E_{2\omega}(\omega)E_{2\omega}(\omega+\delta\omega)\right|\cos\left[\varphi_{2\omega}(\omega+\delta\omega)-\varphi_{2\omega}(\omega)+\omega\tau\right] \end{aligned} \tag{3.16}$$

where:

$E_{2\omega}(\omega)$ is the Fourier image of the pulsed electric field after SHG

The first two terms on the right-hand side of the above equation are the spectra of the second harmonics of the first two replica pulses and they can be obtained from independent measurements. The third term contains information about $\varphi(\omega)$ that can be found by a noniterative algorithm. Note that the interferogram $S(\omega)$ is not the Fourier image of the pulsed electric field $E(t)$. To obtain the Fourier image $E(t)$ from $S(\omega)$, a complicated processing procedure (including two Fourier transformations, filtration, and some independent measurements) is required. Figure 3.23 shows an example of a measured interferogram $S(\omega)$. Figures 3.24 and 3.25 show examples of reconstructions using the SPIDER technique.

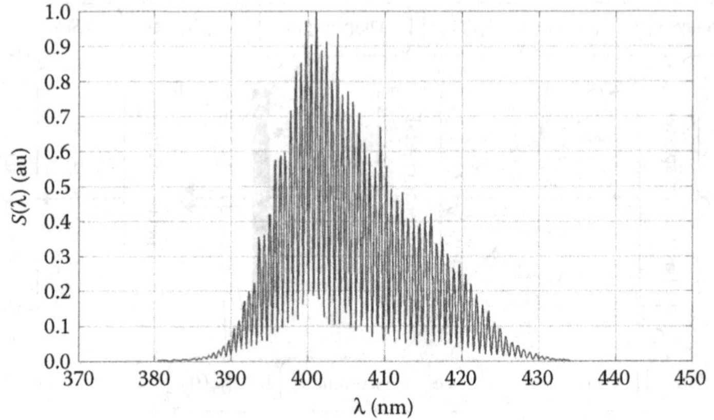

FIGURE 3.23 An example of a measured interferogram $S(\omega)$.

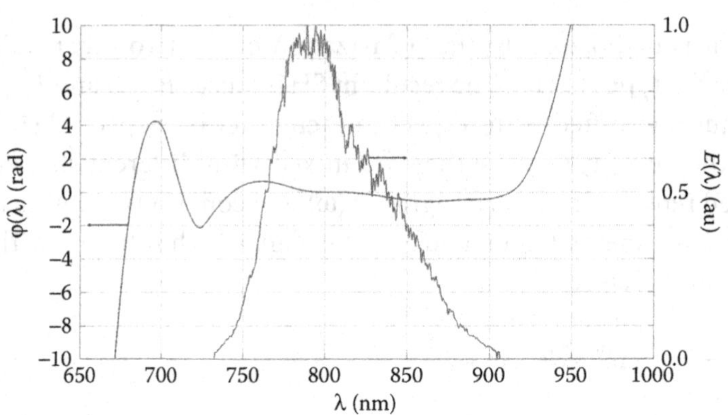

FIGURE 3.24 Measured spectrum of the input pulse and reconstructed phase $\varphi(\lambda)$.

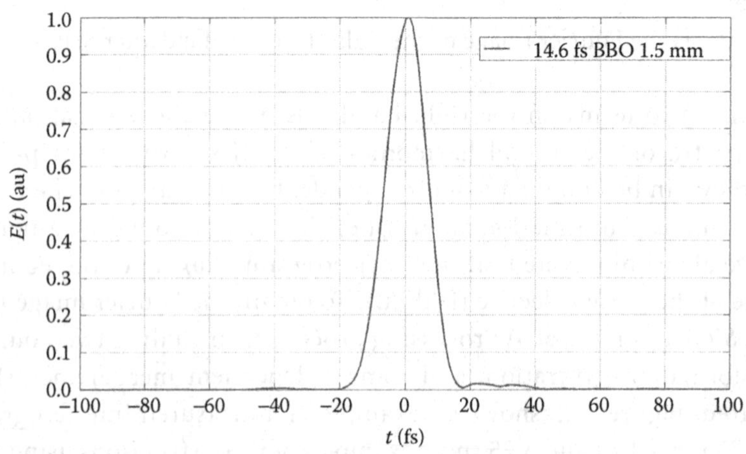

FIGURE 3.25 Reconstructed amplitude of the input pulse $E(t)$.

REFERENCES

1. Gallmann, L., Sutter, D.H., Matuschek, N. et al. 1999. Characterization for direct electric-field reconstruction. *Optics Letters* 24(18):1314–1316.
2. Iaconis, C. and Walmsley, I.A. 1998. Spectral phase interferometry for direct electric-field reconstruction of ultrashort optical pulses. *Optics Letters* 23(10):792–794.
3. Trebino, R. 2002. *Frequency-Resolved Optical Gating: The Measurement of Ultrashort Laser Pulses*. Boston, MA: Kluwer Academic Publishers.

CHAPTER 4

Photographic Methods of Experimental Research and High-Speed Photography

According to a well-known saying, a picture is worth a thousand words. In the context of this book, this statement provides an intuitive comparison between the data transfer rates of two different information channels: optical and acoustic. In addition, the primary information about physical objects and phenomena is often conveyed by light. These circumstances account for the widespread use of optical, and in particular photographic, techniques in modern experimental physics. However, optical images recorded during an experiment most often defy immediate visual interpretation, representing rather some sort of primordial data about the spatial distribution of the parameters of an object under investigation. What is more, in general these data nowadays do not even take the form of a traditional image, but rather take the form of an array of numbers, reflecting a luminosity distribution. The output data (photographic images) have to meet the requirements of the experiment to obtain the reliability and accuracy of a reconstructed set of functions of space and time, describing the state of the object that is the final goal of each experiment. Meeting these requirements can be either possible or impossible, depending on certain specific parameters of the measuring and recording devices. The latter devices usually come in a sequence of individual units, each one transforming the original information contained by the light. This often results in a significant discrepancy between the input and output images, that is, a loss of accuracy. Trying to predict the output as well as to explore the limits of the measuring and recording devices, which in addition to optical components include a variety of radio-electronic devices with built-in communication channels, automatic microphotometers, analog-to-digital converters, memory units, and so on, calls for a unified and systematic approach. That is why, for the purpose of describing and analyzing these systems, it is natural to apply well-developed methods of information theory. Indeed, information theory was designed to define and optimize the transmission rate of the information channels as well as the capacity of information storage systems. Generally speaking, all the experimental equipment listed earlier is used to transmit, receive, transform, and record information. As was mentioned earlier, there is no essential difference whether the information is encoded in the spatial or temporal signal structure in different devices. All these devices, to an equal extent, have a fundamental set of characteristics in common, the first and foremost being linearity and stability (spatial or temporal invariance). This was explored in detail in Chapter 2, taking advantage of the almost complete similarity of the mathematical tools to extend the well-developed analysis and methods of communication theory to applications well beyond the original scope of this theory, such as to optical devices.

However, to put the ideas for such an approach into practice, we will have to reconsider the parameter system that is in use today, characterizing the optical devices exactly on the basis of the visually perceptible image quality. This parameter system is deeply rooted in the photographic tradition, whose main objective has always been to produce images of high quality for visual perception. This objective had determined a widely accepted system of quality criteria for photographic images and devices, which is still extant and dominating. We are not trying to pass judgment here, whether this system is good or bad in principle. The point is that it is extremely difficult and often impossible to use it in order to arrive at the set of required functions. This is the problem. To illustrate this issue in its essence, consider, for instance, the commonly accepted concept of the graininess of photographic film, that is, film noise. The graininess G describes a measure of photographic noise, its value being defined as $G = 100/M_{\text{lim}}$, where M_{lim} is the maximum magnification that allows an observer to distinguish the image pixel structure with the naked eye. The basis of this concept is clear, as it determines the maximum possible print size, and it is understandable and useful for a photographer. However, it does not allow an experimenter to derive from it either a signal-to-noise ratio or an ultimate value of the error of the measured illumination.

Another parameter is the so-called spatial resolution, which traditionally "is determined by the fragment of the resolution target with the thinnest lines, which can be visually resolved separately in all directions." For instance, the image of a resolution target taken with an electron-optical imaging camera is shown in Figure 4.1. As one can see, the resolution target used in this measurement contains 25 segments (fields) with different orientations of black lines. Both the spacing and the thicknesses of the lines in the various fields are not only different but strongly determined. The limiting segment, where "lines can be visually resolved separately in all directions" is represented by field #10 in Figure 4.1. The same figure also contains the microdensitometer traces obtained by scanning across the lines of the images at the input and the output of the camera. This shows the dramatic difference between the input and output images. This figure also shows that in this particular case a brightness modulation of 10% is sufficient to provide the visual resolution. What is exactly implied by the statement that such a small brightness modulation enables

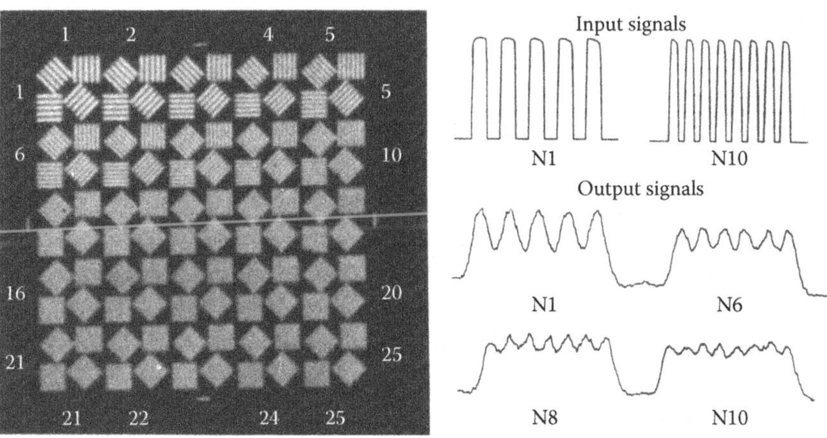

FIGURE 4.1 The image of a resolution target (left) taken with an electron-optical imaging camera. The numbering, which is duplicated in frame corners, identifies the resolution target segments. The microdensitometer traces (right) were obtained by scanning across the lines of the images at the input and the output of the camera.

one to "resolve the lines separately in all directions"? Overall, it reflects the brain's immense capacity for pattern recognition and nothing more!

Is it possible to make use of visual measurements of this kind and to derive the criteria for the evaluation of the input signal measurement errors? Unfortunately, not! Quite a few attempts have been made, trying to patch these ostensible gaps by using the Raleigh criterion. This numerical criterion may give a correct threshold value of visual resolution, but only in a rather specific case when the transfer function results from diffraction at a slit or circular aperture for point radiation sources with nearly equal intensities. Let us repeat that in the above case, the use of the criterion is justified and efficient. However, for the purpose of creating a general formalism for visual perception, this approach is just as irrelevant as the use of photometric units, pertaining to visual light units (lumen, lux, etc.) to describe the sensitivity of instruments in the infrared (IR) or ultraviolet (UV) regions of the spectrum. Indeed, in the latter case, the visibility function and thus the illumination are, strictly speaking, identically 0, unless one comes up with some formal artifice in defining them.

It seemed for a while that the frequency–contrast characteristic method, if properly put into practice, might bring some significant positive changes to the issue. Once again, being a slave to tradition prevented us from taking advantage of existing opportunities. Indeed, it would hardly occur to a radio engineer to use a periodic rectangular pulse generator to record an amplitude–frequency characteristic. But open any reference book in optics and it will tell you that frequency–contrast characteristics can be measured with a rectangular line test grating! The point is not only that the use of a *sinusoidal test grating*, that is, one whose spatial brightness varies as $I(x) = I_0[1 + \sin(\omega x)]/2$, brings a significant simplification mathematically, but more importantly that measurements with a line test grating just cannot give the necessary accuracy in a high spatial frequency range. We have listed by now practically all the major parameters required to characterize present-day optical instruments, that is, all those rudiments that are still being widely addressed within the mass of modern reference literature. The reasons mentioned earlier should make it sufficiently clear that the traditional frame of reference is suitable neither for the purposes of the design, analysis, and synthesis of even slightly complex optical systems, nor for making *a priori* estimates of measurement errors, regarding the input illumination $E_{in}(x,y)$ on the output image basis, as well as estimates of the accuracy of the input image reconstruction. The estimates and computations in question are simply not feasible if they are based on the parameter system that is commonly accepted today. For this reason, we shall attempt to reach these goals by different means, namely, those provided by the methods of information theory. These methods were discussed in detail in Chapter 2.

4.1 OPTICAL-MECHANICAL CAMERAS

Optical-mechanical cameras, creating an image on a photographic film, are simple and reliable. Using them is beneficial in all those cases, when neither an ultrafast time resolution nor an extremely high sensitivity is required. The extra care that has to be taken in developing the film and further processing does not generate enthusiasm, but the effort is usually rewarded with a high-quality image. Optical-mechanical cameras come in two types: (1) the so-called streak cameras, which are designed to record one-dimensional images with time resolution, that is, those images whose illumination $E(x,y,t)$ is constant along one of the axes, say the Y-axis, so that $E = E(x,t)$, and (2) the framing cameras, whose objective is the instantaneous recording of two-dimensional images with illumination $E(x,y)$ in a series of discrete moments of time, separated by equal intervals Δt that correspond to a photographing frequency $N = 1/\Delta t$ frames per second. The quantity Δt is naturally referred to as the temporal resolution. What should be understood as the time resolution for streak cameras is discussed in more detail later in this section.

Figure 4.2 shows the principle of the optical arrangement of a streak camera. Let us describe how this instrument actually works. The objective lens L_1 creates an image of the object O in the plane S that contains a slit. This selects a streak of characteristic size Δ, which is of the same order

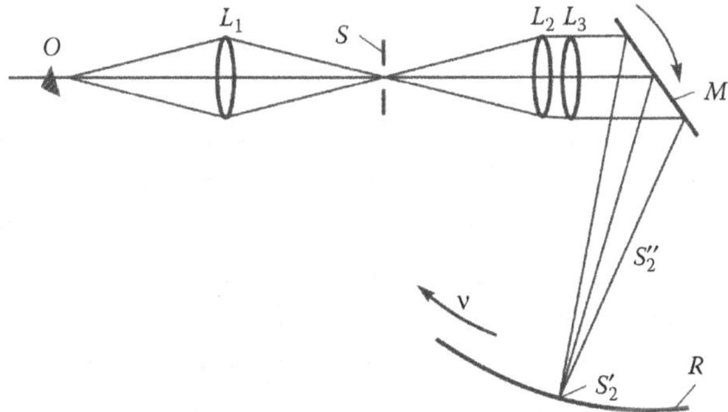

FIGURE 4.2 Schematic showing the principle of the optical arrangement of a streak camera.

as the transfer function full width at half maximum (FWHM) β of the camera. Naturally, the time-varying image $E(x,t)$ of the object being studied should be projected onto the slit in such a way that the X-axis is directed along the slit. The selected image streak is projected by means of a pair of lenses L_2 and L_3 as well as a rotating mirror M onto the photographic film that is pressed against a cylindrical surface R. Note that the consecutive lens pair L_2 and L_3 represents a so-called reproducing objective. The slit is situated in the focal plane of the first lens and hence the second lens operates in parallel light, thus ensuring a high spatial resolution. If one uses objectives with rather small relative f-number, as is usually the case with streak cameras, it is possible to gain a spatial resolution that is close to the diffraction limit. To estimate the temporal resolution, one can use a simple relationship $\Delta t \approx \beta/\nu$, where β is the FWHM of the transfer function and ν is the linear velocity of the image of the slit along the film. From Figure 4.2, one can see that $\nu = \omega R$, where R is the distance between the film and the axis of rotation of the mirror. Let us pay attention to the following two circumstances. It is not only the linear velocity ν that grows proportionally to R, but also unfortunately the FWHM β, because the optical magnification (which also reduces the film illumination) also increases with R. Therefore, a naive increase of R will not result in the improvement of the time resolution. An apparently more serious way of doing it is to increase the mirror angular velocity ω, which is helpful up to a certain limit. This limit is due to the effect of centrifugal forces proportional to $r\omega^2$, which simply tend to rip the mirror into smithereens, unless it is possible to decrease the mirror radius r, in order that the stress due to the centrifugal forces does not reach its critical value. Obviously, in this near-critical regime, in order to increase ω by a factor of two, r should be decreased by a factor of four. At the same time, β in such systems is primarily determined by diffraction at the mirror aperture and this is roughly proportional to λ/r, where λ is the wavelength of the light. Hence, the possibilities here are also limited. In the reference literature, one can usually find values $N_{\text{max}} = \omega_{\text{max}}/2\pi$, usually of the order 100,000–500,000 rpm, with maximum velocity v_{max} well within the limits of 3–15 mm per μs^{-1}. However, from the above discussion, it follows that these figures do not allow one to pass judgment about the time resolution, unless β or at least r is known. The real temporal resolution of optical-mechanical cameras is usually not higher than 10^{-8} s. The f-number of this type of device is not very high with $D/f \approx 1/10 - 1/20$.

Let us return to Figure 4.2. The operating angle γ over which it is possible to record an image is not very large (being typically in the

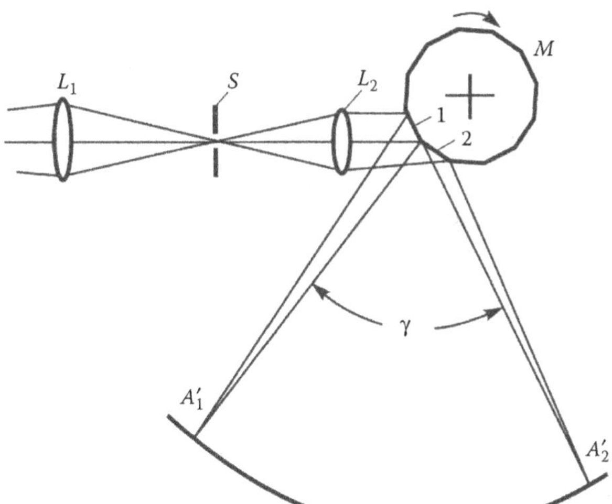

FIGURE 4.3 Schematic showing the principle of a streak camera with a continuously rotating multifaceted mirror.

range 60°–90°). This requires an accurate synchronization between the initial time of the physical phenomenon that is being studied and the start of the photographic recording process. This is not at all easy with mechanical systems, and a widely used solution is to use a continuously rotating, multifaceted mirror. This ensures that the object is always projected onto the film, so that it can be recorded at any desired moment. The principle of the optical arrangement of such a device is shown in Figure 4.3. As one can see, a light beam, emanating from the object at some time instant, is incident onto two different facets of the mirror at the same time. After being reflected from these facets, it creates two images of the object at the same time, located at the points A'_1 and A'_2 with the angular separation γ. Further rotation of the mirror causes one of the images (A'_1) to leave the operating angle limits, whereas the other moves well within it. Thus, at any particular time, there is at least one image of the object available for recording.

Note that a circle of radius R (as shown in Figure 4.2) would be the geometric locus of the slit image points only in the case when the axis of revolution lies in the plane of the reflecting surface of the mirror M. Otherwise, if the reflecting surface of the mirror is displaced with respect to the axis of revolution, as it is the case with multifaceted mirrors (Figure 4.3), the locus of the image of the slit represents a part of the inner arc of the Pascal snail. However, by changing the radius R and simultaneously shifting the center of the circle, it is possible to approximate the part of the Pascal snail arc by a circle sufficiently well that any related loss of spatial resolution would be negligible.

As already mentioned, the focusing quality of the lenses L_2 and L_3 is close to the diffraction limit. Therefore, the two-dimensional transfer function of a streak camera can be represented as follows:

$$g(x,y) = g(r) = \left(\frac{4\lambda}{\pi D^2}\right)^2 \left[\frac{2J_1((\pi D/q\lambda f)r)}{(\pi D/q\lambda f)r}\right]^2 \tag{4.1}$$

where:

J_1 is a Bessel function
D is the light beam diameter in the objective L_3 output pupil plane
f is the focal length of the objective
λ is the light wavelength
r is a coordinate in the film plane
q is a coefficient of order unity that corresponds to the amount by which the diffraction limit is exceeded

Hence, the transfer function FWHM

$$\beta = \frac{q\lambda f}{D} \tag{4.2}$$

To conclude this discussion, we show an example of an image of a periodic axially symmetric plasma compression by a weakly dissipating alternating current with a period of about 8 μs (Figure 4.4). The direction of the current in the plasma is parallel to the tube axis and the diameter of the tube is about 8 cm. The image was taken with an optical-mechanical streak camera of the type AWREC5.* The full optical scanning duration was 11.4 μs. As one can see in the schematic (Figure 4.4b), the slit of the streak camera selects a streak image that is perpendicular to the tube axis. Therefore, the vertical coordinate of the picture shown in Figure 4.4a is the radial direction through the plasma and the horizontal coordinate is the time; in other words, this picture is determined by radius- and time-dependent plasma radiation, that is, by the function $E(r,t)$. The figure shows that the

* Originally published in the *Proceedings of the SMPTE International Congress on High-Speed Photography* (1962), Application of the type C5 rotating mirror framing camera, by K.R. Coleman and A. Skinner, p. 364, Figure 7. Copyright © SMPTE.

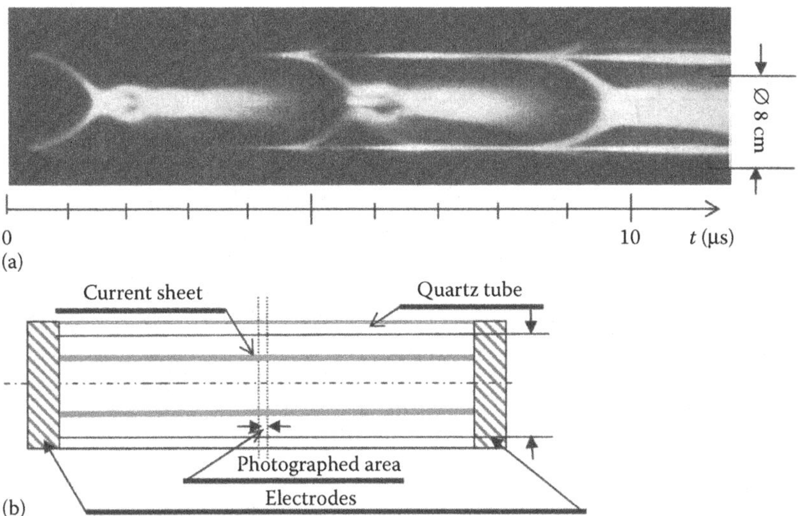

FIGURE 4.4 (a) An optically scanned streak image of a discharge in deuterium and (b) a schematic of the experimental arrangement.

process starts near the inner wall of the quartz tube where the discharge is initiated; then a thin cylindrical plasma layer is accelerated by the Lorentz force $\mathbf{j} \times \mathbf{H}$ toward the tube axis, which is confined for some time by the pressure of the magnetic field. At the end of the first half-cycle, as the current falls the plasma is no longer confined and it escapes to the walls. (The experimental arrangement is shown schematically in Figure 4.4b 1 μs after the start of the current.) The sequence is repeated every current half-cycle, but intense wall plasma radiation accompanies the discharge during later cycles.

The principle of the optical arrangement of an optical-mechanical framing camera is shown in Figure 4.5. An image $a'b'$ of the object ab that is to be photographed is projected by a pair of objective lenses L_1 and L_2 onto a plane that is perpendicular to the optical axis of the objectives and that contains the axis of rotation of a rotating mirror M. The mirror rotates the image, creating an image $a''b''$. (This would be the position in which the photographed object ab would be seen, if viewed in the direction of the arrow A). Finally, the image $a''b''$ is projected onto a fixed film P, via a small diameter lens L_3 (creating the image $a'''b'''$). The latter lens is an element of an array of identical objectives, located on an arc of a circle, centered at the point C of intersection of the optical axis of the objectives L_1 and L_2 with the axis of rotation of the mirror. The output pupil

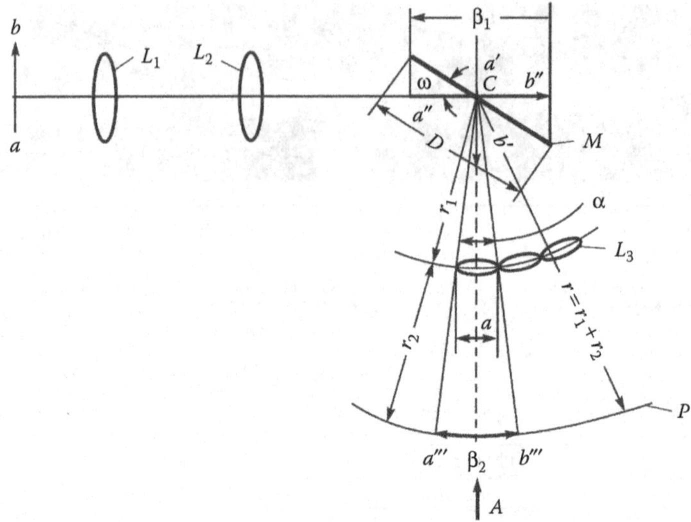

FIGURE 4.5 Principle of the optical scheme of a framing optical-mechanical camera.

plane of the objective L_1 is optically conjugated via the objective L_2 with the surface on which lie the input pupils of the objective array L_3. Thus, a light beam leaving the objective L_1 is directed toward the input pupil of L_3. The optical magnification is chosen such that all the rays emanating from L_1 would eventually fit into the aperture of L_3. There is no image that could be seen in the output pupil plane, as far as L_1 is concerned. Thus, the same is true with respect to the input pupil plane of L_3. The optical commutation is fulfilled via the rotating mirror M, which transfers just a fuzzy blob of light rather than an image from one small aperture lens L_3 onto another. Naturally, the frequency of recording the image is determined by the angular rotation velocity of the mirror and the diameter of each single objective L_3. Modern high-speed framing cameras have the capability of operating at a frequency of some 10^5–10^7 frames per second. Their principal flaw is that they have relatively low f-numbers ($D/f \approx 1/20 - 1/40$). Also there are many difficulties related to the synchronization of the start of the recording with the physical phenomena, though, to some extent, this is compensated by the large number of frames. Finally, the vast majority of optical-mechanical systems require the use of a film, which does not really kindle the passion of an experimental physicist, due to the considerable headache that is related to its further processing.

4.2 ELECTRON–OPTICAL CONVERTERS AND CAMERAS

There is no exaggeration in saying that the advent of electron–optical converters has caused a virtual revolution in the methods used to investigate high-speed processes. Indeed, their real-time high resolution of nearly 10^{-12} s (which is four orders of magnitude better than what one can achieve with optical-mechanical cameras), their near-limit* sensitivity, the simplicity of image scanning, and the possibility to record radiation with wavelengths ranging from the IR through the soft X-ray make these unique devices simply irreplaceable. Electron–optical converters are vacuum devices, in which an optical image is transformed into an electron analog, that is, an electron image, which is further amplified and projected onto a luminescent screen, where it becomes again an optical image. Currently, there are several dozen varieties of electron–optical converters available on the experimental physics market. However, each one of them would fit into one of the two following categories: time-analyzing devices and light amplifiers. Let us first become acquainted with the latter class of device, as it is the simpler one.

The structure of a typical electron-optical light amplifier is shown in Figure 4.6, where all of its components are itemized. Let us go on to discuss each component in more detail. The image that is to be studied is

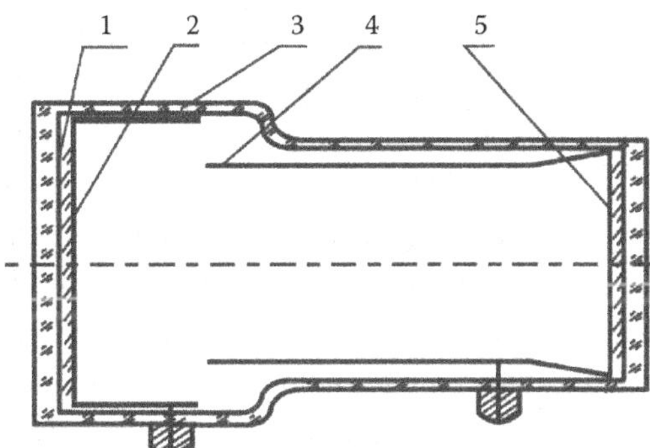

FIGURE 4.6 Typical structure of an electron-optical light amplifier: 1, the semitransparent metal substrate; 2, the photocathode; 3 and 4, the electrodes; 5, the phosphor screen. The electrode system creates the electric field for electron acceleration and for projecting the electron image onto the phosphor screen.

* The limit sensitivity is that allowing one to detect a single photon or electron. Clearly, there is no point in trying to exceed the limit sensitivity, as nature does not provide us with a half of an electron or a photon.

projected by an external lens onto a semitransparent photocathode (2), which usually sits on a semitransparent metal substrate (1). The substrate is necessary to avoid a voltage drop across the photocathode when illuminated by an intense light pulse, as the inherent resistance of the majority of photocathodes is excessively high. The electron emission from a given point on the photocathode surface is proportional to the incident light intensity at this point. In view of this, the spatial distribution of emitted electrons is

$$n_e(x, y) = \mu n_p(x, y) \tag{4.3}$$

where:

n_e is the spatial density of the emitted electrons*

μ is the quantum yield of the photocathode

n_p is the spatial density of the photons that illuminate the photocathode

Thus, apart from random fluctuations, the electrons emitted from the photocathode carry all the information about the projected optical image. The dynamic range of a photocathode is in excess of several millions. This means that the number of electrons emitted from a spatially resolvable region of the photocathode, whose area is roughly of the order β^2, is linearly proportional to the number of incident photons over a range that extends from a few emitted electrons up to several millions. Note that if the number of emitted electrons is small, the proportionality discussed earlier is determined by the photoelectron statistics and is valid only in the mean sense. Figures 4.7 and 4.8 show the spectral characteristics of various photocathodes, designed for operation in the IR and visible ranges. The curves in Figure 4.7 give the quantum yield of a silver–oxygen–cesium photocathode and are normalized with respect to the maximum sensitivity. The curves in Figure 4.8 give the dependence of the quantum yield of a multialkaline photoemission layer on the photon wavelength, without taking into account the light absorption in the conducting substrate. The light absorption of the conducting substrate is usually of the order 10%–20%. Thus, in order to find the quantum efficiency of a multialkaline photocathode with a conducting substrate, one should multiply the data on Figure 4.8 by a factor of 0.8–0.9. Hence, it is not hard to convince oneself that at the

* Strictly speaking, the generation of photoelectrons is a random process, and one should understand Equation 4.3 in the mean sense, that is, with n_e being a mathematical expectation.

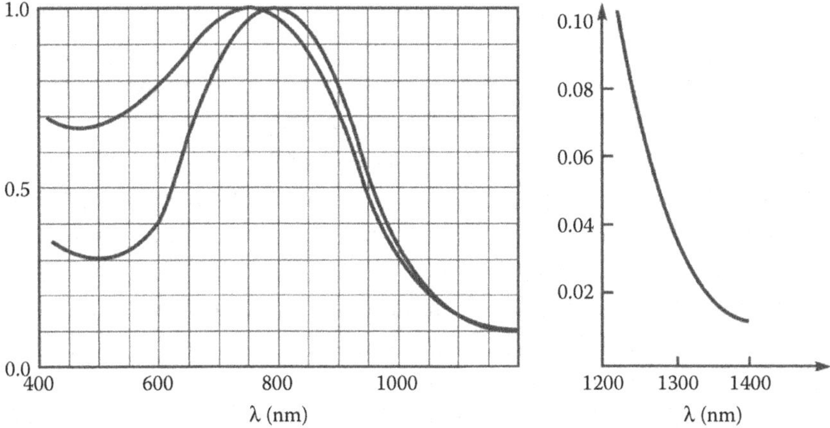

FIGURE 4.7 Quantum yields for a silver–oxygen–cesium photocathode (left – range 400–1200 nm) and (right – range 1200–1400 nm).

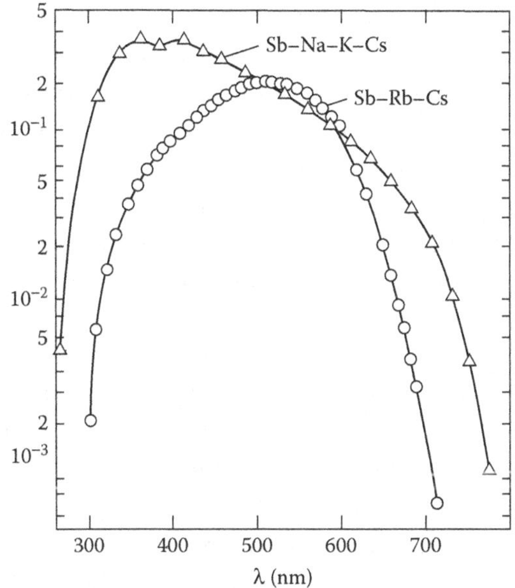

FIGURE 4.8 Yields for a multialkaline photocathodes.

maximum spectral sensitivity, it would take typically three photons to produce a single electron. This is the best that one can possibly get, while in reality the photocathode in an average commercially available tube is considered good if its quantum yield is in the range 0.15–0.2. In this case, one needs typically five or seven photons to knock out a single electron.

The screen of an electron–optical converter (5 in Figure 4.6) is a composite part of the anode and sits on the wall of the device's vacuum vessel, opposite to the cathode. It serves to convert the electron image into an optical image. Screens of electron–optical converters, apart from being characterized by a higher resolution ($\beta \approx 1.25–2 \times 10^{-2}$ mm), are not very different from those of television sets and computer displays. These screens consist of a prompt luminophore emitting blue–violet radiation ($\lambda_{max} \approx 450$nm; $\Delta\lambda \approx 60$nm), which corresponds to the maximum sensitivity of the majority of radiation detectors. A 50–100-nm-thick aluminum film is vacuum deposited over the luminophore. A film with this thickness is essentially transparent for an electron with energy around 10 keV, but it is practically opaque to light. Its role is twofold: First, it creates an electric field in the screen's vicinity, and second, it reflects the light of the luminophore toward the radiation detector, thus enhancing the efficiency on the one hand and screening the photocathode from parasite secondary illumination on the other. The screen's energy efficiency* is of the order of 10%–15% in the electron energy range 10–20 keV. Incidentally, a potential difference of 10–20 keV is just about the right operational value for a single-cascade electron–optical converter. Thus, every electron whose energy lies within the above range gives rise to 300–1000 photons, the exact number being dependent on the supply voltage and the efficiency of the screen. The average energy of these photons turns out to be about 2.75 eV, which corresponds to the blue–violet visible light range, coinciding with the maximum of the spectral density distribution of the luminophore radiation.

To summarize, an electron analog of the original optical image that was projected onto the photocathode is formed on the photocathode surface. The next task is to carry it all the way to the screen of the electron–optical converter. This is accomplished by means of electron optics, that is, a special electric field configuration, which ensures that all the electrons that are emitted from a single point on the photocathode are delivered to a single conjugate point on the screen. Note that the equipotential surfaces of the electric field play the same role in electron imaging as do the surfaces of a lens in traditional optics in changing the direction of propagation of a beam of light. Indeed, if the velocity vector of an electron forms

* The screen's energy efficiency is defined as the ratio between the power of the photons emitted by the screen and the power of the electron beam incident onto the screen [W/W]. In the pulsed operation regime, the same quantity is also defined as the ratio of the total energy of the emitted photons to the total energy of the incident electrons over the duration of the light pulse, which is being recorded [J/J].

a nonzero angle with the direction of the electric field, then there exists a force proportional to the transverse component of the electric field, which changes the direction of the electron's trajectory. For a given potential, the larger the angle, the greater is this force. The electric field in question, necessary to carry the electrons from the photocathode to the screen, is produced by a series of electrodes, numbering only two in the simplest case. These electrodes are marked as 3 and 4 in Figure 4.6.

An obvious question is whether it is possible to do without complex geometric electric field configurations, being simply content with a uniform longitudinal electric field, which is naturally produced by applying an accelerating voltage between a flat photocathode and the screen. Let us estimate in this case the characteristic size of the circular spot on the screen that results from the emission of electrons from a single point of the photocathode. It is well known that the photocathode emits the electrons isotropically in all directions. For a typical multialkaline photocathode, bombarded by visible light, the maximum of the energy probability distribution occurs around $\mathcal{E}_0 \approx 0.5$ eV. Let an "average" electron, while being accelerated by a longitudinal electric field directed from the photocathode toward the anode (screen), move uniformly in a transverse direction with a velocity corresponding to the above probability distribution maximum. Based on these assumptions, it is not difficult to estimate the radius of the circular spot $r = 2S\sqrt{\mathcal{E}_0/\mathcal{E}}$, where S is the distance between the cathode and the anode, $\mathcal{E}_0$ is the initial energy of the emitted electron, and $\mathcal{E}$ is its final energy. The distance S is typically of the order of 10 cm, and thus for $\mathcal{E}_0 = 0.5\,\text{eV}$ and $\mathcal{E} = 10$ keV, $r \approx 1.4$ mm, and this is absolutely unacceptable. What can we do to improve the situation? First, we can decrease S down to, say, 5 mm. This reduces the radius of the spot down to about 0.14 mm. This is yet not a stunning achievement, but it already allows one to use the device, and indeed such devices, usually referred to as biplaners, exist. Despite their alluring simplicity, they have not gained extreme popularity due to their small amplification factor, a relatively high parasite noise level, and a modest spatial resolving power. Another possibility is to use a longitudinal magnetic field parallel to the electric one. This would cause the electrons to follow spiral trajectories, completing one step of a helix over a time interval $\Delta t = 2\pi/\omega$, with $\omega = eH/mc$ being the well-known Langmuir frequency. Note that the latter does not depend on the electrons' velocity. Thus, by choosing the right ratio between the electric and magnetic fields to ensure that it takes an electron exactly the time Δt (equal to the helix period) to travel from the cathode to the screen, one can guarantee that all the electrons emitted at a specific point on the photocathode will also arrive at a single point on the screen, their trajectories having completed a full circle in the transverse direction. Such devices used to be widely used, despite their elephantine dimensions, their excessive weight, and the necessity to stabilize the accelerating voltage and the magnetic coils current to an accuracy of 0.1%. At the time when these devices were in vogue, it did not appear possible to achieve a comparable focusing quality in purely electrostatic devices. However, as electrostatic focusing technologies evolved, and their spatial resolution was improved considerably, the magnetic focusing devices became extinct like dinosaurs. A decisive role in the improved quality of electrostatic focusing devices was played by the arrival and proliferation of fiber optics technology. The heart of the matter is that in simple electrostatic devices, similar to that shown in Figure 4.6, it is not possible to have equally good spatial resolution both in the center of the field of vision and at the periphery. In other words, the planes of the photocathode and the screen are conjugate only in the region close to the axis. However, there always exist surfaces (unfortunately not planar), which can be optically conjugated via an electric field of a relatively simple configuration. This is, of course, not a new invention; even as long as 70 years ago, the photocathode of the device RCA C73435 was formed on the surface of a sphere. However, there was the problem of how to project the optical image onto the spherical surface without getting geometrical distortions and inhomogeneity of the spatial resolution within the field of vision. A general solution to all the problems listed earlier was the invention of the so-called fiber-optical pack. This is used as an input and/or output window of an electron–optical converter, and it consists of a bundle of millions of extremely densely packed optical fibers, aligned along the optical axis of the device (Figure 4.9). A single step d of this matrix structure is essentially the outer diameter of a separate double-layered optical fiber. In modern devices, d is typically 5–10 μm. It is clear that an image, carried by such a structure, will consist of separate pixels, whose brightness diameter is 3–8 μm and the pixel density is of the order of $1–4 \times 10^6$ cm^{-2}. That is to say, there are some 2000 × 2000 pixels per square centimeter. Let us try to get a feel for this number. To compare, note that the number of pixels in a single high-resolution television frame is 10 times smaller. However, for our purposes, it is in fact much more important that $d \ll \beta$. After packing the fibers into a bundle and baking them together, the pack becomes similar to ordinary glass, at least as far as grinding and polishing are concerned. Hence, one of the pack's surfaces, onto which one is going to project the optical image, remains plane, whereas the other, onto which

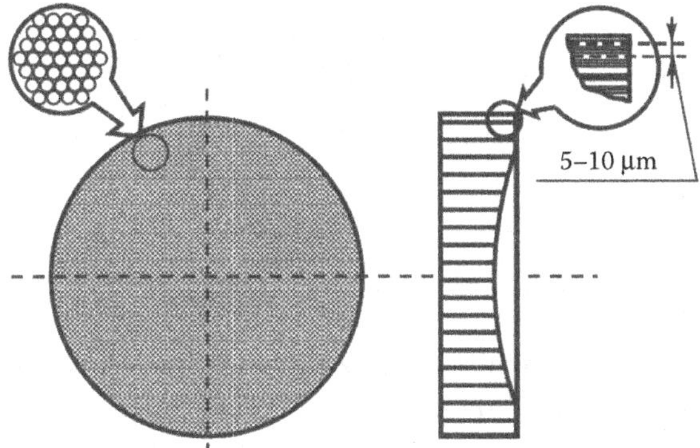

FIGURE 4.9 Fiber-optical pack.

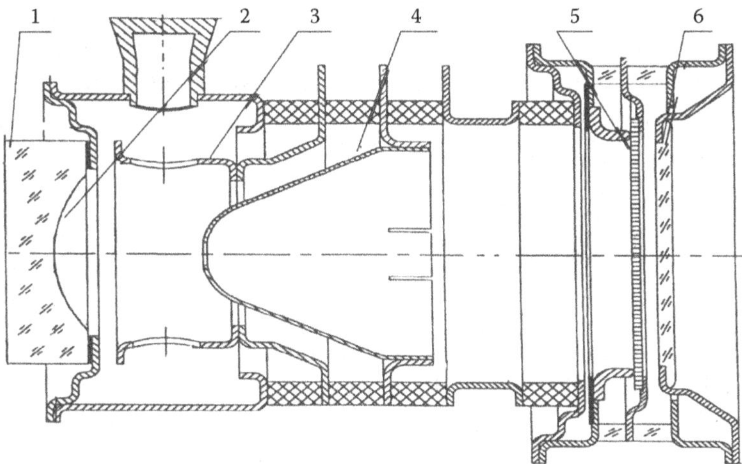

FIGURE 4.10 Light amplifier PMU 12 × 18: 1, the fiber-optical input window; 2, the photocathode; 3, the focusing electrode; 4, the anode; 5, the microchannel amplifier; 6, the output window.

the photocathode will be formed, is made spherical. Of course, the construction of modern light amplifiers has evolved far beyond the primitive scheme shown in Figure 4.6, as can be easily seen after even a casual look at Figure 4.10, which illustrates a real light amplifier PMU 12 × 18, manufactured by the company BIFO Industries Ltd. Moscow, Russia, (BIFO), and which contains this type of fiber-optical input window. Figure 4.11 shows the basic vacuum device and the same device when packaged in a container. The container is filled with a high-voltage insulating compound, and it contains a high-voltage transformer, a voltage divider, and other electronics

FIGURE 4.11 The light amplifier PMU 12 × 18: (left) the whole commercial unit housed in a container, (right) the electro–optical converter.

components as well as the electron–optical converter. Ultimately, the device operates in such an easy way that usually it is necessary only to connect a single pair of wires to a three-volt battery.

The amplification factor of an electron-optical image intensifier (which is the ratio of the number of photons emitted by the screen luminophore to the number of photons incident upon the photocathode) can be computed by the following equation:

$$\eta = \mu\varphi \frac{eU}{h\bar{\nu}_{\text{out}}} \tag{4.4}$$

where:

- μ is the quantum yield of the photocathode
- φ is the energy efficiency of the screen
- e is the charge of an electron
- U is the potential difference between the photocathode and the anode
- $h\bar{\nu}_{\text{out}}$ is the photon energy averaged over the spectral distribution of the photons that are emitted by the screen

Taking $\mu = 0.2$, $\varphi = 0.125$, $U = 10$ kV, and $h\bar{\nu}_{\text{out}} = 4.4 \times 10^{-19}$ J, one finds $\eta \approx 90$. This would be excellent, but this quantity usually lies within the range 30–50 in a typical single-cascade light amplifier. This would not be too bad either, but we have to relay the image from the screen onto a photographic recorder, for example, a film. As was mentioned already, the screen radiates isotropically in all directions, and it is easy to see that a reproduction objective lens with a typical *f*-number $D/f = 1/3$ would intercept approximately one-hundredth of this light stream! In other words, the net result is not the amplification, but rather the reduction of the light. What should one do? Today, there are two possibilities. The first is to use multicascade devices. In these devices, there are several identical amplification stages, separated from one another by a thin film (usually made of mica and some 5–10 μm thick), within a single vacuum vessel. A screen is placed on one side of the transparent layer with a photocathode on the other so that in the whole sandwich-like construction the screen for one amplification stage is combined with the photocathode for the next stage. Owing to its extreme thinness, there are practically no losses of light or spatial resolution within the mica film. Certainly, this does not mean that the transfer function or the transfer ratio of a multicascade device would eventually be the same as they are for a single stage, because an image passing through the device is formed not just by one, but by many consecutive electron lenses. Thus, if the transfer ratio of a single amplification stage is $K_i(\omega)$, then the overall transfer ratio of the n-cascade device will be $K_\Sigma(\omega) = K_1(\omega), K_2(\omega),\ldots, K_n(\omega)$ In the same way, the overall amplification factor will also equal the product of the individual amplification factors for each stage. Then, for a two-stage amplifier, the amplification coefficient $\eta_\Sigma = \eta_1 \times \eta_2$ will be approximately 2.5×10^3.

The other way to increase the amplification factor is to use microchannel plates to multiply the electrons inside the device. A microchannel plate consists of millions of individual electron multipliers that are densely packed together with their longitudinal axes perpendicular to the plane of the assembly. A single microchannel electron multiplier is shown schematically in Figure 4.12. Such a device is often referred to as an electron multiplier with evenly distributed dynodes. It consists of a hollow tube with an outer diameter of 5–10 μm and a length of 0.3–0.5 mm. The inner surface of the tube is coated with a special chemical in order to make it weakly conducting. As a result, a potential applied to the ends of the tube distributes itself evenly along the length, and this ensures the uniformity of the corresponding electric field that is directed along the axis of the tube.

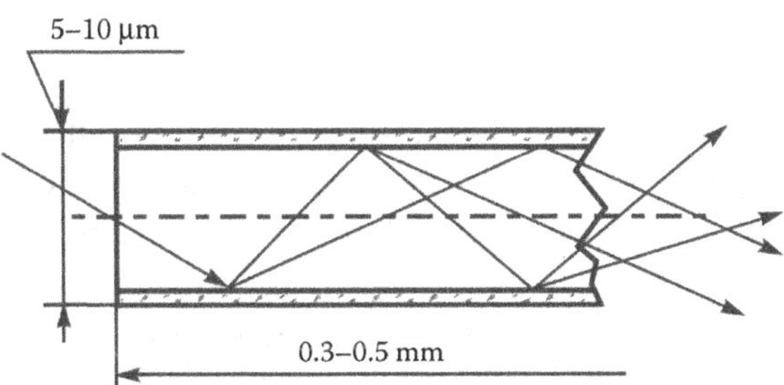

FIGURE 4.12 A microchannel electron multiplier.

Electrons inside the tube will be accelerated by the electric field along the direction of the axis. However, the electrons will frequently collide with the inner surface of the walls of the tube due to their transverse component of velocity and the small diameter of the tube. The applied accelerating electric field can be chosen so that during a single step between two consecutive collisions with the wall, an electron will acquire enough energy to cause secondary electron emission. If the secondary electron emission coefficient is greater than one, the number of electrons inside the tube will grow. The value of the electron multiplication factor, or the magnification coefficient, depends on the applied exterior voltage. For applied voltages in the range 0.5–1 kV, the electron multiplication factor will be in the range from 1 to 10^4. Using the whole packed plate structure rather than a single microchannel multiplier, the electron image is amplified before it reaches the screen of the electron-optical image intensifier. As one can see from Figure 4.10, which shows the arrangement in the PMU 12 × 18 device, a microchannel plate is installed just in front of the screen. A voltage is applied between the output face of the microchannel plate and the aluminum coating of the screen to accelerate the electrons leaving the plate up to between 5 and 7 keV; this ensures high-energy efficiency on the screen. Clearly, in this case the electron image is projected onto the input face of the multichannel plate, rather than onto the screen itself. The light amplifier PMU 12 × 18 is a typical example of this class of modern devices. Its characteristic parameters are rather high: $\beta < 20\ \mu m$ all along the input pupil of diameter 12 mm and amplification factor $\eta = 2 \times 10^4$. Once again, the fact that the amplification factor $\eta = 2 \times 10^4$ implies only that the luminosity of the output screen is 2×10^4 higher than the input photocathode illumination. However, the information content of the screen luminescence is certainly independent of the amplification factor. Instead, it is determined by the number of photons that fall on the photocathode throughout the duration of the exposure, its quantum yield, and the statistics of the primary electrons. All these issues will be discussed later.

Time-analyzing electron–optical converters differ from light amplifiers only by the presence of deflecting plates, whose purpose is to scan the electron image. A time-varying electric field between these plates is essentially an analog of the revolving mirror in optical-mechanical devices. In electron-optical analogs of streak cameras, the voltage on the deflecting plates has a linear time dependence, similar to that in oscilloscopes. It produces the linear motion of the image strip along the screen of the electron–optical converter. In electron-optical framing cameras,

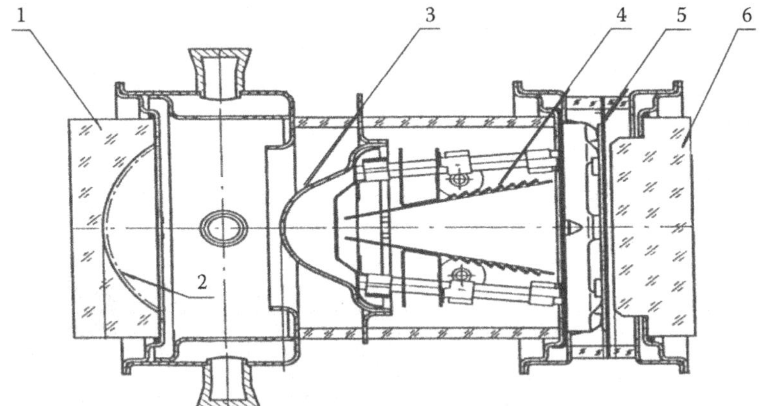

FIGURE 4.13 Time-analyzing electron–optical converter PV201: 1, the fiber-optical input window; 2, the photocathode; 3, the anode; 4, the deflecting plates; 5, the microchannel plate; 6, the fiber-optical output window.

the pulses applied to the plates have a rectangular shape, which makes the electron image jump from one location on the screen to another. The total number of frames in devices of this sort is not extremely high and rarely exceeds 16. However, it is possible to synchronize the images with the recorded phenomena in such a precise way that it generally compensates for the small number of frames.

As an example, let us discuss the principles of the time-analyzing electron–optical converter PV201, whose main components are shown in Figure 4.13. It is not hard to see that the general features of the design of this device, apart from the presence of the deflecting plates, are in general the same as in the light amplifier that was described already (Figure 4.10). The main characteristic parameters of the device PV201 are summarized in Table 4.1. Here one should note that this device does not have the highest possible time resolution. Today, the electron–optical camera K005, which incorporates an electron–optical converter PV-001, has maximum possible time resolution. The time resolution of this camera approaches 0.3 ps when a voltage pulse is applied to increase the electric field near the photocathode surface* up to 24 kV cm^{-1}. It does not appear possible to maintain permanently an electric field with such intensity near the photocathode. That is why the voltage to produce the above electric field is applied as a short pulse with duration shorter than the time that would

* It will be shown in Chapter 4 that the time resolution of an electron–optical converter is determined by the electric field intensity near the photocathode surface.

TABLE 4.1 Principal Parameters of the Electron–Optical Converter PV201

Parameter	Value
Time resolution	10 ps
Spectral sensitivity range	0.4–0.8 μm
Operating photocathode size	20 × 15 mm
Transfer function FWHM β	<50 μm
Amplification factor	10^4

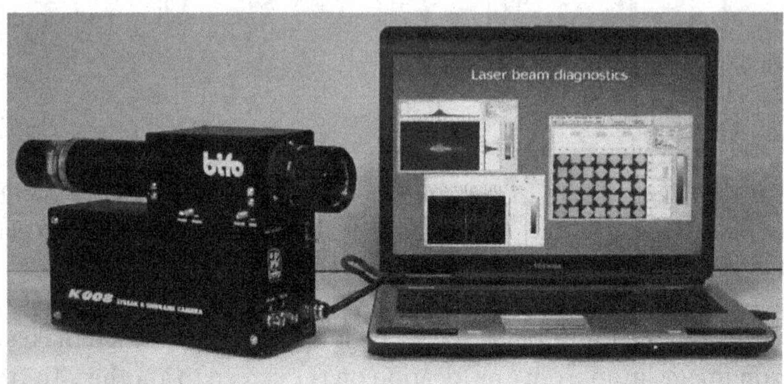

FIGURE 4.14 Universal electron-optical camera K005.

allow an electric breakdown to develop. The exterior view of the new camera K008 is shown in Figure 4.14. This universal camera can operate both in the streak mode with a time resolution of up to 50 ps and in the single-frame mode with a frame exposure of 640 ms–10 ns. The camera is equipped with a light amplifier with a microchannel plate, which is especially important for dealing with low-illumination objects. This light amplifier has an aperture of 40 × 40 mm². The output image is read by a charge-coupled device (CCD) camera* connected to a computer, which avoids the inconvenience of using a photographic film and considerably speeds up the image processing.

It is well known that the transfer function of an electron-optical converter can be well approximated by a Gaussian:

$$g(x)=\frac{\alpha\sqrt{2}}{\sqrt{\pi}}\exp\left[-(\alpha x)^2\right] \tag{4.5}$$

The FWHM for such a transfer function is $\beta=2\sqrt{\ln 2/\alpha^2}$.

* The principles of the design and operation of CCD cameras will be discussed in detail in Section 4.3.

Using the relation (2.25), we get the transfer ratio

$$K(w) = \exp\left[-(\omega/2\alpha)^2\right] \tag{4.6}$$

and the frequency–contrast characteristics*

$$K(\nu) = \exp\left[-(\pi\nu/\alpha)^2\right] \tag{4.7}$$

Now, according to Equation 2.57, it is possible to compute the power loss coefficient:

$$\mathcal{K} = \exp\left(-\pi^2\nu^2/3\alpha^2\right) \tag{4.8}$$

In the two-dimensional case with no astigmatism, in view of Equation 2.67,

$$g(r) = \frac{\alpha^2}{\pi}\exp\left[-(r\alpha)^2\right] \tag{4.9}$$

and hence the transfer ratio

$$K_2(\omega_x,\omega_y) = \exp\left(-\frac{\omega_x^2+\omega_y^2}{4\alpha^2}\right) \tag{4.10}$$

the frequency–contrast characteristics

$$K_2(\nu_x,\nu_y) = \exp\left[-\frac{\pi^2\left(\nu_x^2+\nu_y^2\right)}{\alpha^2}\right] \tag{4.11}$$

and the power loss coefficient:

$$\mathcal{K}_2 = \exp\left(-\frac{\pi^2\nu_c^2}{2\alpha^2}\right) \tag{4.12}$$

It has been already mentioned that with low illumination the main noise component of an electron-optical device arises from quantum noise on the input photocathode. Given an average photocathode illumination $\bar{E}$ and quantum efficiency μ, the average number of electrons emitted from an element of the photocathode surface area $\Delta^2 = \Delta x\Delta y$ during an exposure time τ will be $N_e = \bar{\mathcal{E}}\tau\Delta^2\mu = \mathcal{E}\Delta^2\mu$. Therefore, if the successive image counts are taken with a step interval of Δ, the signal-to-noise ratio

* It should not be difficult for the reader to convince himself or herself that the FWHM of the functions $K(\omega)$, $K(\nu)$, and $g(x)$ are related by $\beta_\omega = 2\pi\beta_\nu = 8\ln 2/\beta_x$. Then $\beta_x \approx 0.8825/\beta_\nu$.

at the system input will be $L = N_e / \sqrt{N_e} = \Delta(\bar{\mathcal{E}}\mu)^{1/2}$.* The contrast at high spatial frequencies will drop at the system output due to the finite width of the transfer function. Hence, the signal-to-noise ratio, denoted by L^*, can be computed using Equations 2.56 and 4.12:

$$L^* = \Delta(\bar{\mathcal{E}}\mu)^{1/2}\mathcal{K}_2 = \Delta(\bar{\mathcal{E}}\mu)^{1/2}\exp\left(\frac{\pi^2\nu_c^2}{2\alpha^2}\right) \tag{4.13}$$

The reading step Δ is bounded on the one hand by the Kotelnikov–Shannon theorem requirement that $\Delta \le 1/2\nu_c$ and on the other hand by a reconstruction error $\delta\bar{\mathcal{E}} / \bar{\mathcal{E}} \ge 1/\Delta(\bar{\mathcal{E}}\mu)^{1/2}\mathcal{K}_2$, whose dependence on Δ is complicated because $L^* = f(\Delta,\beta)$. Thus, it is reasonable, after the substitution of $\nu_c^2 = 1/4\Delta^2$ and $\alpha^2 = 4\ln 2/\beta^2$, to represent the relationship indicated by Equation 4.13 as

$$L^* = \Delta\sqrt{\bar{\mathcal{E}}\mu}\exp\left[-\frac{1}{2\ln 2}\left(\frac{\pi\beta}{4\Delta}\right)^2\right] \tag{4.14}$$

As a result, the density of the information flow transmitted through the electron-optical channel is as follows:

$$R = \frac{1}{\Delta^2}Q_n\log_2\left\{\Delta\sqrt{\bar{\mathcal{E}}\mu}\exp\left[-\frac{1}{2\ln 2}\left(\frac{\pi\beta}{4\Delta}\right)^2\right]\right\}\left[\text{bit cm}^{-2}\right] \tag{4.15}$$

[Recall that $Q_n = Q_n(L^*)$ is a slowly varying function, depending on the whole ensemble of signals and noise, which in this case are Poisson noise and the $E_{in}(x,y)$, characterized by the uniform probability density function $p(X) = 1/E_{max}$ = const.] Hence, in one dimension,

$$L^* = \sqrt{\bar{\mathcal{E}}\mu\Delta x\Delta y}\exp\left[-\frac{1}{3\ln 2}\left(\frac{\pi\beta}{4\Delta}\right)^2\right] \tag{4.16}$$

$$R = \frac{1}{\Delta}Q_n\log_2\left\{\sqrt{\bar{\mathcal{E}}\mu\Delta x\Delta y}\exp\left[-\frac{1}{3\ln 2}\left(\frac{\pi\beta}{4\Delta}\right)^2\right]\right\}\left[\text{bit cm}^{-1}\right] \tag{4.17}$$

Let us use the above concepts and equations to analyze a real-life situation, arising in connection with recording an image under the circumstances of weak light fluxes. Suppose, as is usually the case, that we know

* A more accurate consideration, taking a variety of additional issues into account and primarily the amplification factor dispersion should use the "generalized quantum yield" $\mu_\Sigma = \mu\eta$, rather than μ, where η is on the order of 1.

the aggregate energy incident upon the photocathode over the time of the exposure, as well as the photocathode's surface area. Let us assume that the amplification coefficient of the electron-optical device is at the limit, that is, the amplification is sufficient to enable one to record a single photoelectron at the output. The transfer function FWHM $\beta \approx 10^{-2}$ cm in such devices, and this makes it clear that the reading step should be taken with the same order of magnitude, that is, let $\Delta \sim 10^{-2}$ cm. What is the value of the ratio Δ/β that would enable us to transfer and read from the screen a maximum amount of information? Consider this problem using the above ratio as a parameter. Let us assume that near a maximum of $R^* = f(\Delta)$, the variable $Q(L^*)$ does not depend on Δ (it will not be difficult to convince ourselves that this is true in a moment, by comparing the curves in Figures 2.4 and 4.15). A maximum of Equation 4.15 is found by requiring that $R'(\Delta) = 0$, which yields

$$\left(\frac{\Delta}{\beta}\right)_{\text{opt}} = \sqrt{\frac{\pi^2}{8\left[2\ln\left(\Delta\sqrt{\bar{\mathcal{E}}\mu}\right)-1\right]\ln 2}} \tag{4.18}$$

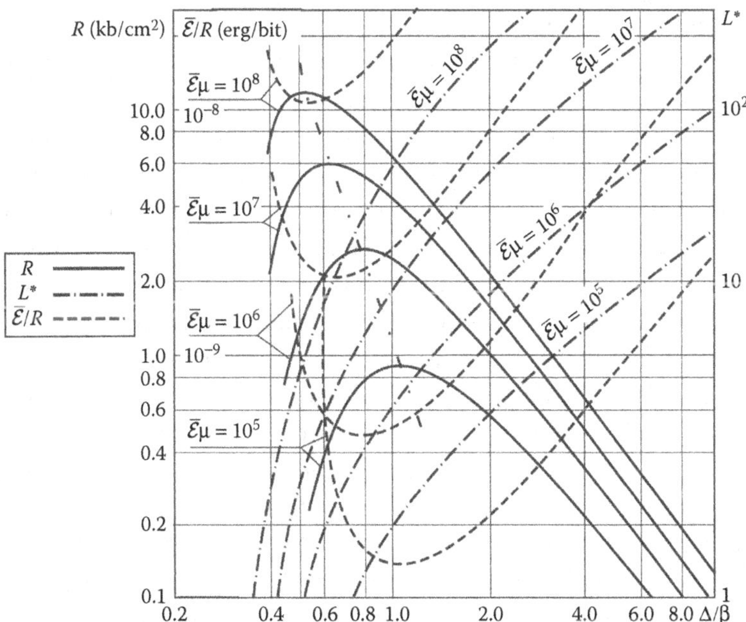

FIGURE 4.15 Dependence on the amount of information transmitted through an electro-optical converter versus the ratio Δ/β.

and this implies, for instance, that if the input signal provides, say, $\mathcal{E}\mu = 10^5\,\text{cm}^{-2}$ primary photoelectrons per square centimeter of the photocathode, then the maximum amount of information can be read off the output screen of the electron-optical image intensifier only if the ratio Δ/β is chosen to be equal to 1.1 (or 0.76 if $\bar{\mathcal{E}}\mu = 10^6\,\text{cm}^{-2}$).

To study the behavior of the other parameters that are characteristic of the image, let us investigate numerically the dependence of the quantities L^*, R, and $\bar{\mathcal{E}}/R$ on the ratio Δ/β. The values of these quantities in the range $0.4 < \Delta/\beta < 10$ are shown in Figure 4.15 for values of $\bar{\mathcal{E}}\mu = 10^5, 10^6, 10^7$, and $10^8\,\text{cm}^{-2}$. The dependence of the energy equivalent $\mathcal{E}/R$ [erg per bit] of the ratio Δ/β given in this figure has been calculated with the assumption that the average photon energy $h\nu = 2$ eV and that the photocathode quantum yield $\mu = 0.3$. As confirmed by the data in question, even under conditions of extremely low illumination, it is possible to transmit 1 kb cm^{-2} through such an electron-optical channel as long as the value of Δ/β is close to the optimum, though a single element of the output image would contain on average approximately only one bit of information. Thus the signal-to-noise ratio at the channel input lies in the range 3–8 and the signal-to-noise ratio at the channel output lies in the range 2–4 because only 100 primary photoelectrons fall on an element of the image surface area Δ^2. It is true also on average. It is worth mentioning that as the input signal increases, the energy equivalent of the information also increases, but a one order of magnitude increase of the input illumination would increase the information contained in the output image by at most a factor of three. Intuitively, this has always been understood, as reflected by a heuristic rule of photographic exposure, prescribing it to continue until the film's optical density reaches $\bar{D} \approx 1$. The fact that a further decrease of Δ after the function $R(\Delta/\beta)$ has reached its maximum results in a catastrophic decline of the channel transmission rate can potentially cause confusion. It might seem that if Δ is a single step size of reading/recording decided by an investigator, it has nothing to do with the information characteristics of the data transmission channel. Of course, the crux of the matter lies not with the man-made quantity Δ, but with the fact that by introducing the ratio Δ/β, we have decided *a priori* that according to Shannon's theorem $\Delta = 1/2\nu_c$. Hence, on the physical level, the quantity Δ/β not only characterizes the ratio of the reading/recording step size Δ to the transfer function FWHM β, but also determines the ratio of the transmitted signal spectrum width $\nu_c = 1/2\Delta$ and the spectral width of the information transmission channel, which is known to be proportional to 1/β. Then it seems natural that if one attempts to squeeze into a channel a signal, whose spectrum is wider than the spectral width of the channel itself, the data transmission rate will drop substantially.

So far, we have obtained the equations that enable us to take into account the transfer function and to compute the density of the information flow that is transmitted through a real electron-optical channel with noise, depending on the input illumination and the method of reading the output signal. These equations are indeed important and useful for practical applications. In concrete situations, they can be used to find optimal conditions for the reception and recording of optical images, as well as to obtain an output-based accuracy estimate for the reconstruction quality of the input image. However, there is still a need for one extra parameter, which would not only give an unambiguous characterization of our measuring–recording devices, but also allow a comparison between different ones. To fulfil this last step, let us return to Figure 4.15 and pay attention to the following two circumstances. Increasing the input signal intensity naturally causes the density of the transmitted information flow to increase. As this happens, the value of $(\Delta/\beta)_{opt} \rightarrow 0.5$. This implies that given the transfer ratio $K(\omega)$ and the noise occurring within the system, which are characterized in terms of their entropy power $\bar{\sigma}^2$ (or variance σ^2 if it is normally distributed), then the highest density of the information transmitted, that is, the *transmission capacity* of the channel, is determined by the limiting value E_{max} of the input illumination. Taking account of the fact that $(\Delta/\beta)_{opt} \approx 0.5$ for a high input illumination, we can use Equation 4.15 to obtain the transmission capacity of an electron-optical image intensifier as follows:

$$C = \frac{4}{\beta^2} Q_n \log_2 \left\{ \frac{E_{max}}{2\sigma} \exp\left[-\frac{1}{2\ln 2} \left(\frac{\pi}{2} \right)^2 \right] \right\} \left[\text{bit cm}^{-2} \right] \tag{4.19}$$

It is essentially this parameter, together with the equivalent information energy, which gives the most complete characteristics of an optical measuring–recording system, whose transmission capacity is determined by the presence of an electron–optical converter. The signal-to-noise ratio $E_{max}/2\sigma$ in electron-optical image intensifiers with microchannel plates is usually smaller than 10^2. However, the ratio $E_{max}/2\sigma$ can be as large as 10^3 in devices that do not use microchannel plates and the transfer function FWHM for such devices is approximately 50 μm. Then it is not difficult to calculate that the transmission capacity of such an electron-optical image intensifier will be 10^6 bit cm^{-2}, which is equivalent to 1.25×10^5 bytes cm^{-2}.

One should notice, however, that in this case the quantity $\bar{\mathcal{E}}$ will increase to ~0.2 erg cm^{-2}, thus approaching a value that is comparable to a photographic film; in the same way, the energy equivalent increases up to ~2×10^{-7} erg per bit. In general, it is not a good idea to exploit any electron-optical device in such regimes where they are fulfilling their utmost transmission capacity. However, such a parameter is irreplaceable for the purpose of drawing a comparison between a pair of different devices, only for the reason that any complex alternative, characterizing the devices in question from the viewpoint of information theory, simply does not exist.

One usually understands the time resolution of an electron-optical device to be the maximum time resolution that can be reached in the streak mode. The time resolution of an electron-optical device is determined by two key factors: First, there is the spatial resolution, primarily established by the FWHM of the transfer function β_c, which determines the positional accuracy and therefore the precision of timing markers on the device's screen. Second, there is the quantity β_t, which is related to the blurring of the bunch of electrons as it travels between the photocathode and the screen. The point is that all the electrons emitted by the same region of the photocathode at the same moment in time will not arrive at the screen simultaneously due to the distribution of their initial velocities. Recall that if, for instance, a multialkaline photocathode is bombarded by visible light photons, the energy distribution function for the electrons has a maximum around $\mathcal{E}_0 \approx 0.5$ eV. Also recall that as the distribution is Poissonian, its variance is of the same order. Let us pay attention to one more factor. While the positional accuracy can be improved by a mere increase of the scanning velocity of the electrons over the screen, nothing can really be done about the quantity β_t that characterizes the original velocity distribution of the electrons. Let us estimate the time resolution $\Delta t \approx \beta_t$ of an electron-optical device, assuming $\beta_c \ll \beta_t$. Thus, the electrons that left the same point of the photocathode at the same instant will require different times to travel the same distance S between the photocathode and the screen. This difference is caused by the spread of initial velocities $v_{01} = v_0 + \sigma_v$ and $v_{02} = v_0 - \sigma_v$, where σ_v^2 is the velocity dispersion. Let $\Delta t = t_2 - t_1 = 2\sigma_t$, $t_1 = t - \sigma_t$, and $t_2 = t + \sigma_t$, where σ_v^2 is the dispersion of the transit time. Then $S_1 = (eE/2m)t_1^2 + v_{01}t_1$ and $S_2 = (eE/2m)t_2^2 + v_{02}t_2$. Equating the right-hand sides of these equalities (since $S_1 = S_2 = S$ is the distance between the photocathode and the screen), we obtain $(eE/2m)(t_2^2 - t_1^2) = v_{01}t_1 - v_{02}t_2$. Taking into account the fact that $t_1 \approx t_2 \approx t$ to a good deal of accuracy, and as we have agreed $\Delta t = t_2 - t_1$, we find that

$(eE/m)t\Delta t = v_{01}t_1 - v_{02}t_2 = v_{01}t_1 - v_{02}t_2 - v_{02}t_1 + v_{02}t_1 = \Delta vt - v_{02}\Delta t$. On the right-hand side of the equation, $\Delta vt >> v_{02}\Delta t$, since Δv is of the order v_{02}, while $t >> \Delta t$. Thus, we finally obtain

$$\Delta t = \frac{m\Delta v}{eE} \approx \alpha \frac{10^{-11}}{E} \tag{4.20}$$

where:

E should be expressed in centimeter–gram–second (CGS) units

This equation was proposed by L. A. Artsimovich. The coefficient α is in the range 1–5 with the precise value determined by the specific type of photocathode used as well as the energy of the incident light photons. Assuming $\alpha = 3$ and $E = 30$ kV cm^{-1}, we find that $\Delta t = 3 \times 10^{-13}$ s. It seems that this is the best time resolution that can be achieved technically in instruments of this kind.

4.3 CCD AND CMOS STRUCTURES CAMERAS

There have been numerous attempts to replace photographic film, which is used in scientific research not so much as a detector but mainly as a long-term memory facility for storing optical information. One example was a relatively recent massive attack against film, attempting to substitute it with a closed-circuit television camera with the image recorded on magnetic tape. However, the processing effort that was required was not significantly less than that in the case of a photographic film, and there existed a number of obstructing factors that diminished the quality of the experimental data. Among these factors were insufficient accuracy, considerable nonlinear distortions, positioning difficulties (i.e., making a connection between the whole image or a fragment of it and the laboratory coordinate frame), the necessity of digitizing the image originally stored in analog form, and a variety of technical difficulties related to interline coding. Not only did it keep all the headaches, but it also eventually reduced the quality. The breakthrough came only when B. Boyle and J. Smith suggested to use solid semiconductor devices, or so-called charge-coupled devices, and further developed the concept of charge transfer devices. The principle of the CCD operation consists in accumulation of electric charges in capacitors with in the metal oxide semiconductor (MOS) structure and transfer of these charges to a device that measures the magnitude of each charge. The charges are guided by special voltage pulses through a sequence of analogous MOS capacitors. By means of modern microelectronic technology, hundreds of thousands of such capacitors can be formed on a single semiconductor substrate or "chip," yielding an orthogonal structure, called a CCD matrix. Simultaneously, a thin (0.1–0.15 μm) dielectric layer (usually an oxide) is formed on the semiconductor substrate (say, of p-type conductivity), where the conducting electrodes are placed (usually made of metal or polycrystalline silicon). A voltage is applied to some of the electrodes creating an electric field within the MOS structure. This causes the main conductivity carriers (holes) to move away from the surface of the semiconductor, thus creating in the near-surface area a potential well for the electrons. A typical orthogonal CCD matrix contains between 10^5 (256×512) and 2×10^6 (1024×2048) cells (pixels). The linear dimension of an individual pixel varies in different CCD arrays from 5 to 15 μm, whereas the spacing between a pair of neighboring pixels is 0.1–1.0 μm. As a result, the characteristic size of the whole CCD matrix rarely exceeds 1 cm. Every cell contains several MOS structures, which detect the charge accumulation due to the photoelectric effect, store the charge, and further transport it. Up to a certain limit, in the linear approximation, the magnitude of the charge accumulated in each cell is proportional to the exposure, that is, to the number of photons that have landed on the light sensor during the exposure time. Thus, the charge distribution within the whole CCD matrix is proportional to the distribution of the illumination of the projected image. Measuring the charge in each cell of the matrix, one obtains a two-dimensional array, which determines the illumination $E(x,y)$ in question.

To understand the principles of CCD matrix operation in more detail, let us first consider the main features of the MOS capacitor. In Figure 4.16, one can see a MOS capacitor formed on a p-type semiconductor substrate (exactly the same reasoning would apply to an n-type semiconductor given the relevant sign changes). The MOS capacitor consists of an electrode (4), which is made of metal or a strongly alloyed polysilicon, a dielectric (3), and a semiconductor substrate (1). To ensure that the charges move in a specific direction along the register during their transport, stop channels are created on both sides of the transfer channel (2). These are regions that are alloyed stronger than the silicon in the transfer channel itself. There is no potential well in the stop channel and the charge package maintains its shape. Immediately after a voltage is applied, the main conductivity carriers leave the electrode very quickly (some picoseconds) causing the formation of a near-surface layer, which is characterized by

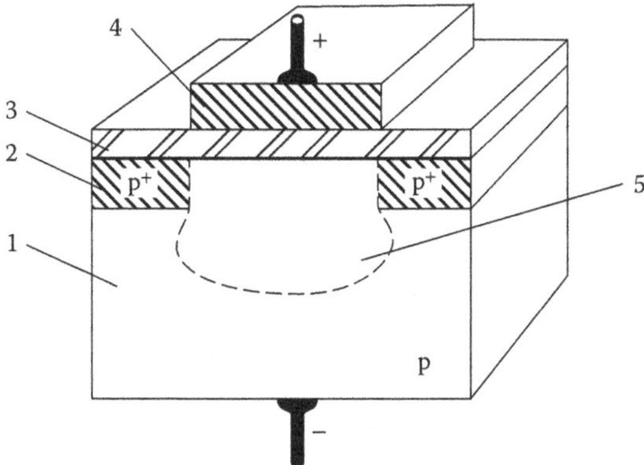

FIGURE 4.16 MOS capacitor.

the deficiency of the main carrier, and hence there is a potential well (5). Note that the CCD operates on the basis of a nonstationary state of the MOS structure. However, as the rate of thermal generation of the carrier is low, the potential well of the MOS structure can be used for accumulation and temporary storage of signal charge packages. As a consequence of thermal generation processes, the signal (informational) charge stored in the potential well acquires noise that is due to the thermally generated parasitic charge. This originates mainly from thermal generation of electron–hole pairs on the surface and within the deficient layer, as well as to a much smaller extent from the diffusion of the secondary conductivity carriers from nearby regions. This obviously causes a distortion of the signal charge. Thus, the maximum accumulation and storage time is dictated by an admissible measurement error and is generally defined by the semiconductor's properties and primarily by the temperature of the CCD. Therefore, high-quality CCD matrices are usually cooled, most often by means of built-in semiconductor refrigerators, although liquid nitrogen is used in some precision cases. All in all, a CCD structure is a temporary warehouse, whose content will deteriorate if it is kept there for too long. Thus, immediately after the accumulation of informational charges, one should read them off the device and measure them, transferring the data to a more permanent storage system.

Let us first consider the informational charge transfer process along a MOS capacitor chain in the form of a shift register. Its structure is depicted in Figure 4.17, each element of the structure being formed by three MOS capacitors. Similar electrodes of all the elements are electrically connected by buses, forming so-called phase electrode schemes, or simply phases.

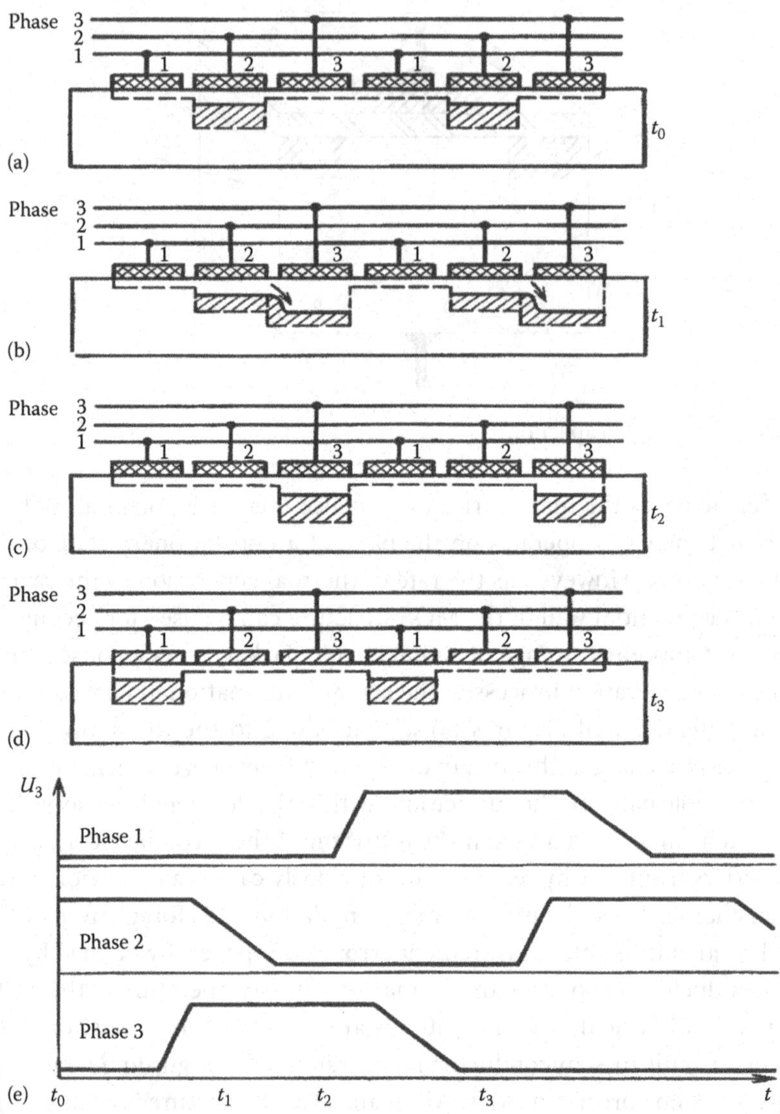

FIGURE 4.17 Transfer of informational charges within a shift register. The situations shown from (a) through (d) correspond with instant times from t_0 to t_3. At the foot of the figure (e) phase voltages are shown.

In the given example, there are three phases and hence the register is referred to as a three-phase register. In general, the number of phases can be 1, 2, 3, and higher.

Suppose, in such a three-phase system, that a high potential is applied at a time instant t_0 (Figure 4.17e) to the second phase, while the potential on the first and third phases is low (Figure 4.17a). After the charge packages have been stored under the electrodes of the second phase, the voltage applied to it is switched to a low level (at the moment t_1) and simultaneously a high voltage is applied to one of the neighboring phases, for example, the third phase, located on the right. As a result, the charges will start flowing into the empty wells on the right (Figure 4.17b). They cannot move to the left, as the potential applied to the electrodes of the first phase is still low. One can see from the figure that the transfer of the charge packages occurs simultaneously in all the elements of the shift register. Furthermore, as the high potential gets switched from the third (Figure 4.17c) to the first phase, all the charges will simultaneously move by one more step to the right (Figure 4.17d). Then the high potential is switched back again onto the second phase and so on. To read the information off the CCD matrix, all of its rows are moved downward simultaneously so that the lowermost row finds itself in the shift register. Then the informational charges are moved along the shift register structure, being measured at the latter's output and sent to the read-only memory (ROM) of the CCD camera or directly to the ROM of a computer that is connected to the CCD. This completes the reading of the lowermost row and the sequence is repeated, until all the rows have been recorded in a number array, reflecting the illumination distribution on the surface of the CCD matrix, and this array has been stored in the computer memory.

When a CCD is illuminated, the photons absorbed in the semiconductor will generate electron–hole pairs. These pairs will become separated due to the electric field in the deficient layer, with the electrons being localized in the potential wells and the holes being carried into the neutral area of the semiconductor. As already discussed, the volume of the charge package that is accumulated in each element is proportional to the photon flux, averaged over the surface area of the photosensor, as well as the time of accumulation and the quantum efficiency, which, as we will see later, is rather large.

A CCD is similar to other semiconductor photodetectors and is characterized by a certain spectral sensitivity band that is the wavelength range within which the conversion of a luminous flux into a charge package is the most efficient. The long-wave boundary is determined by the semiconductor's forbidden zone width, which for silicon is equal to 1.1 μm. The short-wave boundary lies between 0.32 and 0.4 μm, and is determined by the strong absorption of short-wave photons within the surface layer, where recombination processes take place simultaneously with photogeneration. The spectral sensitivity (quantum efficiency) of a CCD is shown

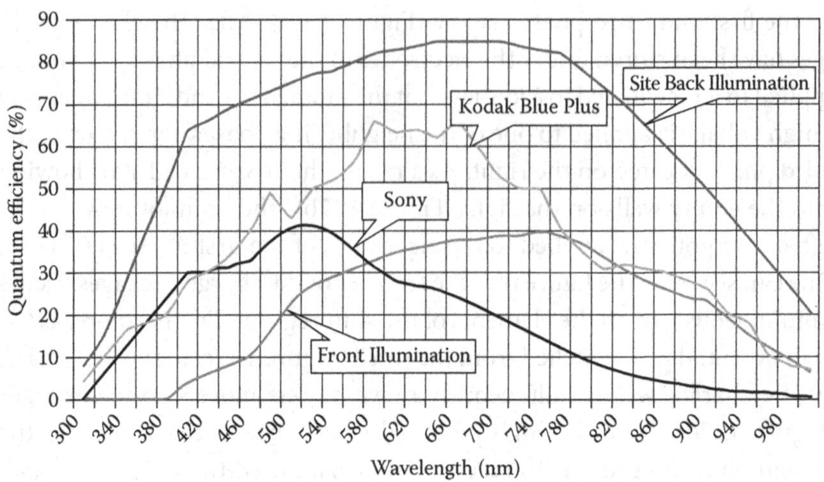

FIGURE 4.18 Spectral sensitivity (quantum efficiency) of typical CCDs.

in Figure 4.18. It should be mentioned that the maximum quantum efficiency can be as high as 85% in the modern generation of CCDs (so-called back-illuminated CCDs where the image is produced at the rear side of the matrix). Such high value of quantum efficiency of a CCD compared to that of the photocathodes of electron-optical devices is due to the fact that in a CCD almost all generated electrons are captured in the potential well. In photocathodes, half of electrons move in the direction opposite to the emitting surface due to the isotropy of the velocity distribution of the photoelectrons.

The FWHM of the CCD matrix transfer function is always larger than the spacing between neighboring cells.* This is a consequence of light scattering in the semiconductor substrate of the matrix, whose characteristics depend on the wavelength of the light. In view of this, not only the FWHM, but also the whole shape of the transfer function $g(x)$, and equally the transfer ratio $K(\omega)$ and the frequency–contrast function $K(\nu)$, turn out to be different in different wavelength bands. Fortunately, in the visible range, except for the red end, there is not a large variation of the scattering characteristic. Figure 4.19 shows the frequency–contrast function of the CCD camera type ST-7E, manufactured by Santa Barbara Instrument Group, Santa Barbara, CA. The CCD matrix of this device (Kodak KAF-1600 + TI TC-211) is equipped with a Peltier refrigerator, allowing its temperature to be reduced to −50°C.

* There is a common misunderstanding that the CCD matrix transfer function can be represented by the size of the sensitive zone of the matrix cell or at least by the distance between pixels. Unfortunately this is not correct.

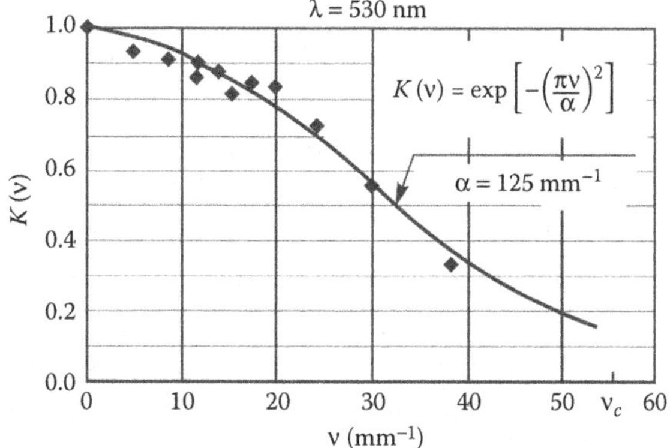

FIGURE 4.19 Frequency–contrast function $K(\nu)$ of the CCD matrix, compared against the experimental data, indicated by the diamonds points on the graph.

It contains 1530 × 1020 pixels and the spacing between a neighboring pair is 9 μm, so that the overall size of the matrix is 1.38 × 0.92 cm². As one can see from the figure, the frequency–contrast characteristic of this matrix is well approximated by a Gaussian, which implies that the transfer function $g(x)$ is also Gaussian. Consequently, all the equations, which have been obtained throughout the analysis of the problems, pertaining to electron–optical converters, in particular the relations starting from Equation 4.5 up to equality (4.19), can be used equally well here.

Let us take further advantage of this situation. First, we take the data from Figure 4.19 and then use Equation 4.7 and footnote on page 127 to find the FWHM β of the transfer function $g(x)$. The value of β in our case is 13.32 μm ≈ 1.5 pixels. Is this good or bad? Basically, it means that after a single cell has been illuminated, the charge that is accumulated in neighboring nonilluminated cells due to the light scattering in the semiconductor base of the CCD matrix will amount to approximately a third (actually 0.28) of the charge in the illuminated cell itself. Clearly, this parasitic charge will reduce the sharpness and the contrast of the stored image, causing information loss in the high spatial frequency band. Let us point out that for cells that are located two cells away from the illuminated cell, the scattered light will not exceed ~0.5% of the intensity that is incident upon the illuminated cell. (These data have been verified by direct measurements, by illuminating the sensitive zone of the matrix cell using a microscope lens.)

The illumination scale of the ST-7E CCD camera contains a maximum (S_{max}) of 4 × 10⁴ scale readings in the linear range. This means that any number of counts $S \leq S_{max}$ can be found by the simple equation $S = k\mu N_{Ph}$, where as usual μ is the quantum yield (in this case, the quantum yield of the semiconductor detector), N_{Ph} is the number of photons incident upon the cell sensor area integrated over the exposure time, and k is a coefficient inversely proportional to the electron equivalence of a single illumination scale count. (In other words, k^{-1} is the number of electrons that would cause an increment of one unit of the quantity S.) As usual, the manufacturers of these types of instruments, including the manufacturer of the ST-7E CCD camera, provide the mean-square informational charge measurement error. Unfortunately, this quantity is not very useful for us, as the error in determining the illumination primarily depends on the cell-to-cell accumulated electron fluctuations, that is, the fluctuation of the quantity μN_{Ph}, rather than the fluctuation of k. Hence, the relative illumination measurement error (which is actually not a constant, but a function of μN_{Ph}) and equally its inverse, that is, the signal-to-noise ratio, have to be determined experimentally. Recall that the theoretical basis for such measurements was established in Section 2.7. Experimentally recorded values for the signal-to-noise ratio, obtained after testing the ST-7E CCD camera, are given in Figure 4.20. As one would expect, the signal-to-noise ratio $L = S/2\sigma$ is proportional to $\sqrt{\mu N_{Ph}}$. It is also not surprising that even with 10^4 readings on the scale of the device, the maximum illumination measurement accuracy hardly reaches 1%. This accuracy is the maximum in the sense that the accuracy estimate $2\sigma/S$ is valid only in the case when, over the whole range of spatial frequencies of the image, the following is true: $K(\nu) \approx 1$. Otherwise, to compute the signal-to-noise ratio or the measurement error, one would have to take into account Shannon's *power loss factor*, which allows one to correct the signal-to-noise ratio for contrast reduction in the high spatial frequency band. This means that instead of the quantity L, one should use $L^* = LK$.

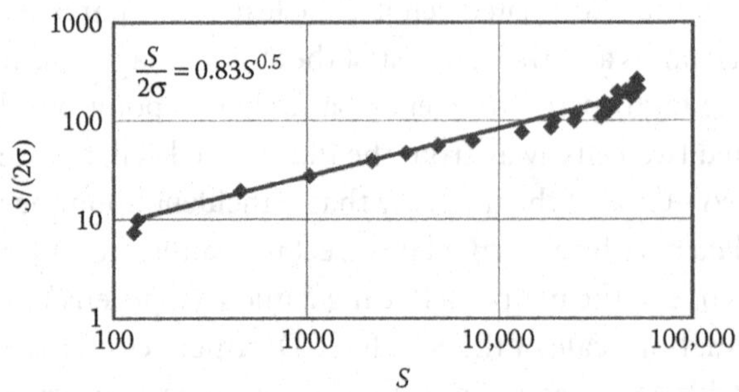

FIGURE 4.20 Signal-to-noise ratio.

In this case, it follows from Equation 4.12 that the power loss coefficient is $\mathcal{K}=\exp(-\pi^2\nu^2\beta^2/8\ln 2)$, and therefore $L^* = L\mathcal{K} = L\exp(-\pi^2\nu^2\beta^2/8\ln 2)$. This means that as ν increases, the measurement error also grows, while the signal-to-noise ratio goes down, doing so very rapidly after ν passes a certain critical value.

It is time to estimate the scale of the problem. Let Δ be the spacing between the pixels in the matrix. According to the Kotelnikov–Shannon theorem, the maximum spatial frequency ν_{max} that can be present in the image stored by the CCD matrix is equal to $1/2\Delta$. To avoid the inevitable and serious consequences, we will take a modest reduction of the admissible spatial frequency ν_c, say, by a factor of 1.5, so that $\nu_c = \nu_{max}/1.5$. In this case, it is not hard to calculate that the power loss coefficient is $\mathcal{K}=0.65$. This means that the measurement error of the illumination distribution averaged over the image field will increase by a factor that exceeds 1.5. In the specific example of the CCD camera under discussion, with 10^4 counts on the instrument's scale, the illumination measurement error will amount to 1.85%, instead of 1.2%. Finally, taking the value $S_{max}/2\sigma = 166$ from Figure 4.20 and taking account of the fact that a single reading step Δ equals the spacing between neighboring pixels, let us compute the maximum information density, which can possibly be recorded and therefore read in this case per square centimeter of the ST-7E CCD camera:

$$H = \frac{1}{\Delta^2}\left(\frac{\nu_c}{\nu_{max}}\right)^2 Q_n \log_2\left[\frac{S_{max}}{2\sigma}\mathcal{K}\right] = 3.1\times 10^6 \text{ bits cm}^{-2} = 389 \text{ kb cm}^{-2}$$

where:

$\Delta = 9\ \mu\text{m}$

$(\nu_c/\nu_{max})^2$ is a coefficient that takes the real frequency bandwidth for the recorded image

$Q = 0.84$

$L^* = 107.4$

Thus, taking the value of Q into account, each pixel contains on average 5.67 bits, despite the image average for the instrument's scale count numbers $\bar{S} = 2\times 10^4$. With such a density, in the whole matrix, whose size slightly exceeds one square centimeter ($1.38 \times 92 = 1.27$ cm^2), one can store a maximum of about 0.5 Mb of information. This is an excellent result! However, in real life, the recorded images usually have a much smaller ν_c than $\nu_{max} = 1/2\Delta$. Therefore, practical recordings usually contain a considerably smaller amount of information.

At this point, it is worth remarking that one should not use the simple and clear equation $H = k\log_2 L$ to estimate the information content of the recorded image. Indeed, why should one consider such "complexities" as light scattering in the semiconductor base of the CCD matrix, cell-to-cell electron fluctuations, and so on? Seemingly, one can just let k be the number of cells, then find L by the equation $L = S_{\max}/2\sigma$, using the manufacturer's data *apropos* of the mean-square informational charge measurement error, and eventually arrive at $H = 2.025$ Mb, that is, four times larger than the value estimated above! This result, however pleasing, has nothing to do with reality.

The parasitic scattering in the semiconductor base of the CCD matrix is especially noticeable in the IR wavelength region. The nature of scattering in this range is different from that in the visible band. It becomes obvious when one compares two different frequency–contrast functions $K(\nu)$, shown in Figures 4.19 and 4.21. It can be seen from the latter figure that the values of $K(\nu)$ in the IR band are just inversely proportional to ν. In this case, the transfer function is as follows:

$$g(x) = F\frac{\sin^2 \pi F x}{(\pi F x)^2} \tag{4.21}$$

where:

F is given implicitly as $K(F) = 0$

In other words, it is at the point where the straight line $K(\nu)$ intersects with the abscissa axis. The FWHM for the above transfer function is $\beta = (1/1.13F) \approx 1/F$, which implies that in the IR region, the transfer function FWHM for the ST-7E CCD camera will be ~20 μm, which is approximately

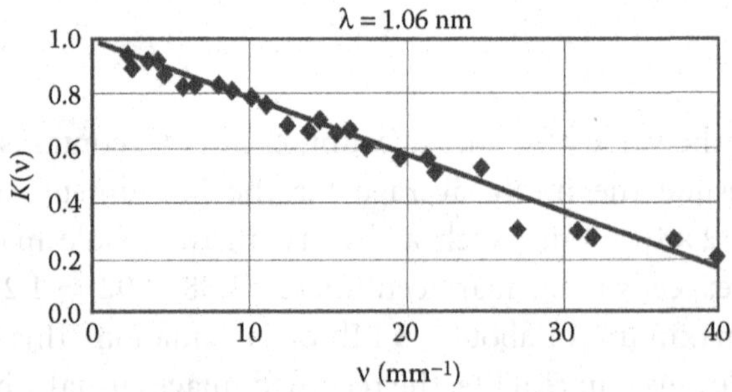

FIGURE 4.21 Frequency–contrast function $K(\nu)$ of the CCD matrix. Diamond points on the graph represent experimental data.

twice the spacing between a pair of neighboring pixels within the CCD matrix. This is almost 1.5 times larger than in the visible light wavelength range. The power loss coefficient in the IR region is as follows:

$$\mathcal{K} = \frac{1}{e}\left(1 - \frac{\nu_c}{F}\right)^{\left(1 - \frac{F}{\nu_c}\right)} \tag{4.22}$$

which means that now $\mathcal{K} = 0.56$, provided one makes an assumption that the highest admissible image spatial frequency in the recording is 1.5 times less than $\nu_{max} = 1/2\Delta$, as has been done earlier.

The further progress in development of solid-state semiconductor detectors for image recording was obtained after creating so-called complementary MOS (CMOS) structures. Similar to CCD matrices, CMOS devices consist of hundreds of thousands of cells (pixels) that form a two-dimensional orthogonal structure. However, in distinction to CCD, in CMOS devices the pixels are connected by a system of orthogonal buses, which make it possible to read out the information charges from arbitrarily selected sets of pixels. As a result, CMOS matrices allow the signals from arbitrarily selected parts of the matrix to be read out and to realize a parallel transfer of data (recall that in CCD matrices it is only possible to read out the data from the whole device). This possibility is extremely advantageous for a number of applications where the arbitrary selection of pixels is of particular interest, such as optical aiming systems and orientation systems for space devices.

Another even more important difference between CCD and CMOS matrices is in the structure of the pixels. In contrast to the CCD, the pixels of CMOS matrix include not only the photosensitive element (photodetector) but also the signal amplifier and analog–digital convertor (ADC). In the so-called active pixel sensors (APSs), the charge-to-voltage conversion is produced inside each pixel, and it is possible to read out the information about the state of any individual pixel by selecting the pixel address (column and row) in a two-dimensional array. It is also important to note that the power consumption of APS devices is ~100 times lower compared to CCD devices, which is of great value in autonomous devices such as notebooks, cell phones, and digital photo cameras. However, all these advantages do not come for free. For instance, due to complexity of CMOS, its pixel size is ~15–20 μm (compared with a pixel size of 5–15 μm for CCD). Accordingly, the CMOS matrices have bigger sizes and their structure is more complicated. Moreover, the fill factor of CMOS

TABLE 4.2 Comparison of CCD and CMOS

Parameter	CCD	CMOS
Output signal of cell (pixel)	Informational charge	Voltage
Matrix output (signal)	Analog	Digital
Fill factor	High	Moderate
Noise	Low	Moderate
Dynamic range	High	Moderate
Uniformity of sensitivity	High	Low to moderate
Complexity of detector	Low	High
Complexity of readout system	High	Low
Readout rate	Low	High
Frame processing options	Limited	Unlimited
Design cost	Low	High

(detector area divided by pixel area) is not high. This last disadvantage can be compensated for by using a microlens array to focus incoming light onto the active area of the photodetectors. It is also worth mentioning that the damage of several pixels in the CMOS matrix does not affect the device performance because this damage would mean just a loss of several counts in an image consisting of million elements. Similar damage in the CCD matrices could result in complete inability to read out the signal. And last but not least—due to standard and well-established technology of CMOS fabrication, their price is an order of magnitude smaller than that of CCD. The parameters of CCD and CMOS devices are compared in Table 4.2.

As one can see from Table 4.2, CCD devices have higher dynamic range and better signal-to-noise ratio. This is why CCD cameras are preferable in scientific applications. At the same time, due to their high readout rate, CMOS matrices are widely used in high-speed digital cameras. For instance, a recently developed camera Phantom V12.1 (Figure 4.22) was reported to provide up to 10^6 frames per second with a minimum exposition time of 300 ps.

4.4 PHOTOLAYERS

Despite all the advantages of CCD cameras, which have tended to push photographic film out of the experimental domain, the latter, notwithstanding the lack of affection by experimenters, will still remain in use for many years to come. What is more, in some of applications, such as holography, one cannot even remotely foresee a substitute. In modern

FIGURE 4.22 Framing camera Phantom V12.1 (Vision Research Company, New Jersey) with CMOS providing up to 10^6 fps.

FIGURE 4.23 Principle structure of a modern photographic film.

experimental physics, the photolayer is used as both a detector and a long-term memory element in optical instruments. The principle structure of a modern photographic film is presented in Figure 4.23. Regardless of their type and functions, all photosensitive materials have one or several emulsion layers as well as intermediate and protective gelatin coatings. The overall thickness of emulsion layers usually lies within the range 10–25 μm. The substrate (2) of the photographic material is normally a glass plate or a synthetic film. To ensure a solid bond with the emulsion layer, the substrate is covered with a thin gelatin coating (3). To prevent halation, that is, halo formation on the images of objects with bright illuminating elements, a varnish or gelatin-absorbing antihalation coating (1) is deposited on the reverse side of the substrate. In the negative types of black-and-white photo and motion picture films, this goal is achieved by incorporating a neutral gray dye directly into the film base. (Sometimes the antihalation coating is placed as a sublayer on the base of the synthetic film or glass plate.) To achieve the desired photographic characteristics, almost all negative black-and-white film types have two emulsion layers, rather than one, the lower layer (4) being less sensitive than the upper one (5). The upper emulsion layer is covered by an extra protective gelatin coating (6) to protect it from mechanical damage.

Photoemulsions contain primarily silver bromide with a small addition of silver iodide. They also may contain silver chloride and some other salts of silver, formed by a variety of organic acids. The grains of all emulsions, made of silver halogenides, possess a crystal structure. The exterior shape of the grains may vary. In most cases, the grains look like polygonal plates, whose thickness is some 10–15 times smaller than either of their other dimensions. More rarely, the crystals have more or less the same characteristic size in all three dimensions. A single grain's diameter varies from some 0.1 to 0.2 μm up to 4–5 μm. The relative size distribution for the emulsion grains depends on how they were synthesized and determines the photographic properties of the emulsion. The grains form multitier structures within the photographic layers. In spite of its small thickness, a single layer can easily contain some 40–50 or more tiers of crystals. Flat emulsion grains are aligned with their larger facet parallel to the photolayer surface. Table 4.3 summarizes some data, showing from left to right the differences in granularity of sample photolayers with high, medium, and low sensitivity.

Photographic properties of photolayers are determined by their sensitometric and structural characteristics. Let us consider them one after another, starting with a key term *optical density*. The optical density D depends on the average size and density of the grains of silver that are produced when the film is developed and defined on a logarithmic scale as follows:

$$D = \lg \frac{I_0}{I}$$

where:

I_0 is the light flux intensity incident upon the developed emulsion layer
I is the intensity of light passing through it

TABLE 4.3 Photolayer Characteristics

Grain density per 1 cm²	3.5×10^8	1.0×10^9	1.6×10^{10}
Average grain radius (μm)	0.65	0.36	0.15
Average spacing between grains (μm)	2.1	0.35	0.56

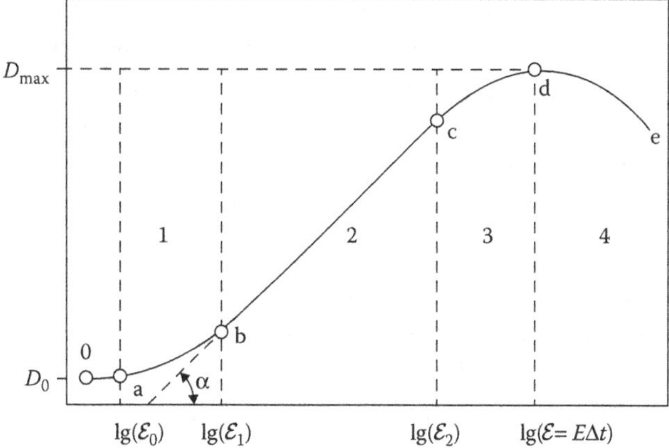

FIGURE 4.24 Characteristic curve of a photolayer: 1, underexposure region; 2, region of normal exposure; 3, overexposure region; 4, solarization region.

Clearly, the number of developed silver grains depends on the number of photons that illuminated the undeveloped emulsion layer during the time interval when the picture was being taken. The energy density incident upon the photolayer during this time interval $\mathcal{E} = E\Delta t$ is called the *exposure.* (In this equation, E is illumination and Δt is the duration of the exposure.) The dependence of the optical density versus the logarithm of exposure is called the *characteristic curve.* A typical characteristic curve is shown in Figure 4.24. One should pay attention to the following four regions of this curve: Region 1 corresponds to underexposure, region 2 to normal exposure, region 3 to overexposure, and finally region 4 to so-called solarization. In this region, a further increase of exposure results in the decrease of the degree of optical density, rather than its increase. It is worth noticing that the optical density starts to increase only after the exposure has reached some threshold nonzero value, marked by the point a in the figure. At lower exposure values $D = D_0$, where D_0 is the optical density of a developed blank photolayer, or the so-called fog. Region 2 of normal exposures is called the linear section* of the characteristic curve. The interval $\left[\lg\mathcal{E}_1, \lg\mathcal{E}_2\right]$ is called the *photographic width,* whereas $D'(\log\mathcal{E}) = \tan\alpha = \gamma$ is called the *contrast coefficient* (Figure 4.24). There is no need for any rescaling of the characteristic curve to perform relative measurements in the linear section.† Indeed, if the only

* "Linear" on the logarithmical scale.

† As one can see in Figure 4.24 within the "photographic width" $D = \gamma(\lg\mathcal{E} - \lg\mathcal{E}_0)$.

quantity of interest is the original intensity ratio, which came, say, to two different points x_1,y_1 and x_2,y_2 of the image when the latter was taken (for instance, it can be an intensity ratio between a pair of close spectral lines), this ratio is easy to find, as $\gamma \lg(I_1/I_2)=\gamma \lg(\mathcal{E}_1/\mathcal{E}_2)=D_1-D_2$, where D_1 and D_2 are the optical densities of the film at the points x_1,y_1 and x_2,y_2, respectively. Unfortunately, the situation is very different as far as absolute measurements are concerned. In fact, all the characteristic curves that one can find in the reference literature result from testing "fresh," that is, brand-new photolayers, which have just been manufactured and then developed in a special, model, perfect, fresh developer, with a certain temperature and a prescribed developing time. However, the storage time and conditions that a film has been through since it was manufactured until a camera's shutter finally opens in front of it can have a dramatic and uncontrollable impact on all its parameters. Moreover, it is hardly feasible to satisfy all the standard development requirements. For instance, a change in the developer's temperature by only 1°C or 2°C would cause a significant change in the characteristic curve's shape. Often, the characteristic curves of two pieces of the same film from the same roll, developed on the same day and in the same container, still turn out to be different. Thus, the only way for absolute measurements appears to be using a test light source to print some fiduciary "calibration marks" on a fragment of the same film, that will be used later on in an experiment. To print these calibration marks, one should place a graded multistage light absorber between the test light source and the film and project its output onto the film's surface. One should take into account additional intensity losses that may occur in the projection objectives as well as the lack of illumination uniformity along the field of vision. Summing up the procedure, one has to (1) print the calibration marks on a film fragment using a test light source and a graded multistage light absorber, (2) use the same film fragment for recording the experimental data, (3) develop it promptly upon completion of the experiment, (4) process the calibration marks, following the above given recommendations, and use them as a basis to generate a characteristic curve, and (5) only then start processing the results of the experiment.

The sensitivity S of photolayers is defined as a quantity, inversely proportional to an exposure, necessary to achieve optical density, which exceeds the fog by some fixed value D. Most frequently, the normalization is used with $D = 0.2$. In this standard situation, the so-called *sensitivity number* is defined as follows:

$$S = \frac{1}{\mathcal{E}_{D=0.2+D_0}} \left[\mathrm{lx}^{-1}\mathrm{s}^{-1}\right]$$

In Germany a sensitivity of photographic materials (S_{DIN}) are marked in a slightly different way:

$$S_{\mathrm{DIN}} = 10\lg\frac{1}{\mathcal{E}_{D=0.1+D_0}} \left[\mathrm{lx}^{-1}\mathrm{s}^{-1}\right]$$

Note that in the latter equation $D = 0.1$. In both cases, the same system of photometric units is used (lumens, luxes, etc.). All these definitions, however, create even more obstructions on the way to using all sorts of reference materials for experiment planning and data analysis as we have already discussed at the beginning of this chapter. From this point of view, the characteristics on the basis of the so-called *energy spectral sensitivity* are much more practical. In the latter case, the data (characteristic curves and sensitivity numbers) are found as a result of measurements, during which a photographic material is exposed by radiation of a succession of different wavelengths. Then the energy spectral sensitivity numbers are defined as follows:

$$S_\lambda = \frac{1}{\mathcal{E}(\lambda)_{D=1.0+D_0}} \left[\mathrm{erg}^{-1}\mathrm{cm}^{2}\right]$$

F, keeping an open eye on the sensitometric properties of photographic materials, one should pay attention to the so-called *noninterchangeability curves* $\lg \mathcal{E}_{D=\mathrm{const}} = f(\lg \Delta t)$, where $lgE_{D=\mathrm{const}}$ is plotted as a function of lgΔt These are gently sloped curves with a minimum around $\Delta t = 10^{-2}$–10^{-4} s. In other words, to achieve the same degree of optical density, one needs slightly more energy if the exposure time falls outside the above range.

Due to the granular structure of the developed image, the optical density can in principle be obtained only as an average with respect to some finite area $\Delta S = \Delta x \Delta y$. The latter area can be quite small; however, the lower limit is always determined by the reconstruction error value $\delta\bar{\mathcal{E}}/\bar{\mathcal{E}} = G(D)/[\Delta\gamma(D)]$, where $\Delta = \sqrt{\Delta S}$, $\gamma = D'(\log\mathcal{E})$ is the so-called *gamma*, or the contrast, D is the optical density, and $G(D)$ is a granularity factor.* Figure 4.25 illustrates the main features of the function $G(D)$.

* In this case, $G(D)$ is normalized to ensure the equality $\delta\bar{\mathcal{E}}/\bar{\mathcal{E}} = G(D)/[\Delta\gamma(D)]$ without any extra numerical factors.

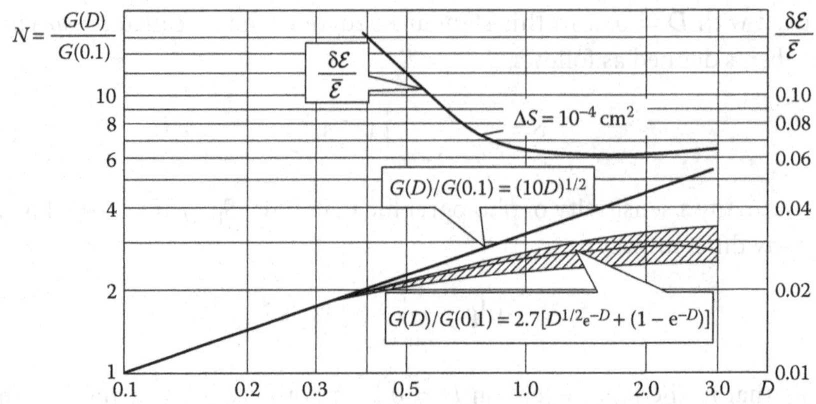

FIGURE 4.25 Noise of films. The experimental data area is hatched.

Two different approximations of the dependence $G(D)/G(0.1)$ and the dependence of the relative error values $\delta\bar{\mathcal{E}}/\bar{\mathcal{E}}$ versus D are also shown in the same figure. This dependence is based on the data for Kodak 4X film, developed with $\gamma = 1$. Let us note that if the optical density $D > 0.8$, both $\delta\bar{\mathcal{E}}/\bar{\mathcal{E}}$ and $G(D)$ are effectively constant. [A sharp increase of $\delta\bar{\mathcal{E}}/\bar{\mathcal{E}} = G(D)/\Delta\gamma(D)$ in the range $D < 0.8$ coinciding with the decline of $G(D)$ can be explained in the following way. When $D \to D_0$,* the function $\gamma(D)$ vanishes faster than $G(D)$.] Also note that in the interval $0.8 \le D \le 2.5$ the contrast $\gamma(D)$ is also approximately constant. Therefore, in the "linear" section of the characteristic curve $D(\lg\mathcal{E})$, that is, in the region where $\gamma \approx$ const, the reconstruction error $\delta\bar{\mathcal{E}}/\bar{\mathcal{E}}$ depends only on ΔS and not on $\bar{\mathcal{E}}$. Hence, if there is enough energy available, one should project an image onto the photolayer with such magnification M that the ratio $\Delta/\beta >> 1$ [accordingly $K(\omega) \approx 1$] for $D > 0.8$. In this case, the signal-to-noise ratio will be

$$L = \bar{\mathcal{E}}/\sigma = \gamma\Delta S/G \tag{4.23}$$

At first sight, according to this equation, one can improve the signal-to-noise ratio just by developing a photolayer up to a larger value of γ. However, the experience of experimental data processing tells that as γ increases with the development duration, the grain size and consequently G also increase. As a result, the ratio γ/G remains more or less

* Recall that D_0 is the optical density of a developed, but not exposed, photolayer, that is, so-called fog.

constant: A gentle maximum that several types of photographic film display in the neighborhood of the recommended value of γ improves the signal-to-noise ratio by only some 10%–15%. It is worth adding that in the spatial frequency range, which is interesting from the practical point of view, the noise of the photolayer is white and normally distributed.

Let us consider the informational capacity of the photolayer as a long-term memory facility for optical systems and find the effective value of the signal-to-noise ratio under the conditions when $K(\omega) \neq 1$, that is, let us compute the quantity L^*. It is well known that the loss of contrast in the high spatial frequency region is caused by light scattering within the emulsion. In the one-dimensional case, the spatial distribution of the scattered light is described in terms of the dispersion curve, which effectively represents the photolayer's transfer function:

$$g(x) = \frac{\beta}{2\pi\left[x^2 + \left(\frac{\beta}{2}\right)^2\right]} \tag{4.24}$$

where:

β is the FWHM

Using this equation, one can get the transfer ratio

$$K(\omega) = \exp(-\beta\omega/2) \tag{4.25}$$

and the frequency–contrast function

$$K(\nu) = \exp(-\pi\beta\nu) \tag{4.26}$$

Now, according to Equation 2.57, compute

$$K_1 = \exp\left(-\frac{\pi\beta\nu_c}{2}\right) = \exp\left(-\frac{\pi\beta}{4\Delta}\right) \tag{4.27}$$

and then

$$L_1^* = \frac{\gamma\Delta}{G}\exp\left(-\frac{\pi\beta}{4\Delta}\right) \tag{4.28}$$

In the two-dimensional case, the transfer function $g(r)$ can be obtained by means of Equation 2.67:

$$g(r)=\frac{\beta}{4\pi\left[r^2+\left(\frac{\beta}{2}\right)^2\right]^{3/2}} \tag{4.29}$$

thence, in a way analogous to the preceding calculation, we find $K(\omega)$, $K(\nu)$, $\mathcal{K}_2$, and L^*:

$$K(\omega_x,\omega_y)=\exp\left(-\frac{\beta}{2}\sqrt{\omega_x^2+\omega_y^2}\right) \tag{4.30}$$

$$K(\nu_x,\nu_y)=\exp\left(-\pi\beta\sqrt{\nu_x^2+\nu_y^2}\right) \tag{4.31}$$

$$\mathcal{K}_2=\exp\left(-\frac{2\pi\beta\nu_c}{3}\right)=\exp\left(-\frac{\pi\beta}{3\Delta}\right) \tag{4.32}$$

$$L_2^*=\frac{\gamma\Delta}{G}\exp\left(-\frac{\pi\beta}{3\Delta}\right) \tag{4.33}$$

The Equations 4.28 and 4.33 enable us to write down the following equations for the photolayer informational density in the one- and two-dimensional cases, respectively:

$$H=\frac{1}{\Delta}Q\log_2\left[\frac{\gamma\Delta}{G}\exp\left(-\frac{\pi\beta}{4\Delta}\right)\right]\left[\text{bit cm}^{-1}\right] \tag{4.34}$$

$$H=\frac{1}{\Delta^2}Q\log_2\left[\frac{\gamma\Delta}{G}\exp\left(-\frac{\pi\beta}{3\Delta}\right)\right]\left[\text{bit cm}^{-2}\right] \tag{4.35}$$

Let us take advantage of these relations to see the effect of the parameter Δ/β on the unit image area entropy H, the information energy equivalent $\bar{\mathcal{E}}/H$, and the information R_1, contained in an *ensemble-averaged* count, obtained with the reading step Δ. The dependencies listed in Figure 4.26 are the Russian manufactured films FOTO-32, FOTO-250T as well as the previously-mentioned film Kodak 4X. These three films have not been selected by chance. First, the comparison of these films will demonstrate the characteristic features of the functions under study. Second, in principle for these films, one can calculate (although not without some

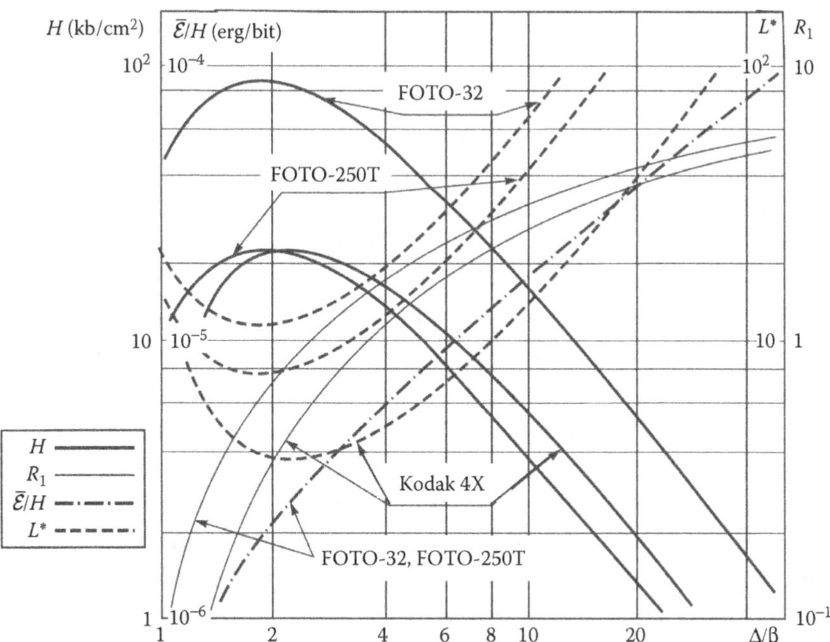

FIGURE 4.26 Information capacities of photolayers.

TABLE 4.4 Main Parameters of Films FOTO-32, FOTO-250T, and Kodak 4X

Film Type	γ_{opt}	$\beta \times 10^4$ cm	$G \times 10^4$ cm	$(G/\gamma) \times 10^4$ cm	$E(D = D_0 + 1.5)$ erg cm^{-2}
FOTO-32	0.8	5.0	2.1	2.6	8.2
FOTO-250T	0.8	10.0	4.2	5.25	1.4
Kodak 4X	0.6	7.5	3.6	6.0	0.68

effort) the values of β, G, and $\bar{\mathcal{E}}$ using the reference literature and verify the trustworthiness of the values in Table 4.4. If three films are developed up to a recommended value γ_{opt}, in the middle of the linear segments of the characteristic curves, the optical density values are the same and correspond to $D = 1.5$. In addition, the ratio of the original illumination values, corresponding to the endpoints of the linear range $0.8 < D < 2.5$, is in excess of two orders of magnitude, which guarantees the existence of the functions that are being studied within the argument range 1–50.

Let us come back to Figure 4.26 and discuss the principal consequences of the numerical investigation. First, let us compare the curves, corresponding to the films FOTO-32 and FOTO-250T. This pair of films is interesting in the sense that with the same contrast factor γ_{opt}, one of them has a transfer function FWHM β and a granularity factor G that are two times smaller than the other film. This implies that the ratio $G\Delta/\gamma\beta$ is the same for both films, and therefore the functions $L^*(\Delta/\beta)$ and $R_1(\Delta/\beta)$, characterizing them, are also the same. Naturally, in view of this, the information density H recording with the same ratio Δ/β is four times greater for the film FOTO-32 than for the film FOTO-250T because in the two-dimensional case the value of H is proportional to $1/\Delta^2$. However, as the first film requires a six times greater exposure than the second film to reach the same optical density, the energy equivalent of recording the same amount of information on the film FOTO-32 is still 1.5 times higher than if one uses the high-sensitive film FOTO-250T. It is worth noting that while the sensitivity numbers of these two films differ by more than an order of magnitude, the energy costs of recording the same amount of information differ only by factor of 1.5! This paradox is well familiar to experimenters used to dealing with weak luminous fluxes: You choose a film, which is 10 times more sensitive, but gain only some 30% in reconstruction accuracy. Of course, this is not the case when the exposure is insufficient to get a degree of optical density $D \geq 0.8$ for the low-sensitive film: In the latter case, the advantage of using photolayers with high sensitivity can be essential.

Let us now compare the films FOTO-250T and Kodak 4X. The former is characterized by a 25% wider transfer function, but a 12% better signal-to-noise ratio. As a result, as can be seen from Figure 4.26, the maximum recording density is approximately the same for the two films and its difference does not exceed 30%–40% within a reasonable range of Δ/β. The fact that Kodak 4X requires a two times smaller exposure to yield the same optical density is an essential point. Thus, the energy equivalent of recording a certain amount of information on the Kodak 4X film will be two to four times lower than if one uses the FOTO-250T film. In the situation discussed here, the minimum energy equivalent of recorded information for the Kodak 4X film amounts to $\sim 4 \times 10^{-6}$ erg per bit. Note that under the condition of the lowest energy equivalent recording, namely, when the information density is maximal, there is no need to expose the photolayer until the optical density reaches $\bar{D} = 1.5$, as in this case the number of optical density gradations is in any case rather low. For $D = 0.8$–1.0, the energy equivalent of recorded information falls to $(1–3) \times 10^{-7}$ erg per bit for the best from this standpoint photolayers. This is still 2.5–3 orders of magnitude higher than in the case of electron-optical systems, the result, generally speaking, is well known and obtained by various investigators in various ways. But on the other hand the informational energy equivalent advantage of electron-optical schemes is realized when the information flux density in these devices does not exceed 1 kb cm^{-2}. It is 1.5 orders of magnitude lower than it is possible to record on the film.

Finally, let us pay attention to the issue (Figure 4.26) that the maximum information recording density for a photolayer is achieved for $\Delta/\beta \approx 2$, while $H\left[(\Delta/\beta)=2\right]$ differs from H_{max} by no more than 1%–2%. This creates a basis for obtaining from Equation 4.35 the following equation for the quantity H_{max}:

$$H_{max} = \frac{1}{4\beta^2} Q\log_2\left[\frac{2\gamma\beta}{G}\exp\left(-\frac{\pi}{6}\right)\right] \tag{4.36}$$

The latter parameter together with the recorded information energy equivalent $(\mathcal{E}/H)_{min} = \mathcal{E}(D = D_0 + 1)/H_{max}$, provide exhaustive characterizations for a photolayer. Naturally, to compute these parameters, one should know β, γ, G, and $\mathcal{E}(D = D_0 + 1)$. Altogether, the data listed above allows us to find the optimal conditions for photorecording, as well as to obtain an estimate for the accuracy of the original image reconstruction.

4.5 ELECTRO-OPTICAL SHUTTERS

From the standpoint of high-speed photography, both films and CCD cameras are only memory facilities in optical systems. No clarification is needed as far as film is concerned. The situation is somewhat more complex with CCD cameras, as in principle one can use electronic controls and turn a CCD camera into a photographic device for taking pictures in the millisecond (rarely in the microsecond) time range. However, one can forget about even approaching nanosecond exposures, and there is no point in talking about the pico- and femtosecond range.* Thus, in either case, no matter whether it is a film or a CCD camera, to record in the range 10^{-4}–10^{-12} s, one needs external fast-speed shutters.

Apparently, electron-optical converters of various types serve quite well for this purpose, for instance, the biplanar devices described above. However, in this section, we are going to address shutters, whose operating principle is based on electro-optical effects. The principle of one such

* At this point, it is worth mentioning that the informational charge read-out time in CCD cameras that are designed for scientific experiments is extremely long. For instance, as far as the camera ST-7E is concerned, which was considered earlier in detail, the reading time for it amounts to some 60 s!

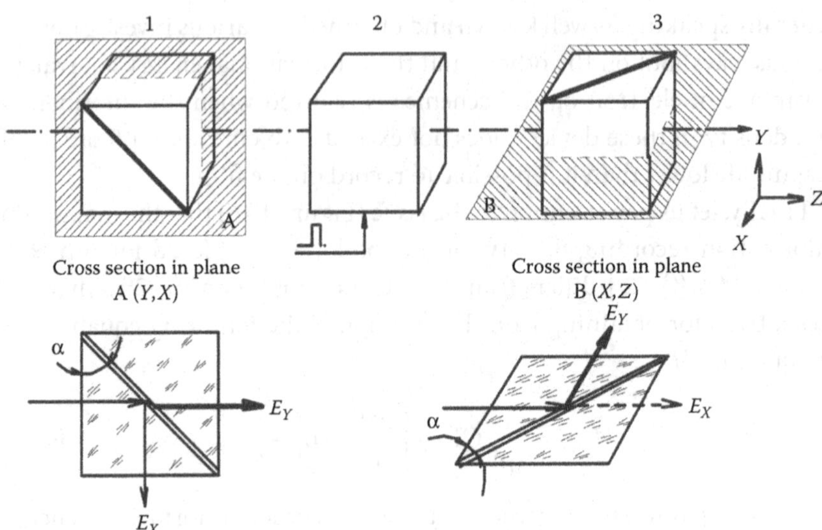

FIGURE 4.27 Electro–optical shutter; 1, Glan prism; 2, polarization rotator; 3, Glan prism.

shutter is shown in Figure 4.27. The shutter consists of two Glan prisms (1) and (3), oriented at a right angle with respect to one another, and a unit (2), whose function is to rotate the electric field vector $\boldsymbol{E}$ by 90° when an external voltage is applied to it. This unit is usually called an electro-optical shutter, and we will henceforth adhere to this convention, although strictly speaking it is not a shutter as a matter of fact, serving as it does, only to change the polarization of light passing through it. It is well known that Glan prisms are made on the basis of single-axis anisotropic crystals, whose refraction coefficient depends on mutual orientation of the electric field $\boldsymbol{E}$ and the crystal's axis. The Glan prism represents a juxtaposition of two separate prisms, joined together in such a way to form a rectangular parallelepiped (Figure 4.27). They are placed together with a narrow air gap in-between. The optical axes of the Glan prism's components are oriented in such a way that the dielectric constants in the directions of the X and Y axes are not equal to each other, namely, $\varepsilon_x \neq \varepsilon_y$. The angle α at the prism vertex is chosen in such a way that at the crystal–air interface, the conditions for total internal reflection will be satisfied for only one direction of polarization of the light wave (in the notation of Figure 4.27, it is the component with polarization E_X). Therefore, at the output of the first (with respect to the direction of light propagation) Glan prism (1), we will have linearly polarized radiation, with the field vector $\boldsymbol{E}$ of this light wave being parallel to the Y-axis. Indeed, a component with the polarization E_X will be reflected at the interface between the crystal and the air gap. The second prism (3) is rotated by 90° about the Z-axis with respect to the first one (1). Accordingly, the optical axes of the underlying crystals will also lie at 90° to one another, and now a wave with the polarization E_Y is reflected off the interface and directed out of the system, just as was previously the case with E_X. This explains why a pair of crossed Glan prisms does not allow practically any light to penetrate through (the contrast, i.e., the ratio $I_{out}/I_{in} \approx 10^{-3}$–$10^{-4}$). However, if one applies an external voltage to the electro–optical shutter (2), the polarization plane of the transmitted radiation will change to an orthogonal direction, which will enable the light to freely pass through the second Glan prism (3).

When a voltage is applied to the electro–optical shutter (2) in this optical system, it turns the optical system into a so-called half-wave plate. Traditional half-wave plates are made of single-axis anisotropic crystals and are intended for rotating the light-wave polarization plane by 90°. The reasons that cause the rotation of the plane of polarization are explained in Figure 4.28. Let the crystal's optical axes be directed along the X and Y axes, and suppose that the refractive indices n_x and n_y in these directions satisfy an inequality $n_x > n_y$. Here, n_x and n_y are the crystal's refractive indices for linearly polarized light waves, whose vectors $\boldsymbol{E}_X$ and $\boldsymbol{E}_Y$ lie parallel to the X and Y axes, respectively. Let us assume that a plane-polarized wave propagates along the Z-axis in such a way that at the entrance to the crystal, the vector $\boldsymbol{E}_{in}$ divides the angle between the X and Y axes

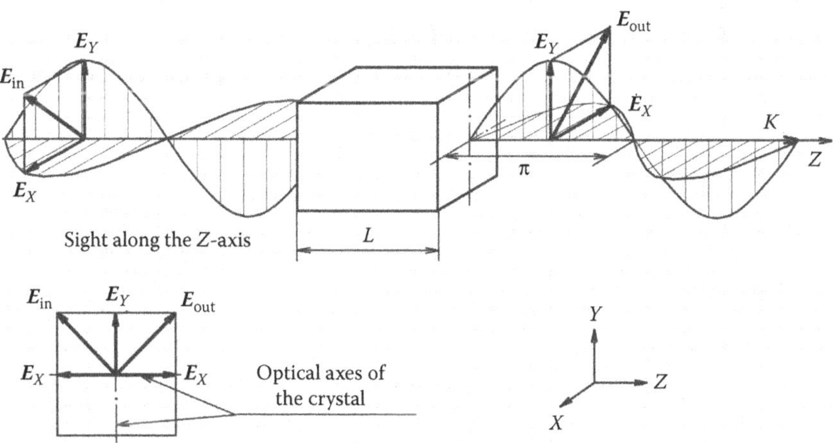

FIGURE 4.28 Half-wave plate.

in half. Clearly, in this case the vector components E_X and E_Y will have the same magnitude, that is, the projections of the vector $\boldsymbol{E}_{\text{in}}$ onto the X and Y axes will be equal to one another. Suppose that the length of the crystal L is chosen in such a way that at the exit from it, the component with the polarization E_X will have been delayed with respect to the component with the polarization E_Y by half a period $\Delta t = (L/c)(n_x - n_y) = (\lambda/2c)$. [In other words, the requirement is $L(n_x - n_y) = \lambda/2$.] With such a delay, the wave with the polarization $\boldsymbol{E}_X$ will be shifted in space with respect to the wave with polarization $\boldsymbol{E}_Y$ (Figure 4.28) by π. Thus, the overall vector $\boldsymbol{E}_{\text{out}}$ will have been rotated by 90° with respect to the vector $\boldsymbol{E}_{\text{in}}$ (see the vector diagram in Figure 4.28).

Now let us return to Figure 4.27. Within the unit (2), which despite an abuse of terminology we have agreed to call an electro-optical shutter, there is an optically active substance (crystal or liquid) placed along the wave's optical path. This substance is isotropic in the absence of an external electric field; however, as the latter is applied, the substance acquires all the properties of a single-axis anisotropic crystal. Hence, as the voltage is applied to the electrodes of the electro-optical shutter (2), the latter effectively becomes a half-wave plate, rotating the wave's polarization plane by 90° and allowing the light to pass through the crossed Glan prisms. Clearly, the exposure time, or the time during which the system is transparent to light, will be determined by the duration of the applied voltage pulse.

Such systems are also used to provide a short-pulse illumination to enable high-speed photography of otherwise dark objects that do not have any internal sources of light. In particular, they are used for multiframe delayed light photography. The principle of the optical scheme is shown in Figure 4.29. In such a system, a ray of light, reflected many times by the mirrors, illuminates the object under investigation for a sequence of short-time

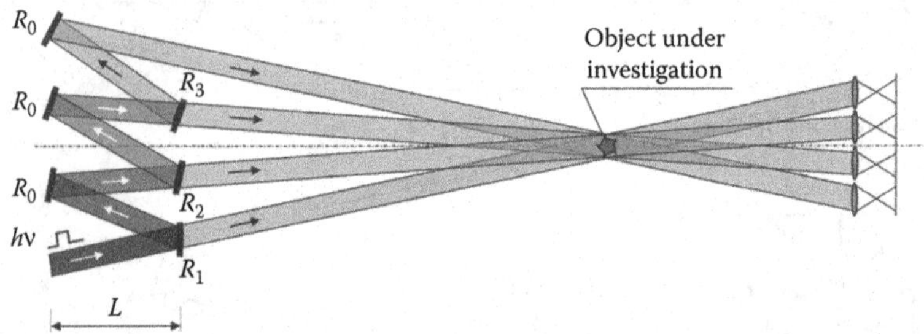

FIGURE 4.29 Principle of the optical scheme of a multiframe delayed light photographic device.

intervals, beginning at instants that are separated from one another by intervals $\Delta t \approx 2L/c$. (Certainly, an obvious condition to be observed is that $\tau < \Delta t$, where τ is the duration of the light pulse allowed by the shutter.) The light delay system uses a set of reflecting mirrors. The reflection coefficient of the mirrors R_0 is ~100%, and there is also an array of dividing mirrors R_i, whose reflection coefficients are $R_i = (n-i)/[n-(i-1)]$, where i is the dividing mirror's ordinal number with respect to the light ray propagation. Finally, n is the number of output rays or the number of frames. When the reflection coefficients are chosen this way, the intensity of the light illuminating the object under investigation will remain the same for a succession of time intervals. Note that in principle one can vary each single distance R_0–R_i and realize any prescribed time sequence for the shots being taken. Clearly, such a system can contain additional units, depending on what kind of objects one is dealing with: holograms, interferograms, and so on. As an illustration, Figure 4.30 shows a series of images of shock-wave formation

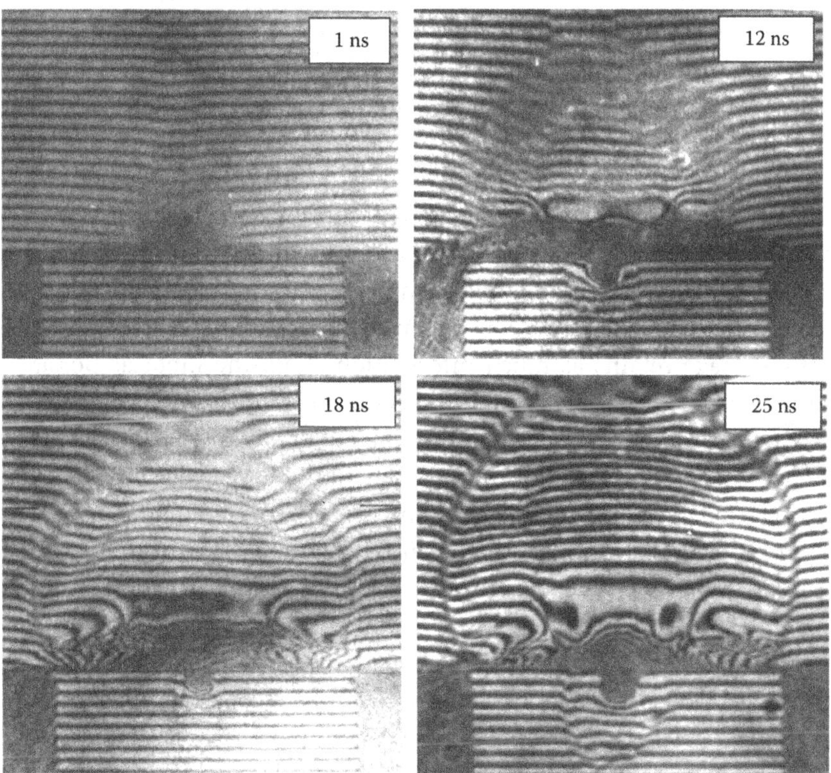

FIGURE 4.30 Shock-wave formation in gas.

in gas at a pressure of 10 torr, containing a flat target, which is hit by a laser pulse of 100 J energy in ~2 ns duration. The interferograms were obtained with a Mach–Zehnder interferometer, the duration of exposure for each frame being 3×10^{-10} s. The short light pulse generator and the delay line, which have been used, are similar to those shown in Figures 4.27 and 4.29.

4.6 SUMMARY

Let us summarize the results. We have considered a series of devices designed for recording optical information and have derived the relationships that enable us to find their most important characteristics. The results obtained enable us to find the rate of information transmission through various optical channels with noise, to determine the information density of its recording at the measurements, and to compute the energy equivalent of a received information unit. These relationships can be instrumental to find the optimal conditions for receiving, transmitting, and recording information, as well as to obtain a precise lower limit for the measurement error. They also allow comparison of various types of optical information systems, using as the criterion a single pair of main parameters: the transmission capacity that is the maximum density of information broadcast and its informational energy equivalent.

If the study of the information broadcast via the optical channels were our main objective, it would have been reached by now. However, our ultimate goal is different: It is to restore an original image from the output signal, that is, to obtain the maximum precision approximation to the input illumination distribution. So far, we have tried to ensure that the output image retains as much information about the input as possible, rather than trying to determine the conditions under which the difference between the two images is minimal. Let us stress that this is not the same thing, and all the effort made so far would have been in vain, if the information maximum in the output image occurred simultaneously with the minimum distortion of the reconstructed image with respect to the original. So far, we have been merely encoding the information and trying to determine how many bits per count or symbols per word results in the maximum transmission capacity of an information channel. Now it is time to get busy with decoding, namely, with the reconstruction of the initial intensity distribution. The main question that arises is as follows: Is it possible to avoid a substantial information loss while doing the decoding?

In other words, how close can we get to the lower limit for the error that is determined by the information content of the output signal?

The second question to be answered is as follows: What is the optimal scale for reading the recorded information? Encoding the information while recording it, or in less ambitious verbiage, matching the signal's spatial structure with the frequency characteristics of the channel, by means of the magnification M of the projecting optical devices, we have used as an optimization parameter the dimensionless ratio Δ/β. This parameter by definition equals the product of the inverse value of two times the critical frequency ν_c and the transfer function FWHM β. At the same time, $\Delta/\beta = 1/2\nu_c\beta = \lambda_c/2\beta$; in other words, Δ/β is the ratio of one-half of the minimum wavelength for the spatial signal harmonics to the transfer function FWHM. This is more natural, as it is not an artificially chosen grid size, but the physical reality described by intrinsic quantities, such as the spectral width and the transfer ratio, that impose the limits on the density of information fluxes being transmitted and recorded. In other words, the above quantity $\Delta = 1/2\nu_c$ simply facilitates the calculations and demonstrates that the parameter Δ/β is convenient for finding the optimum of information and recording. Still, the answer to the question, whether the above value of Δ contains any physical sense, is certainly affirmative. First, $\Delta_c \leq 1/2\nu_c$ gives a maximum possible step size, which determines the minimal number of counts per unit length $1/\Delta_c$ or area $1/\Delta_c^2$ of an optical image, which according to the Kotelnikov–Shannon theorem is necessary for unambiguous interpretation of signals, whose spectrum is bounded by the maximum frequency ν_c.

Second, $1/\Delta_c^2$ is the maximum possible reading aperture, realizing the maximum signal-to-noise ratio. Note that the maximum signal-to-noise ratio does not really correspond to the minimum possible value of the error, because by increasing Δ we reduce the random error component but enhance the systematic one, due to averaging the values over a larger interval Δ. This already implies the fact that the optimal value for the reading step Δ should be sought in the region $\Delta < 1/2\nu_c$; more supporting evidence in its favor will be provided in Chapter 12.

CHAPTER 5

Fourier Optics and Fourier Spectroscopy

"THE LENS IS AN optical device that fulfils the Fourier transform" is typical of the kind of statement in many books dealing with Fourier optics. Let us notice right away that no lens itself does or in principle can possibly perform the Fourier transform. Its role in Fourier optical systems is merely to focus parallel beams. Really, any wave with a flat wavefront incident on the aperture of the lens at some angle β with respect to the optical axis will be sent to a single point in the focal plane of the lens at the distance $r = \beta f$ away from the axis, where f is the focal length of the lens*. Nonetheless, one can easily verify the following fact on a purely formal level, using the Fresnel diffraction formula [1]: If any transparent mask, with amplitude transmission coefficient $t(x, y)$, is placed in the front focal plane of a lens with focal length f, perpendicular to the optical axis (this focal plane is the X–Y plane; Figure 5.1), it is illuminated by a flat monochromatic light wave with wavelength λ and constant amplitude $a_0(x, y) = const$ traveling along the Z-axis. [The amplitude transmission coefficient is defined so that the amplitude of the wave field immediately behind the transparent mask is $a(x, y) = a_0(x, y)t(x, y) = a_0 t(x, y)$.] Then the spatial distribution $U_f(x_f, y_f)$ of the wave field amplitude in the rear focal plane of the lens X_f, Y_f in accordance with Fresnel diffraction is given by the following:

* For large angles, one has $r = f\tan\beta$.

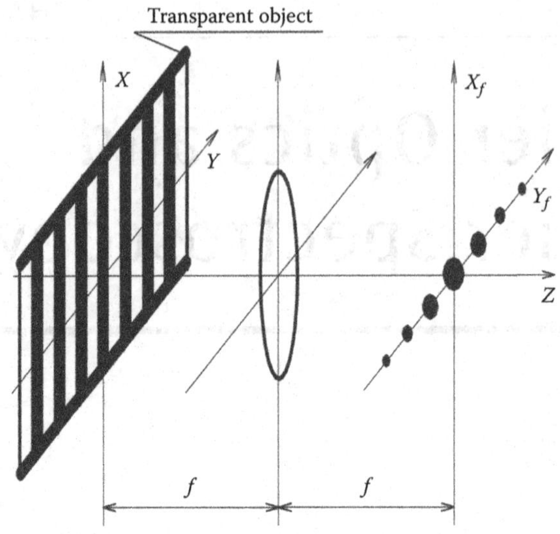

FIGURE 5.1 Optical system.

$$U_f(x_f, y_f) = \frac{a}{i\lambda f} \iint_{\infty} t(x, y) \exp\left[-i\frac{2\pi}{\lambda f}(xx_f + yy_f)\right] dxdy \tag{5.1}$$

It follows that $U_f = (x_f, y_f)$ is indeed the Fourier transform of the function $(a/\lambda f)t(x, y)$.

To understand the physical meaning of this maxim, we shall consider a variety of physical phenomena, which constitute the foundations of Fourier optics.

5.1 DIFFRACTION BY OPTICAL TRANSPARENCIES

Let us begin by "removing" the lens and consider the light diffraction by a sinusoidal transparent object (transparent mask). Suppose a monochromatic wave $a_0(x, y) = const$, with flat wavefront and propagating in the direction of the Z-axis, is incident on the transparent object, which has been placed in the X–Y plane. Suppose that this object has an amplitude transmission $t(x) = [1 + \cos(\omega_0 x)]/2$, where ω_0 is the spatial modulation frequency satisfying $\omega_0 << 2\pi c/\lambda$ [graphs of the functions $t(x)$ and $a(x) = a_0(x)t(x) = at(x, y)$ are shown in Figure 5.2].* It follows that in

* Clearly if $a(x, y) = \text{const}$, then if $a(x, y) \equiv 1$, the functions $t(x)$ and $a(x) = at(x)$ coincide, since the spatial frequency has the same value ω_0.

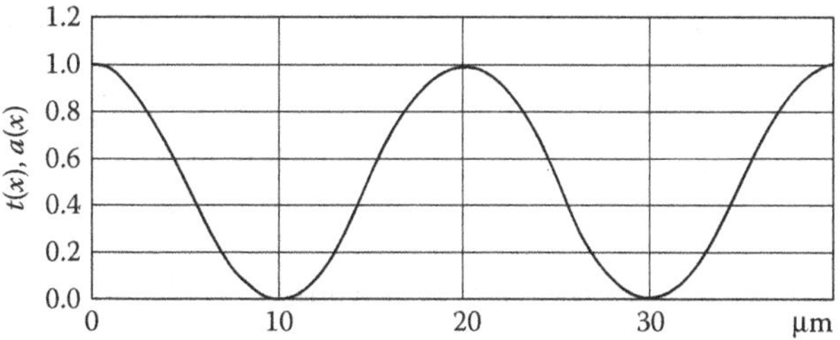

FIGURE 5.2 Amplitude transmission coefficient $t(x)$ and wave amplitude $a(x)$ behind a transparent object with modulation index $m = 1$.

the X–Y plane immediately after the transparent object, the amplitude distribution has the following form:

$$a(x) = at(x) = a\left[\frac{1+\cos(\omega_0 x)}{2}\right] \tag{5.2}$$

According to the Huygens–Fresnel principle, the propagation of the electromagnetic field further "downstream" will be uniquely determined by the electromagnetic field distribution in the X–Y plane. Let us point out some important features of this field:

- The vectors $\boldsymbol{E}$ and $\boldsymbol{H}$ of the electromagnetic wave $a_0(x, y) = \text{const}$, incident on the transparent object, both lie in the X–Y plane, where the object itself also lies and the wave vector $\boldsymbol{k}$ is parallel to the Z-axis.
- In the case of a linearly polarized wave, the vectors $\boldsymbol{E}$ are collinear in the whole X–Y plane and point in the same direction throughout every half-period of the electromagnetic wave.
- The modulation coefficient $m = \dfrac{E_{\max} - E_{\min}}{E_{\max}} = \dfrac{t_{\max} - t_{\min}}{t_{\max}} = 1$ (100%).

Let us now remove the transparent object and try to find a combination of waves incident on the X–Y plane such that in the absence of the transparent object would yield exactly the same field in the X–Y plane, characterized by the same amplitude distribution $U(x) = a[1 + \cos(\omega_0 x)]/2$, which is indeed the field equal to the field behind the transparent object when it is illuminated by the electromagnetic wave traveling along the Z-axis. It does not take long to convince oneself that the outcome can be obtained by using three waves. Indeed, the sum of three waves is as follows*:

$$\frac{a}{2}+\frac{a}{4}e^{i\omega_0 x}+\frac{a}{4}e^{-i\omega_0 x}=\frac{a[1+\cos(\omega_0 x)]}{2} \tag{5.3}$$

The first wave with amplitude $a/2$ is traveling along the Z-axis, whereas the second and third waves $\left[(a/4)e^{i\omega_0 x}\text{and}(a/4)e^{-i\omega_0 x}\right]$ are skewed to the X–Y plane, propagating at angles, respectively, $\pm\beta$ relative to the Z-axis, with the phase $\Delta\Phi(x)=2\pi\beta x/\lambda=\omega_0 x$.† Since both of these scenarios result in equivalent electromagnetic fields, by the Huygens–Fresnel principle, the electromagnetic field further along the Z-axis will be the same in both cases. Therefore, the result of the light diffraction on a sinusoidal transparent mask is given by superposition of three waves, one of which travels in the same direction as the incident wave, and the other two are, respectively, at some angles $\pm\beta$ to the former wave. It is worth mentioning that if $2\pi\beta x/\lambda=\omega_0 x$, then $\beta=\lambda\omega_0/2\pi=\lambda/\Lambda$, where $\Lambda=2\pi/\omega_0$ is the spatial wavelength of the transparent mask's sinusoidal structure. In plain terms, β is just the diffraction angle.

If we now take a lens, whose optical axis is the Z-axis and whose focal length equals f, and place it in the path of the three plane waves, they will be brought to a focus at three points in the focal plane.‡ The first wave is focused on the system's axis and the other two at distances $\pm x=\pm\beta f=\pm(\lambda/\Lambda)f=\pm(\lambda f/2\pi)\omega_0$ from it. Note that the factor $2\pi/\lambda f$ determines the scale factor ω in the focal plane of the lens, namely, $\omega=(2\pi/\lambda f)x$. On this scale, the light incident onto the lens will be focused at three points with $\omega = 0$, $\omega=+\omega_0$, and $\omega=-\omega_0$. The field amplitudes at these points relative to one another are, respectively, in the proportions $(1/2):(1/4):(1/4)$. Note that the consideration so far has made no mention of any Fourier transform: We have just dealt with diffraction of a plane light wave on a transparent mask, whose amplitude transmission is $t(x)=[1+\cos(\omega_0 x)]/2$, and it has turned out that the diffraction effect causes three different plane waves to arise. Then the lens was used to bring these waves to foci at three different points in the focal plane.

* It follows, in particular, that a standing wave arises as a result of wave interference in the X–Y plane.

† Note that this is valid for small angles β only, when one can approximate $\sin\beta\cong\beta$.

‡ The width of these spots, due to the diffraction on the aperture of the lens, will be of the order $2\lambda f/D$, where D is the diameter of the lens.

Now, using the Euler formula, let us represent Equation 5.2 as follows:

$$a(x) = a\left[\frac{1+\cos(\omega_0 x)}{2}\right] = \frac{a}{2} + \frac{a}{4}e^{i\omega_0 x} + \frac{a}{4}e^{-i\omega_0 x}$$

and calculate its Fourier transform as follows:

$$\begin{aligned}\Phi_a(\omega) &= \int_{-\infty}^{\infty}\left(\frac{a}{2} + \frac{a}{4}e^{i\omega_0 x} + \frac{a}{4}e^{-i\omega_0 x}\right)e^{i\omega x}dx \\ &= \frac{a}{2}\delta(\omega) + \frac{a}{4}\delta(\omega+\omega_0) + \frac{a}{4}\delta(\omega-\omega_0)\end{aligned} \tag{5.4}$$

so that the quantity $\Phi_a(\omega)$ is nonzero only if $\omega = 0$, $+\omega_0$, and $-\omega_0$, moreover the coefficients before members of series are in the proportions $(1/2):(1/4):(1/4)$. Note that in the same way one can obtain the Fourier image of the function $t(x)$. An equally straightforward calculation yields $\Phi_t(\omega) = (1/2)\delta(\omega) + (1/4)\delta(\omega+\omega_0) + (1/4)\delta(\omega-\omega_0)$. We see that the quantity $\Phi_t(\omega)$ is also nonzero only if $\omega = 0$, $+\omega_0$, and $-\omega_0$, the values $\Phi_t(\omega)$ at these points being related to one another in the proportions mentioned earlier. Therefore, the electromagnetic field distribution in the focal plane of the lens, which is focusing a plane wave that has been diffracted by a sinusoidal transparent mask, *coincides*, up to a multiplicative constant, with the formally computed Fourier transform of the transparency transmission function $t(x)$. Equivalently, the electromagnetic wave field distribution $A(x)$ [or $A(\omega)$] in the focal plane of the lens *equals* the Fourier transform of the electromagnetic wave in the plane of the transparent mask $at(x)$.*

We have so far investigated the range of phenomena arising when a sinusoidal transparent object, with 100% modulation, is illuminated by a plane light wave. Let us now pass on to the more general case where the modulation index $m \neq 1$. Then,

$$t(x) = 1 - \frac{m}{2} + \frac{m}{2}\cos(\omega_0 x) \tag{5.5}$$

* It should be clear that the ratio of amplitudes A/a will be proportional to the D^2/d^2 ratio, where D is the lens diameter and d is the so-called "circle of dispersion," which is approximately $2\lambda f/D$, if aberrations are disregarded.

where:

$m = [(t_{max} - t_{min})/t_{max}] = 1 - (t_{min}/t_{max})$ so that the variation t_{min} in the interval between 0 and 1 causes the modulation index m vary in the range between 1 and 0

A graph of the $t(x)$ dependence is shown in Figure 5.3. Consider now a similar scenario, namely, when a plane monochromatic wave with amplitude $a_0(x, y) = \text{const}$ travels in the Z-direction and shines upon a transparent object with modulation index $m \neq 1$ placed in the X–Y plane. In this case, the spatial distribution of the electromagnetic wave field in the X–Y plane immediately behind the transparent object will be

$$a(x) = a_0 t(x) = a_0 - \frac{a_0 m}{2} + \frac{a_0 m}{2}\cos(\omega_0 x) \tag{5.6}$$

The function $a(x)$ is also plotted in Figure 5.3. Let us use the same thought process as earlier: Remove the transparent mask and try to find a superposition of plane waves, which, if incident to the same plane in the absence of the mask, would yield in this plane exactly the same field, with the same amplitude distribution, as if the mask were in the wave's path along the Z-axis. It is easy to verify that the above amplitude distribution in the X–Y plane can be obtained by superposing three waves. The first one will have amplitude $a_0(1 - m/2)$ and propagate along the Z-axis, whereas the second and third waves will have amplitudes $(a_0 m/4)e^{i\omega_0 x}$ and $(a_0 m/4)e^{-i\omega_0 x}$, respectively, and will propagate at angles $\pm\beta$ with the Z-axis (and thus with the first wave). Indeed, we have the following:

$$U(x) = a_0\left(\frac{1-m}{2}\right) + \frac{a_0 m}{4}e^{i\omega_0 x} + \frac{a_0 m}{4}e^{-i\omega_0 x} = a_0 - \frac{a_0 m}{2} + \frac{a_0 m}{2}\cos\omega_0 x \tag{5.7}$$

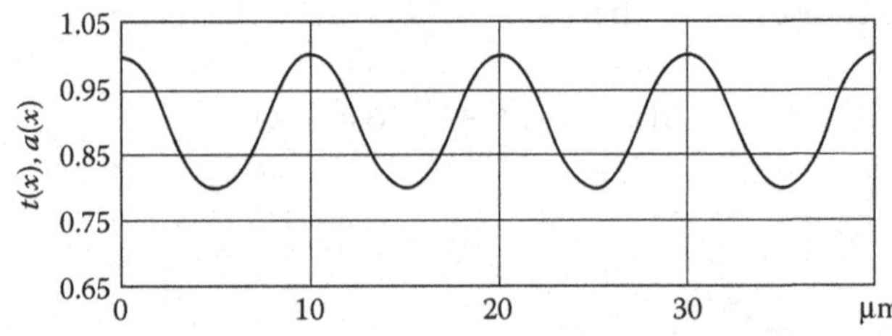

FIGURE 5.3 The functions $t(x)$ and $a(x)$ with modulation index $m \neq 1$.

The same outcome can be obtained in a different way. Let us take explicit advantage of the superposition principle and seek the field distribution behind the transparent object as a sum $a_0 = a_1 + a_2$, where clearly $a_1 = a_0 - a_2 = a_0(1-m)$ and $a_2 = a_0 m$. Here $a_1(x,y) = \text{const}$ represents an unmodulated component, whereas $a_2(x) = a_2[1+\cos(\omega_0 x)]/2$ represents a component with 100% spatial modulation, which, as it follows from the equation presented above, is going to produce a standing wave in the X–Y plane and give rise to three waves: $a_2/2 = a_0 m$, $(a_2/4)\mathrm{e}^{\mathrm{i}\omega_0 x} = (a_0 m/4)\mathrm{e}^{\mathrm{i}\omega_0 x}$, and $(a_2/4)\mathrm{e}^{-\mathrm{i}\omega_0 x} = (a_0 m/4)\mathrm{e}^{-\mathrm{i}\omega_0 x}$. Naturally, the sum of the two components $(a_1 + a_2)$ will be

$$a(x) = a_1 + \frac{a_2}{2} + \frac{a_2}{2}\cos\omega_0 x = a_0 - \frac{a_0 m}{2} + \frac{a_0 m}{2}\cos\omega_0 x$$

which is nothing more than the electromagnetic field distribution immediately behind the transparent mask. Note that in this case, when $m \neq 1$, the amplitude distribution of the electromagnetic wave field in the focal plane of the lens will be different from the distribution considered earlier (for $m = 1$) only in the sense that the zero component on the Fourier side ($\omega = 0$) will be bigger and the plus or minus first orders of diffraction ($\omega = +\omega_0$) and ($\omega = -\omega_0$) will be smaller than in the $m = 1$ case. A particularly important detail is that the amplitudes of the waves diffracting at angles $\pm\beta$ are directly proportional to m.

Let us go one step further as follows:

- Suppose we have manufactured transparent masks for the first and second harmonics $t_1(x) = 1-(m_1/2)+(m_1/2)\cos\omega_0 x$ and $t_2(x) = 1-(m_2/2)+(m_2/2)\cos 2\omega_0 x$ so that $m_1 + m_2 < 1$.
- Now we place them with one immediately following the other along the Z-axis. It is clear that we have $t_\Sigma(x) = t_1(x)t_2(x)$ as the net transmission of such a combination of transparent masks.
- Let us now illuminate this sandwich with a homogeneous $[a_0(x,y) = \text{const}]$ plane monochromatic wave traveling along the Z-axis.
- Finally, we collect the diffracted light with a lens and consider the amplitude distribution in the focal plane of the lens.

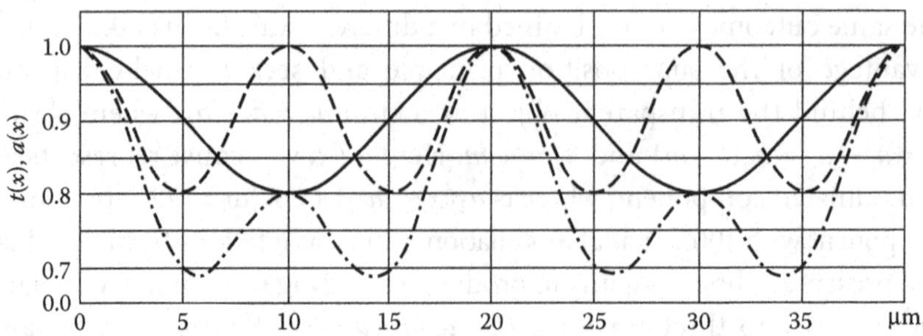

FIGURE 5.4 Functions $t_1(x)$ (—), $t_2(x)$ (– – –), and $t_\Sigma(x)$(–·–·–).

Omitting the second-order terms $m_1 \times m_2/4$, we have the following*:

$$t_\Sigma(x) \approx \left(1 - \frac{m_1 + m_2}{2}\right) + \frac{m_1}{2}\cos\omega_0 x + \frac{m_2}{2}\cos 2\omega_0 x$$

The graphs of the functions $t_1(x)$, $t_2(x)$, as well as $t_\Sigma(x)$ are shown in Figure 5.4. Without repeating the argument and omitting the accompanying calculations, the final diffraction outcome is as follows:

$$a(x) = a_0\left(1 - \frac{m_1 + m_2}{2}\right) + \frac{a_0 m_1}{4}e^{i\omega_0 x} + \frac{a_0 m_1}{4}e^{-i\omega_0 x}$$
$$+ \frac{a_0 m_2}{4}e^{i2\omega_0 x} + \frac{a_0 m_2}{4}e^{i2\omega_0 x}$$

We now see that each wave, that is, each of the harmonics, will be assigned to its own point in the focal plane of the lens, the coordinates of the points being proportional to ω and the wave amplitudes, respectively, to m_1 and m_2. The logic of further generalization should be clear by now; for any finitely supported function $t(x, y)$, the wave $a_0 t(x, y)$ can be represented as a superposition of plane waves. To each term in the superposition, that is, to each of the harmonics, there will correspond a point with coordinates $x_k = (\lambda f/2\pi)k\omega_{0x}$, $y_k = (\lambda f/2\pi)k\omega_{0y}$ in the focal plane X_f, Y_f of the lens, with the amplitude at this point being proportional to the relative contribution of the amplitude of the corresponding plane wave. In other words, given a transparent object with transmission coefficient $t(x, y)$, the

* The latter approximation is rather accurate. For instance, if we have $m_1 = 0.3$ and $m_2 = 0.15$, the maximum relative error pertaining to only some values $t(x_i)$ versus the true ones is ~2.5%; if $m_1 = 0.2$ and $m_2 = 0.1$, this error decreases to ~1%. In principle, a completely error-free transparency can also be made.

amplitudes of the waves that diffract on the object determine the amplitude distribution $U(x_f, y_f)$ of the electromagnetic wave field in the focal plane X_f, Y_f, which is proportional to the Fourier expansion of $at(x, y)$.* In essence, the phenomenon of light diffraction on a transparent subject constitutes the physical basis for the formal corollary (5.1) of the Fresnel diffraction formula:

$$U_f(x_f, y_f) = \frac{a}{\mathrm{i}\lambda f} \iint_{\infty} t(x, y) \exp\left[-\mathrm{i}\frac{2\pi}{\lambda f}(xx_f + yy_f)\right] \mathrm{d}x\mathrm{d}y$$

After a change of variables $x_f = (\lambda f/2\pi)\omega_x$ and $y_f = (\lambda f/2\pi)\omega_y$, the formula acquires a more familiar form:

$$\Phi(\omega_x, \omega_y) = \frac{a}{\lambda f} \iint_{\infty} t(x, y) \exp[-\mathrm{i}(\omega_x x + \omega_y y)] \mathrm{d}x\mathrm{d}y$$

Therefore, if the Fourier transform is caused by the phenomenon of diffraction, then what is the role played by the lens? To answer this question, let us consider diffraction on a slit aligned along the X-axis in the X–Y plane, so that the wave vectors $\boldsymbol{k}$ of the diffracted light waves lie in the Y–Z plane. The amplitude transmission of such a "transparent object" or "transparent mask" can be expressed as follows:

$$t(x) = \begin{cases} 1 \text{ if } & |x| \leq \dfrac{d}{2} \\ 0 \text{ if } & x > \dfrac{d}{2} \text{ or } x < -\dfrac{d}{2} \end{cases} \quad (5.8)$$

where:
 d is the slit width

It is well known that diffraction can actually be observed on a screen without any lens, provided that the screen is positioned far enough away from the slit at a distance $z >> [k(d/2)^2/2] = [\pi(d/2)^2/\lambda]$ (in the so-called far-field zone). This distance gives the scale at which the Fraunhofer approximation becomes valid: $z_F >> [k(x^2 + y^2)_{\max}/2]$. It is also well

* This is why the rear focal plane of the lens is often referred to as the "Fourier plane."

known that the angular distribution of the diffracted wave amplitudes will be as follows:

$$a(\varphi) = a_0 \frac{\sin(d\pi\varphi/\lambda)}{d\pi\varphi/\lambda} \tag{5.9}$$

Replacing the variable φ by $\varphi = x/z$,* we arrive at the field amplitude in the x-direction on a screen positioned at distance z from the slit:

$$a(x) = a_0 \frac{\sin\left(d\pi x/z\lambda\right)}{d\pi x/z\lambda} \tag{5.10}$$

The Fourier transform of the characteristic function of the slit (Equation 5.8) is $\Phi_t(\omega) = d[\sin(\omega d/2)/(\omega d/2)]$. In other words, if along the X-axis on the screen we mark a scale for values of ω with the scale $\omega = 2\pi x/z\lambda$, then the wave amplitude distribution on this scale in the x-direction will correspond to the Fourier transform of the function apart from a constant factor (Equation 5.8). Therefore, the lens has nothing to do with it, we did not use it to arrive at the final answer. Why then are we dealing with the lens and what for? The point is that in real Fourier optical systems, the quantity z_F that satisfies the Fraunhofer condition is extremely large. It is only for a slit with $d = 200$ microns illuminated by, say, a He–Ne laser with wavelength equal to 632.8 nm that the screen can be positioned behind it at a distance of a little over 5 cm. However, this is absolutely not the case for real objects of Fourier optics. For instance, if we want to use the methods of Fourier optics to process a standard frame of a developed film with dimensions 3.6 × 2.4 cm^2, the distance to the far-field zone will end up being in excess of 2.3 km. It may be rather inconvenient having a detector positioned kilometers away from the measured object, but the fact is that the intensity of the signal in the vicinity of such a remote screen will be so weak that, in practice, measuring or even seeing anything there will be simply impossible. The lens is necessary first of all in order to satisfy the Fraunhofer condition without having to deal with kilometer length scales. Indeed, the focal plane of the lens is precisely the far-field zone for incident radiation. Second, the lens concentrates the incident radiation by aligning together diverging rays. Indeed, all the rays incident on the lens at angle φ will be concentrated at a point in the focal plane at the distance $\varphi = x/f$

* This and the preceding equalities are valid, provided that the approximations $\sin\varphi \approx \varphi$ and $\tan\varphi \approx \varphi$ are valid.

from the axis, f being the focal length. Hence, we are allowed to substitute in Equation 5.9 $\varphi = x/f$ and then find the field amplitude distribution in the x-direction in the focal plane of the lens:

$$a(x) = a_0 \frac{\sin(d\pi x/f\lambda)}{d\pi x/f\lambda} \tag{5.11}$$

As one can see, this distribution is only different from Equation 5.10, which describes the amplitude distribution in the x-direction on the screen positioned at distance z from the transparency, by the scale factors ω, which are, respectively, $\omega = 2\pi x/f\lambda$ and $\omega = 2\pi x/z\lambda$.

5.2 FILTRATION IN THE FOURIER PLANE

The principle of a Fourier plane filtration device is shown in Figure 5.5. It can be seen in the figure that the initial image $f(x, y)$, which is to undergo filtration, is placed in the front focal plane of the first lens that the light rays encounter. The image is illuminated by a plane monochromatic wave. Hence, in the rear focal plane of this lens, alias the Fourier plane, one gets the Fourier transform of this image $\Phi(\omega_x, \omega_y)$. The second lens in the direction of the light rays adds up the Fourier components, reproducing in its focal plane the original image $f(x, y)$. In essence, we are dealing with a confocal reproduction lens system. The difference is that the system shown in Figure 5.5 provides an intermediate access to the Fourier plane, thus enabling one to interfere with the Fourier spectrum of the original image.

This scheme was first realized over a century ago in 1873 in the so-called Abbe experiment. Ernst Abbe placed a rectangular mesh in the front focal plane of the first lens and a slit in the rear one. From one experiment to

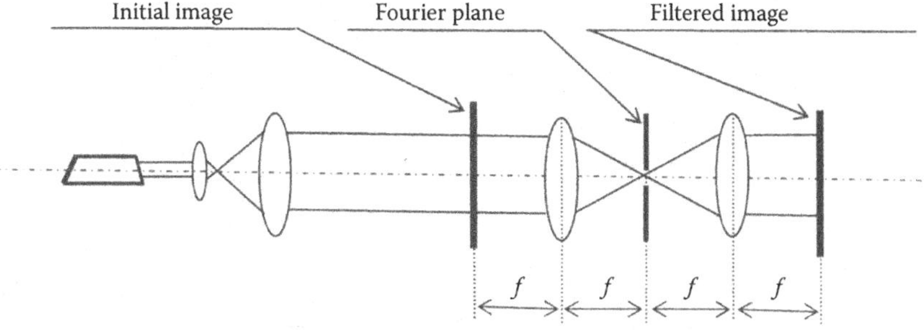

FIGURE 5.5 Schematic illustrating the principle of a Fourier filtration device.

another, he would turn the slit by 90°, filtering out either horizontal or vertical components of the Fourier spectrum, and observed that in the restored image in the focal plane of the second lens either vertical or, respectively, horizontal strips of the mesh are left (Figure 5.6). Based on these experiments in the Fourier plane, Abbe succeeded in developing the diffraction theory of the resolution of optical devices. We shall consider it in Section 5.3.

Filtration in the Fourier plane is often used to single out waves that are intended for further processing from waves that are diffracted in various directions. This procedure is important first and foremost in holography, where throughout the wavefront restoration procedure one has to face three, six, or more waves diffracting on the hologram, while for the purpose of further processing one is interested in only one

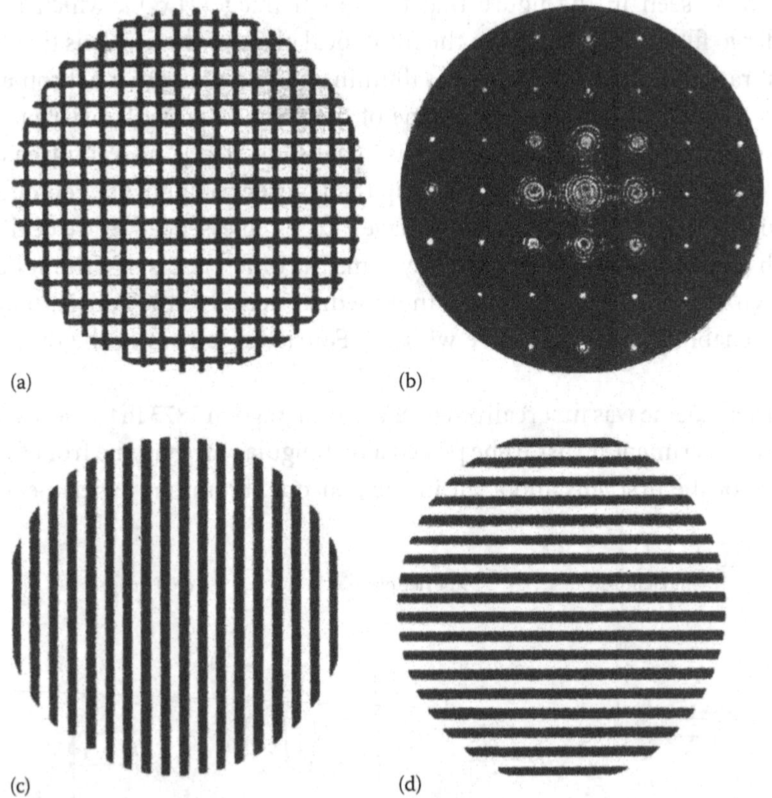

FIGURE 5.6 Abbe experiment: (a) transparency; (b) Fourier spectrum; (c) the image when the horizontal components of Fourier spectrum are filtered out; (d) the image when the vertical components of Fourier spectrum are filtered out.

or two of those waves.* By changing the incidence angle of the restoring wave (or waves) onto the hologram, it becomes possible to select the required wave via an aperture positioned in the Fourier plane by selecting the necessary order of diffraction. As a result, at the device output, one obtains only one wave (or two waves) that is needed. The same procedure is used to process measurement results when one deals with the Hartmann version of the three-wave shift interferometer, as well as in a number of other optical devices.

The other benefit of filtration in the Fourier plane is suppressing noise that may come from the film grain or multielectron component of the electron optical image,† as well as speckle structures. The corresponding procedure is realized by placing an iris diaphragm in the Fourier plane, designed to cut off higher orders of the Fourier spectrum. This is especially beneficial when the ranges of the image and noise spectra are well separated; then filtration will not have a damaging effect on the image. Otherwise one has to compromise, not always successfully. An example illustrating the iris diaphragm filtration technique is given in Figure 5.7. It is worth noticing that as of today, computers have become a natural vehicle for this kind of filtration [2] (Figure 5.8). What are then today's (for yesterday it was still impossible) benefits of analog filtration? Perhaps they consist exclusively in the fact that in this case filtration is done in the interactive regime; we can simultaneously (online) adjust the iris diaphragm and observe the filtered image. This helps us to find the desired compromise between the image quality and

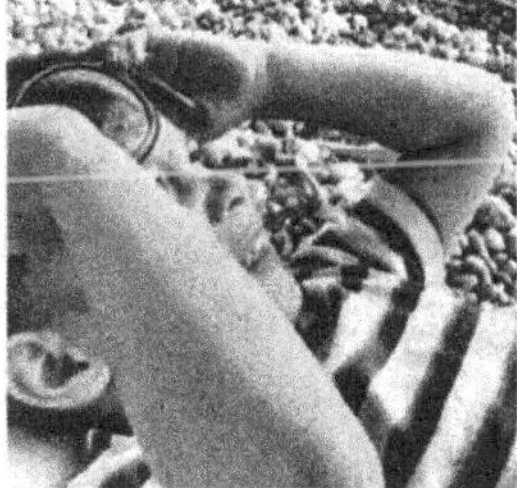

FIGURE 5.7 Filtration in Fourier plane: (left) the original image; (middle) its Fourier spectrum; (right) the image after filtration.

* See Chapter 8 for further detail.

† The multielectron component of electron optical devices displays itself in numerous bright spots appearing on the image. This effect is caused by photoemission fluctuations, when several, rather than one electron, are born.

FIGURE 5.8 Example showing the result of software-based filtration. (From Gruzman, I. S. et al., *Digital Image Processing of Information Systems*, NGTU Publisher, Novosibirsk, Russia, 2002. With permission.)

the remaining noise. (By the way, the electronic analog of the image is also not always available.) Let us note yet another opportunity: If one covers the zero order (in the so-called dark-field microscopy), then in the focal plane of the second lens there will be an image with inverted contrast. This may often be useful for better identifying the contours of the image's contour parts and details.

Let us finally analyze the range of possibilities provided by analog methods to solve convolution-type equations. Abbe had already shown that the image could be influenced by placing an amplitude-phase transparency in the focal plane of the optical device. Essentially, this constituted the first practical realization of optical analog computation. Much later in the 1950s, Maréchal and Françon [3] began an active development of coherent space filtration tools aimed at improving the quality of photographic images. Unfortunately, in our opinion, the computer revolution has, in effect, precluded this elegant method of image restoration from progressing as far as it merits. It turned out that just scanning the image into a computer and then processing it with correspondingly designed software was much easier than manufacturing physical filtration devices. How can one then characterize today's status of the method in question? At the very least, it is undoubtedly very useful for correcting the transfer function in complex optical devices.

What is the core of the method? As we know, restoration of an input signal from its output value formally boils down to solving the equation

$\Phi_{\text{in}}(\omega) = \Phi_{\text{out}}(\omega)/K(\omega)$,* where $K(\omega) \leq 1$. The essence of this procedure is to restore the frequency components of the signal by means of multiplying them by the factor $1/K(\omega)$, that is, exactly compensating the extent of their attenuation while traveling through the optical device. There are two main difficulties arising in an attempt at an analog solution of this equation. One is that, generally speaking, the function $K(\omega)$ is complex valued, and manufacturing phase filters is a difficult task. However, if the spread function of the device producing the image is real and even (which is usually the case), we have $K(\omega) \equiv A(\omega)$, that is, the transfer function has zero imaginary part, and therefore, there is no need for a phase filter. The second problem with filtration, as opposed to numerical processing, is that the former has no power to amplify the previously weakened high-frequency spectral components. Well, we cannot do that, but what we can do is to suppress the low-frequency part of the spectrum over a reasonably wide spectral range to ensure that $A_g(\omega)L(\omega) \approx \text{const}$, where $A_g(\omega)$ is the Fourier image of the spread function $g(x)$, alias the transfer function of the photographic device, and $L(\omega) = A_g^{-1}(\omega)$ is the filter. (Note that if the transfer function of the photographic device is a δ-function, there are no distortions on the spectral side, since we have $K(\omega) \equiv 1$. In other words, the relationship $A_g(\omega)L(\omega) \approx \text{const}$ is precisely the condition for the absence of frequency distortions in the corrected image.) Naturally, the correction filter $L(\omega)$, or more precisely its physical realization, which is a transparency with amplitude transmission $t(x_f)$ or $t(\omega)$, where $\omega = (2\pi/\lambda f)x_f$, must be positioned in the Fourier plane. (It is worth recalling at this point that the role of the multiplier $2\pi/\lambda f$ is to determine the scale factor of the quantity ω in the rear focal plane of the lens.)

To make such a transparency, one should proceed as follows:

- Obtain a negative image of the spread function by, say, shining monochromatic light onto a spectrograph slit, whose width ($\delta \ll \beta$) is smaller than the "normal one."
- Develop this image with inversion or print the positive image, that is, obtain the physical realization of the function $g(x)$.

* In order to see the principal aspect of the matter more clearly and not to overload our explanation with technical details, let us assume that the solution's stability is provided via a cutoff at frequency ω_c such that the noise $N(\omega_c)$ is still smaller than the output signal $\Phi_{\text{out}}(\omega_c)$, while $K(\omega_c) > 0$.

- Place the positive image in the front focal plane of the input lens and illuminate it with a flat wave with amplitude $a_0(x,y) = \text{const}$ to obtain the Fourier image $g(x)$ in the rear focal plane.
- Place a photosensitive layer in the Fourier plane. Upon developing it, we will get a negative with the physical realization of the quantity $A_g^{-1}(\omega) = L(\omega)$ that is our desired filter.

The high-frequency Fourier spectral components with $\omega \geq \omega_c$ will be cut off by a diaphragm. The transformation is corrected by choosing the adequate magnitude of the contrast γ throughout the processes of photographic development involved in the procedure for obtaining $L(\omega)$.

In this context, we would like to mention some experiments of Yu. P. Dontsov (which were carried out in 1965 in the Kurchatov Institute and regrettably unpublished). Dontsov succeeded in obtaining images with clear, well-defined multiplets from spectrograms with vague, practically indistinct spectral lines by implementing the above-described procedure in a device similar to the one in Figure 5.5.

5.3 DIFFRACTION THEORY OF THE RESOLUTION OF OPTICAL DEVICES

We shall consider in this section the range of possibilities provided by the diffraction theory of the resolution of optical devices. Although the theory was founded by Abbe over 100 years ago, it has failed to be applied in practice by engineers, who mostly adhere to the geometric optics that they are used to. This should not be surprising, first because most of the practical challenges that faced researchers in the field of optics throughout the last century could be successfully met within the limits of geometric optics, and second because as the result of many years of practicing and perfecting the geometric approximation, researchers have developed numerous recipes and experimental data processing tricks, which enable them to model and design the majority of optical devices. One of the few fruitful applications of diffraction theory was the application to the resolution of microscopes. Once again, this should not be surprising as this is a topic where geometric optics would be powerless. Our mission here, however, is not to develop alternatives to geometric optics, but rather to show that the diffraction theory of resolution opens a new range of opportunity. In addition, it will allow us to embark on and to explore alternative ways of investigation and design of optical devices—and first and foremost those applications that are used in experimental physics.

Abbe discovered in his experiments that if one narrows the size of the aperture positioned in the focal plane of the lens, first the image starts losing small details, and then sharp contours of image objects begin to smooth out until eventually one is left only with some washed out periodic structure, if any was present in the original image. He used the term "primary image" for the intensity distribution in the focal plane; today we call it the Fourier spectrum. He found out that it was the components of the primary image most remote from the optical axis that determined the resolution. No doubt, he understood that their presence in the primary image was caused by small-scale elements of the image itself on the one hand, which accounted for the maximum diffraction angles, as well as by cutoff lens apertures on the other. Of course, today after over a hundred years, it is easy for us to say that by cutting off the Fourier spectrum of the image, one smoothes out the image that is projected by the lens; removing high-frequency components causes small spatial details to disappear.

We shall actually refrain from expounding the diffraction resolution theory of optical devices *per se*, but rather we will assume that the contents of this chapter so far have been read with attention and we will confine ourselves to describing two examples illustrating its methods and scope. Let us start by calculating the transfer function $K(\omega)$ of a reproduction lens, having corrected spherical and chromatic aberrations. Recall that there are two ways of exploring linear systems: the frequency method and the pulse method (Chapter 2). In the former case (in optics), one deals with the input signal of the system in the form $\mathcal{E}_{\text{in}}(x) = \mathcal{E}_0(1 + \sin\omega x)/2$ and all the output signals together, obtained for different ω and $\mathcal{E}_{\text{out}}(x) = \mathcal{E}_0[1 + A(\omega)\sin\omega x]/2$, are used to determine the quantity $A(\omega)$.* The second method uses input signals reminiscent of the δ-function and the output signal represents the spread function $g(x)$. In principle, there is no difference between the two methods. Indeed, the Fourier transform of the δ-function is identically equal to unity $\mathcal{F}[\delta(x)] = \Phi(\omega) \equiv 1$, which means that the input signal of the system, similar to the frequency method, contains all harmonics within the limits of the system's bandwidth. The only significant difference is that in the frequency method, the harmonics

* It follows by symmetry that the spread function $g(x)$ of the lens is even, and thus, $K(\omega) \equiv A(\omega)$, that is, the transmission coefficient has zero imaginary part.

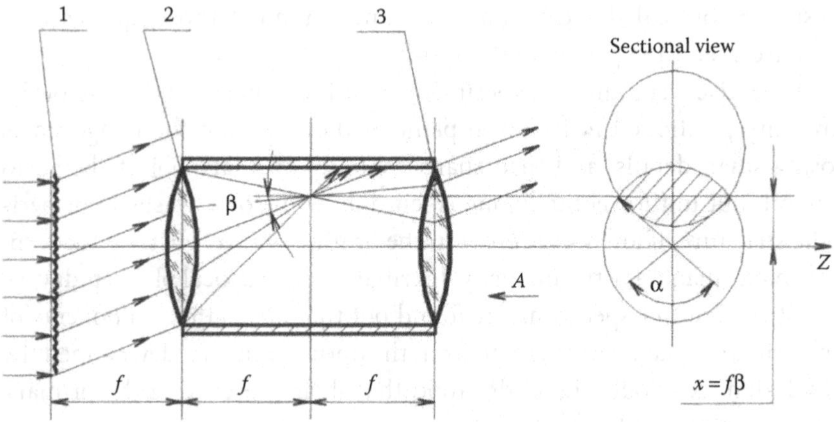

FIGURE 5.9 On calculating the transfer function $K(\omega)$ of a reproduction lens (1) transparent mask, (2) input lens, (3) output lens.

are applied as input in a consecutive way, whereas in the pulse method, they come in simultaneously.

Let us place in the front focal plane of our lens a transparent mask with amplitude transmission $t(x) = (1 + \cos\omega x)/2$ and illuminate it with a plane wave propagating along the Z-axis (Figure 5.9). [The wavelength is λ and the amplitude $a_0(x, y) = \text{const.}$] As we have already mentioned in the beginning of this chapter, the diffraction angle at such a transparent object is $\beta = \lambda\omega/2\pi$. The diffracted radiation will be focused in the joint focal plane of our confocal system at a distance $x = f\beta$ from the axis,* forming a round light spot with radius $r = D/2$ (D is the diameter of the aperture of the entrance lens), and be centered at the point with coordinates $x = 2f\beta$, $y = 0$ in the entrance of the second lens. The resulting scenario will be clear from Figure 5.9, which shows that the radiation will pass through the dashed region in the figure due to the diffracted rays being cut off by the finite size of both lenses. The area of this region, consisting of two equal segments, is $S = r^2(\alpha - \sin\alpha)$, where α is the angle from the center of the circle, which can be found from the equation $f\beta = r\cos(\alpha/2)$: $f\beta/r = \cos(\alpha/2)$, that is, $\alpha = 2\arccos(f\beta/r)$. The value of $\sin\alpha$ follows by dividing the half-chord by the radius: $\sin\alpha = (2f\beta/r)\sqrt{1-(f\beta/r)^2}$. It follows that

$$S = r^2(\alpha - \sin\alpha) = 2r^2\left[\arccos\left(\frac{f\beta}{r}\right) - \frac{f\beta}{r}\sqrt{1-\left(\frac{f\beta}{r}\right)^2}\right] \tag{5.12}$$

* The relative apertures D/f of reproduction lenses are usually not very large (~1/3), and therefore, the approximation $f\tan\beta \approx f\beta$ causes a relative error that is less than 1%.

Note that if $\beta = 0$ (i.e., if there is no cutoff due to the finite size of the lenses), we have $S = \pi r^2$, while the other extreme case when the radiation simply cannot propagate through the reproduction lens is $f\beta = r$ $(\beta_c = r/f)$. It is then easy to determine the bandwidth ω_c, since on the one hand $\beta_c = r/f$ and on the other $\beta_c = \lambda\omega_c/2\pi$, so that $\omega_c = 2\pi r/f\lambda = \pi D/f\lambda$. It follows that the bandwidth is proportional to the value of the relative aperture D/f. Thus, one practical way of increasing ω, and hence the resolution power, consists in increasing the diameter of the second lens, and this indeed is sometimes pursued in designing optical systems.

Making the change $\beta = \lambda\omega/2\pi$ in Equation 5.12 yields

$$S = 2r^2\left\{\arccos\left(\frac{f\lambda\omega}{2\pi r}\right) - \left[\frac{f\lambda\omega}{2\pi r}\sqrt{1-\left(\frac{f\lambda\omega}{2\pi r}\right)^2}\right]\right\}$$

Normalizing by unity, that is, dividing by πr^2, we find

$$K(\omega) = \frac{2}{\pi}\left\{\arccos\left(\frac{f\lambda\omega}{2\pi r}\right) - \left[\frac{f\lambda\omega}{2\pi r}\sqrt{1-\left(\frac{f\lambda\omega}{2\pi r}\right)^2}\right]\right\} \tag{5.13}$$

The function $K(\omega)$ is plotted in Figure 5.10. Since the function $K(\omega)$ is strictly bounded by the frequency ω_c, the spread function $g(x)$ [i.e., the inverse Fourier transform of $K(\omega)$] can be represented as a sum, using, for instance, Equation 2.31.

Let us analyze the diffraction at a transparent diffraction grating from the standpoint of Fourier optics. Suppose, as we had earlier, that a plane

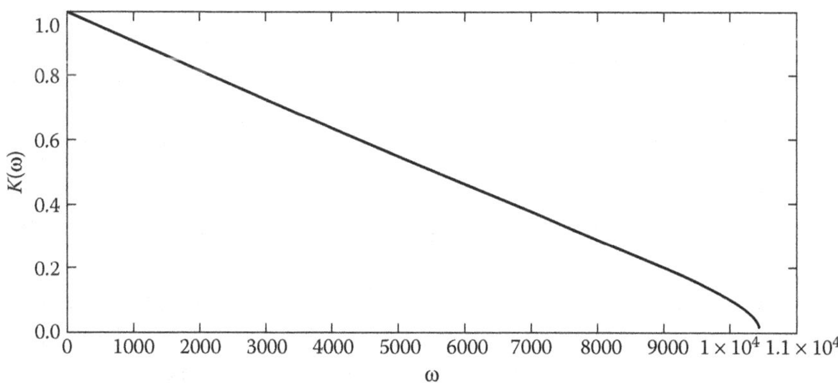

FIGURE 5.10 Function $K(\omega)$. Lens diameter $D = 5$ cm, focal length $f = 15$ cm, and radiation wavelength $\lambda = 10^{-4}$ cm. The dimension of ω is cm^{-1}.

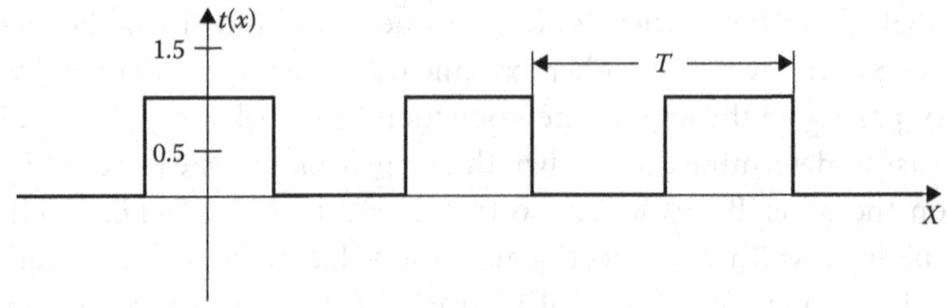

FIGURE 5.11 Transmission function of a diffraction grating.

monochromatic wave with constant amplitude $a_0(x, y) = \text{const}$ is traveling along the Z-axis, illuminating the grating positioned in the X–Y plane. Now the transparent object is a grid with period T, consisting of alternating segments with amplitude transmission $t = 0$ and $t = 1$ [see the quantity $t(x)$ in Figure 5.11]. We have already established that to find the field value distribution in each of the diffraction directions, it suffices to calculate the Fourier transform of the function $a(x) = a_0 t(x)$. Let us do this and calculate the Fourier series for the function $a(x)$, assuming that the widths A of the nontransparent and B of the transparent rulings of the grating are equal to one-half of its period, $A = B = T/2$. The latter assumption ensures that even modes are not present in the expansion. Therefore, we have

$$\mathcal{F}[a(x)] = \Phi(\omega) = a_0 \sum_{n=-\infty}^{\infty} b_n e^{in\omega_0 x} \tag{5.14}$$

where:

$\omega_0 = 2\pi/T$, T being the periodicity of the grating
$|n| = 0, 1, 2, 3, \ldots$

In this specific expansion, that is, when $A = B = T/2$, the Fourier coefficients turn out to have the following values: $b_0 = a_0/2$, $b_n = 0$ where $n = \pm 2k$, and $b_n = a_0(-1)^k/\pi(2k+1)$ if $n = \pm(2k+1)$, when $k = (|n|-1)/2$. Therefore, the intensities of the rays that are diffracted in the zero, first, and third orders of diffraction are, respectively, as follows: $I_0 = a_0^2/4$, $I_{+1} = I_{-1} = a_0^2/\pi^2(2k+1)^2 = a_0^2/n^2\pi^2 = a_0^2/\pi^2$, and $I_{+3} = I_{-3} = a_0^2/\pi^2(2k+1)^2 = a_0^2/n^2\pi^2 = a_0^2/9\pi^2$. (The same result could be obtained in a different and perhaps more traditional way [although this calculation might be a bit more complicated] by first considering the diffraction by a single slit of the grating and then summing the diffracted beams.)[*] It follows that the electromagnetic field intensity in the first order of diffraction, most commonly used in spectrometry, constitutes ~40% of the zero-order intensity and is merely ~10% of the overall radiation intensity in question. This account shows why, in spite of patent advantages, the diffraction grating spectrometer was at a disadvantage in comparison with prism spectrometers. Grating spectrometers were not widely used for many years because they were inferior to prism spectrometers in terms of their luminosity. The breakthrough was due to Rayleigh, who came up with the idea of using profiled (or blazed) gratings,[†] which will be considered in Chapter 6. For the moment, let us mention that the first blazed diffraction gratings were manufactured by the brilliant experimental physicist Robert Wood.

5.4 FOURIER SPECTROSCOPY

The objective of Fourier spectroscopy is to reconstruct the spectra of radiation sources under investigation using recorded interferograms. The principle of a typical Fourier spectrometer is shown in Figure 5.12. As can be seen from the figure, the key element in the arrangement is a Michelson

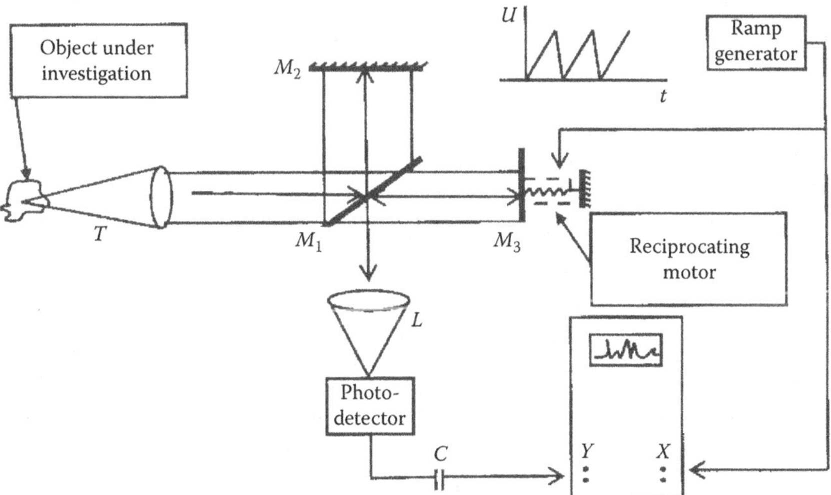

FIGURE 5.12 Schematic illustrating the principle of a typical Fourier spectrometer.

[*] See Chapter 6.

[†] They are also known as phase diffraction gratings.

interferometer,* one of the mirrors (M_3) of which can be moved in the direction of the incident radiation. The source of radiation in question is placed in the focal plane of the input telescope lens T, creating a plane light wave, which is directed onto the interferometer's semitransparent mirror M_1. This splits the incident radiation into two parts: One is reflected by the fixed mirror M_2 and the other is reflected by the movable mirror M_3; both are then focused by the lens L onto a photodetector. A *ramp generator* produces a voltage that varies linearly with time and is applied to the reciprocating motor that moves the mirror M_3. The ramp voltage is also applied to the X-plates of the oscilloscope creating a correspondence between the position x of the mirror M_3 and the oscilloscope's sweep. The signal from the photodetector, after passing through an isolating capacitor C, is applied to the oscilloscope's Y-plates. The reciprocating motor, which moves the mirror M_3, is driven on a piezoelectric, magnetostriction, or simply mechanical basis.

Let us assume that we are dealing with a monochromatic source of radiation. If the radiation frequency is ω and the amplitudes of the two interfering waves that arrive at the detector after having been reflected from the mirrors M_2 and M_3 are, respectively, a_1 and a_2, then, as will be shown later in Chapter 7, their net intensity at the input of the photodetector is

$$I_{\Sigma}(x) = a_1^2 + a_2^2 + 2a_1 a_2 \cos(\omega x) \tag{5.15}$$

where:

$x = L_2 - L_1$ is the optical path difference between the two waves

This implies that $L_1 = L_2$ for $x = 0$ (which is usually taken for the origin). Clearly, if the reflection coefficient R of the mirror M_1 is exactly equal to 50%, that is, $a_1 = a_2 = a$, then $I_{\Sigma}(x) = 2a^2 + 2a^2\cos(\omega x)$. The constant term in the summation does not depend on the time-varying quantity $x(t)$ and therefore will be blocked by the capacitor. In other words, the input signal applied to the oscilloscope will be $I(x) = 2a^2\mu_1\cos(\omega x)$, where μ_1 is the quantum yield efficiency of the detector at a frequency ω. Before moving on, let us draw the reader's attention to the fact that the scale along the

* Generally speaking, one does not specifically have to use the Michelson interferometer in Fourier spectrographs. Any interference scheme, allowing for a change of the optical path of one of the rays involved, would do the job.

X-axis of the actual device is different from the scale along the X^*-axis of the oscilloscope's sweep. This fact has to be taken into account when the measurement results are processed. It is clear that the scale factor involved is $M = (x_j - x_i)/(x_j^* - x_i^*)$ and we will repeat that there is an unambiguous assignment of the coordinate x_k^* in the oscilloscope's sweep to any given value of the coordinate x_k, which describes the position of the mirror M_3 at some instant in time t_k.

We have already established that if the input to the Fourier spectrometer is a monochromatic wave with frequency ω and amplitude a, then the oscilloscope will record the signal $I(x) = 2a^2\mu_1\cos(\omega x)$. Clearly, if the input signal is characterized by frequency 2ω and amplitude b, then the oscilloscope will record the signal $I(x) = 2b^2\mu_2\cos(2\omega x)$. If the two input signals are applied simultaneously to the input of the Fourier spectrometer, then the oscilloscope will record the signal $I(x) = 2a^2\mu(\omega)\cos(\omega x) + 2b^2\mu(\omega)\cos(2\omega x)$. Rather than summing separate harmonics, let us ask how one can calculate the signal applied to the oscilloscope if the spectrometer's input signal is described in terms of the spectral radiation density $S(\omega)$. The answer is immediate: We integrate over the range of frequencies to get

$$I(x) = \frac{1}{\pi}\int_0^\infty S(\omega)\mu(\omega)\cos(\omega x)\mathrm{d}\omega \tag{5.16}$$

Moreover, if the function $I(x)$ is real valued and even, which indeed applies to the situation in question, then the integral on the right-hand side of the above equation is simply the Fourier integral. It can therefore be inverted and the spectral density $S(\omega)$ of the radiation source under investigation can be obtained from the measured values of $I(x)$:

$$S(\omega) = \frac{2}{\mu(\omega)}\int_0^\infty I(x)\cos(\omega x)\mathrm{d}x \tag{5.17}$$

In real-life conditions, unless extremely fast phenomena are being investigated, the data recording and control of the Fourier spectrometer can be computer based. Of course, the basic optical elements of the Fourier spectrometer remain as depicted in Figure 5.12, but there is a variety of options for the recording and control systems. For example, the linear periodic voltage generator in Figure 5.12 can be replaced by a clock rate counter that periodically, upon reaching a predetermined number of counts, sends a digital signal corresponding to the total number of elapsed time pulses to a digital-to-analog converter (DAC) that moves the mirror M_3. The same signal can simultaneously trigger an analog-to-digital converter (ADC) to measure the output signal of the photodetector. Note that if one uses digital technology, then the following equation is used to calculate the spectral density of the input radiation:

$$S(\omega_j) = \frac{2\Delta x}{\mu(\omega_j)} \sum_{k=0}^{k=n} I(x_k)\cos\omega_j x_k \tag{5.18}$$

where:

Δx is a discrete step of the displacement of the mirror M_3

$n = (x_{max} - x_{min})/\Delta x$ is the number of recorded values of the function $I(x)$, $(x_{max} - x_{min})$ being the overall displacement of the mirror M_3 during the measurement process.

REFERENCES

1. Goodman, J.W. 1968 *Introduction to Fourier Optics*. New York: McGraw-Hill Book Co.
2. Gruzman, I.S., Kirichuk, V.S., Kosykh, V.P., Peretyagin, G.I., and Spector, A.A. 2002. *Digital Image Processing of Information Systems*. Novosibirsk, Russia: NGTU Publisher.
3. Maréchal, A. and Françon, M. 1960 *Diffraction Structure des Images*. Paris, France: Editions de la Revue d'Optique Théorique et Instrumentale.

CHAPTER 6

Methods of Spectroscopy

PHYSICS OWES ITS UNDERSTANDING of atomic and molecular structure to spectroscopy. Spectroscopy also once enabled physicists to determine the numerical values of a number of fundamental constants. Today this is history. Modern spectroscopy is an important area of applied physics aimed at solving practical problems in a vast range of scientific and technological endeavor, from metallurgy to microbiology and ecology. To meet these modern challenges, we no longer use classical spectrographs. Instead, they have formed the basis for the development of numerous advanced special devices. We shall not go into all the countless detail concerning these devices but rather concentrate on the set of fundamental principles underlying spectroscopy, and how they are revealed in some key design features of spectral devices.

6.1 SPECTRAL DEVICES AND THEIR MAIN CHARACTERISTICS

Spectral devices are intended to explore the spectral composition of radiation. Accordingly, a key role therein is played by their components that fulfill wavelength decomposition of radiation, as well as measurement of spectra. The principal schemes of generic classical spectrographs are shown in Figure 6.1. The radiation to be analyzed first enters the device through a slit (1) positioned in the focal plane of a collimating lens (2) that provides for illumination of the dispersive elements (3) with a wave with plane wavefront. Upon passing through a prism or being scattered by a diffraction grating, rays of different wavelengths are deflected by different angles and then focused by a lens (4) to different loci on a

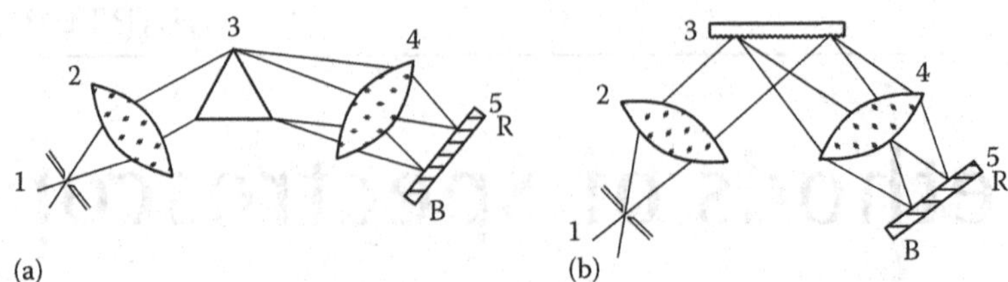

FIGURE 6.1 The principles of classical spectrographs: prism spectrograph (a) and diffraction grating spectrograph (b). 1, entrance slit; 2, collimator lens; 3, dispersive element (prism, diffraction grating); 4, camera lens; 5, photoreceiver; B, blue side of spectrum; R, red side of spectrum.

photoreceiver (5). The latter can be a photographic film, a charge-coupled device (CCD) linear array, or a CCD or photodiode matrix. There are two principal features of the spectrograph design: (1) The dispersive elements are illuminated by a plane wave and (2) the planes of the slit and photoreceiver are optically conjugated. This means that the so-called *spectral lines* that we see on spectrograms are simply multiple images of the entrance slit provided by light rays of different wavelengths. The same principles underlie the construction of the Czerny–Turner type of reflecting optics spectrograph (Figure 6.2), as well as autocollimation schemes. The operational principles of the latter schemes are illustrated in Figure 6.3. As the figure suggests, the lens in such a device acts simultaneously as a collimator generating a wave with flat wavefront, as well as a lens *per se*, focusing the diffracted radiation onto the photoreceiver. Both the entrance slit and the receiver are located in the same plane, and equidistant from the *Z*-axis. Rotating the diffraction grating enables one to obtain different segments of the spectrum appearing on

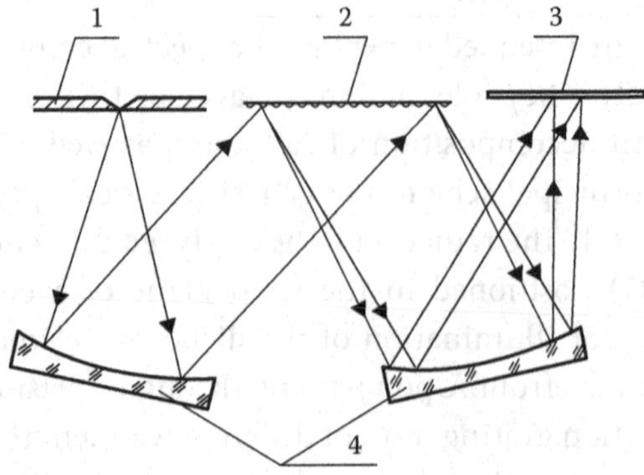

FIGURE 6.2 A reflective optics spectrograph (Czerny–Turner type): 1, slit; 2, diffraction grating; 3, photoreceiver; 4, mirrors.

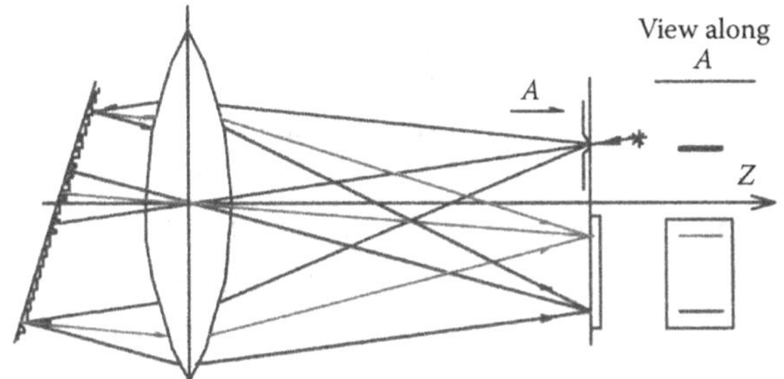

FIGURE 6.3 Light propagation autocollimation spectrograph.

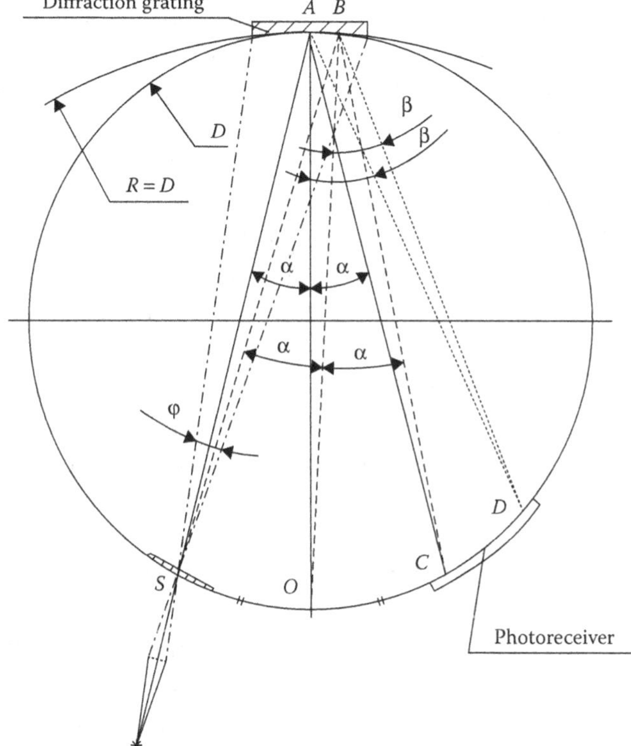

FIGURE 6.4 Spectrograph with concave grating.

the photoreceiver. Quite often, autocollimation schemes also make use of reflecting optics.

Concave grating devices have a special place among spectrographs (Figure 6.4). It can be seen that a concave focusing (often cylindrical shaped) grating with radius of curvature R is tangent at its middle part to a virtual circle with diameter $D = R$ (the so-called *Rowland circle*). Both the entrance slit and the photoreceiver are located on the Rowland circle. The receiver has a cylindrical surface that also lies on the Rowland circle. Let us follow the rays traveling through this device. Suppose that the entrance slit is given by some point S on the circle. The ray SA, incident on the grating at an angle α, will be reflected in zero order at the same angle and arrive at the point C on the Rowland circle (the line OA is perpendicular to the grating, since it originates from its center of curvature). Some other ray SB is incident on the grating at the same angle α, since both angles SAO and SBO lie on the same arc SO. Since OB is normal to the grating, the latter zero-order ray will be reflected at the same angle α and will arrive at the same point C, for the angles SAC and SBC lie on the same arc SC. It follows that the zero-order rays, originally propagating within the angle φ in the figure, will all congregate at the point C. Let AD be some ray in the first order of diffraction, which leaves the point A at the angle β and reaches the Rowland circle at some point D. The ray BD with the same wavelength, diffracted from the point B (clearly at the same angle β), will also reach the Rowland circle at the same point D, simply because the angles OAD and OBD are equal and must therefore lie on the same arc OD. Hence, it follows that all the first-order diffracted rays, given their common wavelength λ, will be focused at D.* In modern designs of this type of spectrographs, one uses holographic gratings, which allow for focusing diffracted rays onto a plane surface tangent to the Rowland circle, rather than a cylindrical surface. This simplifies the design of modern photoreceivers.

Monochromators differ from spectrometers by possessing an exit slit, which transmits radiation in a certain, usually narrow, spectral interval $\Delta\lambda$. The exit slit is positioned in a plane that is optically conjugate with the entrance slit position and is immediately followed by a photoreceiver (a photodiode [PMT]). Monochromators are used in two principal applications. If a process in question is stationary or slow enough in time, the spectrum is scanned through the exit slit by rotating the diffraction grating, the data from the receiver being recorded as a function of the corresponding rotation angle, which in turn depends on the wavelength. If the process is fast, however, the photoreceiver output becomes indicative of the time evolution of a chosen spectral line.

* The formal treatment of focusing properties of concave gratings can be found in the work of Sivukhin [1] (2002, 320–323).

Polychromators are multichannel spectrometers that differ from monochromators by having several, rather than a single, exit slits, optically conjugate to the entrance slit. This type of measurement device is useful if, for instance, one wishes to observe simultaneously several time-dependent spectral lines or continuous spectral intervals and relate them to one another. Such measurements are essential, in particular, for the Thomson scattering method for determining plasma temperature (see Chapter 9). In fact, in the latter case, it becomes feasible to design a polychromator using reflecting band-pass filters, replacing the dispersive elements. A schematic showing the principle of this type of device, used for determining the plasma temperature by the Thomson scattering method in the Joint European Torus (JET) experiment, is illustrated in Figure 6.5. The diagram explains the operational principle of the polychromator. Note that polychromators of this type are especially useful for dealing with extended radiation sources.

A list of the main characteristics of spectral devices is as follows:

Angular dispersion is the quantity $D_\varphi = \mathrm{d}\varphi/\mathrm{d}\lambda$, where $\mathrm{d}\varphi$ is the change of the ray deflection angle by a dispersive element, incurred by the wavelength change $\mathrm{d}\lambda$.

Linear dispersion is defined as $D_l = (\mathrm{d}l/\mathrm{d}\lambda) = D_\varphi f$, where f is the focal length of the camera lens. *Inverse dispersion* is the quantity D_l^{-1}, which is usually expressed in angstroms per millimeter.

Resolution power is characterized by the spread function $g(\lambda)$, and sometimes by the quantity $\Delta\lambda/\lambda$ or its inverse $\lambda/\Delta\lambda$, where $\Delta\lambda$ is the FWHM of the function $g(\lambda)$.

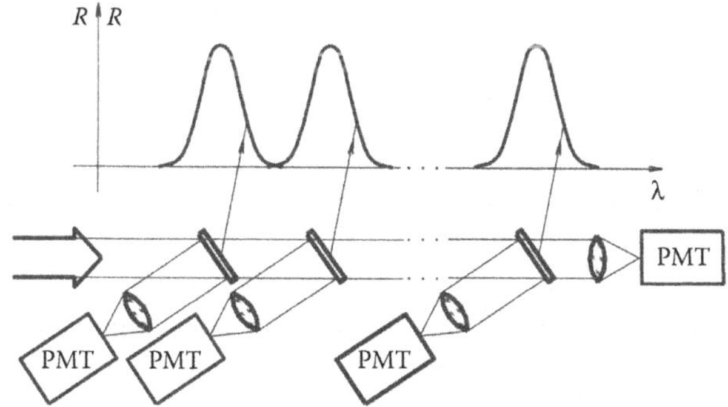

FIGURE 6.5 The principle of a polychromator based on reflecting band-pass filters and photomultipliers (PMT) used to determine plasma temperature by the Thomson scattering method in the JET device. R is reflection coefficient.

Spectrometer throughput (energy efficiency) is the ratio of the output energy of the spectrometer to the input energy of the source.

Dispersion region is described by the spectral interval $\lambda_{max} - \lambda_{min}$, which can be measured by the device. In some designs, the dispersion region can be varied by rotating the diffraction grating; it can also be increased or decreased by using gratings with higher or lower groove densities.

6.2 DISPERSIVE ELEMENTS

Diffraction gratings constitute an indispensable component of the majority of modern spectral devices. Spectrographs and monochromators based on prisms are extremely rare these days. The first diffraction grating was made and investigated in around 1786 by an American inventor David Rittenhouse and consisted of an array of thin hairs strung on a frame.* The theory of the diffraction grating was developed by Fraunhofer as early as 1821. He also suggested a technique for manufacturing both transparent (i.e., transmission) and reflective gratings, which is still being used today. In Figure 6.6, one can see how diffraction occurs on such gratings. It is well known that the maximum intensity occurs in the directions satisfying the condition that the difference Δ between the optical paths of light rays propagating through different slits of the grating is an integer multiple k of the wavelength λ, that is, $\Delta = k\lambda$. (This relation in itself can be viewed as a necessary condition for formation of the diffraction radiation wavefront, i.e., a surface on which the

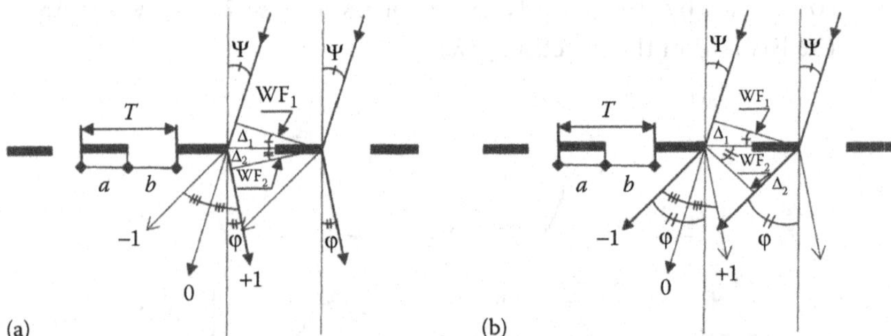

FIGURE 6.6 Light diffraction on transparent grating: diffraction orders +1 ($\Delta = \Delta_1 + \Delta_2$) (a) and −1 ($\Delta = \Delta_2 - \Delta_1$) (b). Directions of rays diffracting in orders +1 or −1 are marked by bold lines. WF_1, incident radiation wavefront; WF_2, diffraction radiation wavefront.

* Clearly, diffraction gratings made this way had a very low angular dispersion.

light wave has a constant phase.) Figure 6.6 shows that for the diffraction orders +1 or −1 on a transmission grating, one should have, respectively, the relations $\Delta = \Delta_1 + \Delta_2$ and $\Delta = \Delta_2 - \Delta_1$, where $\Delta_1 = T\sin\psi$ and $\Delta_2 = T\sin\varphi$, with T being a grating period. (In Figure 6.6, ψ represents the angle of incidence between the radiation and a perpendicular to the grating and φ is the diffraction angle.) Thus, for instance, the condition for diffraction of order +1 on a transmission grating is

$$\Delta = \Delta_1 + \Delta_2 = T(\sin\varphi + \sin\psi) = k\lambda \tag{6.1}$$

Observe that the zero-order wave is traveling in the direction ψ, that is, along the same direction as the incident radiation, while the diffraction angles of the orders +1 and −1 are symmetric with respect to ψ (Figure 6.6). Let us remark that in the case of a reflecting grating, we have, conversely, $\Delta = \Delta_1 + \Delta_2$ for the diffraction angle of the order −1 and $\Delta = \Delta_2 - \Delta_1$ for the diffraction angle of the order +1 (Figure 6.7). Note that Equation 6.1 is satisfied for any λ as long as $\sin\varphi = -\sin\psi$. This is the condition for the zero-order wave formation, which behaves as if it were a mirror reflection from the surface of the grating ($\varphi = -\psi$). The diffraction angles of the orders +1 and −1 are once again symmetric with respect to ψ.

Taking the derivative in Equation 6.1 yields the following equation for the angular dispersion:

$$D_\varphi = \frac{d\varphi}{d\lambda} = \frac{k}{T\cos\varphi} \tag{6.2}$$

Let us analyze the diffraction process of the grating using the traditional approach, where diffraction at a single slit is considered first and then the radiation from each individual slit is superimposed [1,2]. To simplify the

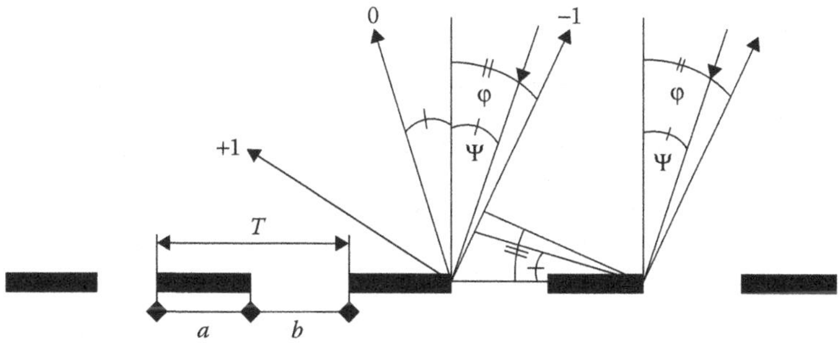

FIGURE 6.7 Diffraction on blazed grating.

calculations and to help the reader to see the forest behind the trees, let us first consider the special case when the incident radiation is normal to the grating. Suppose that we have a transmitting grating with parameters $a = b = T/2$ and a monochromatic plane wave with amplitude $a_0(x, y) = \text{const}$ falling perpendicularly onto the grating. The difference between the optical paths of two waves traveling through two neighboring slits of the grating will, due to the perpendicular direction of the incident radiation relative to the plane of the grating, be equal to $T \sin\varphi$; hence, the phase difference $\delta = (2\pi/\lambda)T \sin\varphi$, where φ denotes, as earlier, the diffraction angle. (Clearly, the phase difference between the waves traveling through, say, the first and third slits will then be 2δ, for the first and fourth slits it will be 3δ and so on.) The field due to radiation traveling through a single slit, let it be the first one, on the observation screen (at infinity or in the focal plane of a converging lens) will be $A_0 \sin u/u$, where $A_0 = a_0/2$ for $a = b = T/2$ and $u = (\pi/\lambda)b \sin\varphi$. Taking summation over all the slits in the grating, we obtain the following result of the interference of the radiation through all individual slits:

$$A_0 \frac{\sin u}{u}[1 + e^{-i\delta} + e^{-i2\delta} + e^{-i3\delta} + \ldots + e^{-i(N-1)\delta}]$$

Using the formula for the summation of a geometric series to evaluate the equation in parentheses, we obtain the following result for the total light wave field at the observation point:

$$U_\Sigma = A_0 \frac{\sin u}{u} \frac{1 - e^{-iN\delta}}{1 - e^{-i\delta}} \tag{6.3}$$

Hence, we have the intensity distribution over the diffraction angle:

$$I = U_\Sigma U_\Sigma^* = A_0^2 \frac{\sin^2 u}{u^2} \frac{1 - (1/2)(e^{iN\delta} + e^{-iN\delta})}{1 - (1/2)(e^{i\delta} + e^{-i\delta})} \tag{6.4}$$

Replacing the complex exponentials by trigonometric functions, we get

$$I(\varphi) = A_0^2 \frac{\sin^2 u}{u^2} \frac{\sin^2 Nv}{\sin^2 v} \tag{6.5}$$

where:

$$v = \frac{\pi}{\lambda} T \sin\varphi$$

We can also rewrite this equation as follows:

$$I(\varphi) = A_0^2 f_1(u) f_2(v) \tag{6.6}$$

In the above equation, the quantity $f_1(u) = (\sin^2 u/u^2)$ characterizes the intensity of harmonics in different orders of diffraction, whereas $f_2(v) = (\sin^2 Nv/\sin^2 v)$ represents the spread function of the diffraction grating $f_2(v) = g(v)$ with minimum possible full width at half maximum (FWHM) β_{min}.* Making the necessary changes for the general case when incident radiation is not normal to the grating surface, Equation 6.5 will remain unchanged, except that the arguments of the two above functions will become

$$\begin{aligned} u &= \frac{\pi}{\lambda} b \cos\psi \sin(\varphi - \psi) \\ v &= \frac{\pi}{\lambda} T(\sin\varphi - \sin\psi) \end{aligned} \tag{6.7}$$

In the case, $\Delta = \Delta_2 - \Delta_1$ and

$$\begin{aligned} u &= \frac{\pi}{\lambda} b \cos\psi \sin(\varphi + \psi) \\ v &= \frac{\pi}{\lambda} T(\sin\varphi + \sin\psi) \end{aligned} \tag{6.8}$$

for $\Delta = \Delta_1 + \Delta_2$.

Figure 6.8 shows the angular intensity $I(\varphi) = A_0^2 f_1(u) f_2(v)$ of diffracted radiation under normal incidence, with parameters $a = b = T/2$, $N = 10$. (Such a small number of interfering rays are chosen to make the illustration more vivid; otherwise, if we dealt with the actual physical number of rays with a chosen angular scale, the graph of the quantity $f_2(v)$ would appear to the viewer as a union of vertical segments.) Since $\psi = 0$, we have $T\sin\varphi = k\lambda$, $u = (\pi/\lambda) b \sin\varphi = \pi k(b/T)$, and $v = (\pi/\lambda) T \sin\varphi = \pi k$, where k, as earlier, is the order of diffraction. Accordingly, $u = v/2$ in the case $a = b = T/2$. Observe that $f_2(v)$ achieves its so-called principal maxima every time that $|v| = k\pi$. Located in the intervals between two principal maxima, there are $N - 2$ additional local maxima of the function $f_2(v)$, as well as $N - 1$ local minima. The angular distance between two neighboring additional local maxima or minima, as well as the distance between

* In practice, the spectrometer's spread function FWHM is determined by the quality of its lens equipment.

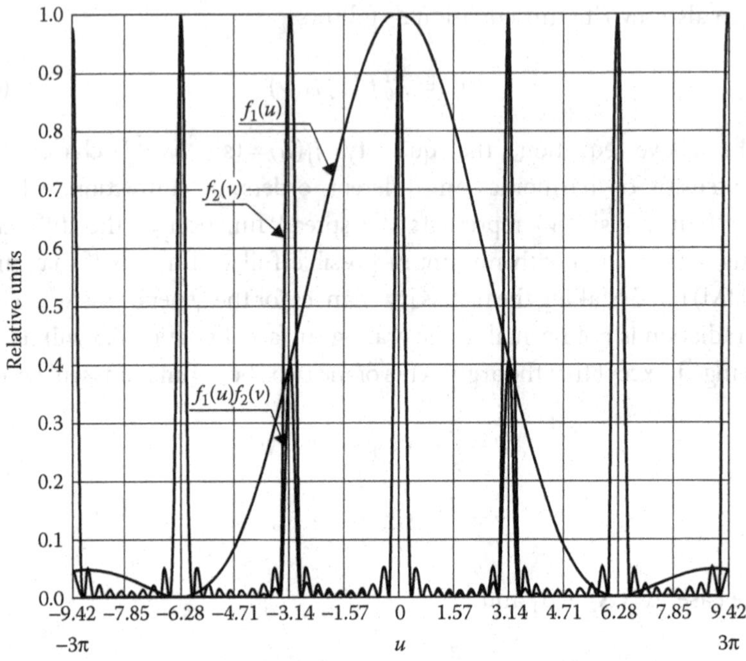

FIGURE 6.8 The angular distribution of diffraction intensity under normal incidence on a grating with parameters $a = b = T/2$. The graphs are normalized to unity.

a principal maximum and a nearby minimum is determined from the condition:

$$T \sin\varphi = \lambda\left(k + \frac{1}{N}\right)$$

Two distinct spectral lines λ_1 and λ_2 are considered resolved in spectroscopy if the principal maximum of the diffraction pattern of the former wave coincides in position with the first diffraction minimum of the latter wave; thus, $\varphi_{\max}(\lambda_1) = \varphi_{\min}(\lambda_2)$. Accordingly, we have

$$T \sin\varphi = \lambda_1 k = T \sin\varphi = \lambda_2 (k + 1/N)$$

and thus $\lambda_1 k = \lambda_2 (k + 1/N)$ or $kN = \lambda_2/(\lambda_1 - \lambda_2)$. Since $\lambda_1 - \lambda_2 = \Delta\lambda$ for the resolution $\mathcal{R}$ of a diffraction grating, we have

$$\mathcal{R} = \frac{\lambda}{\Delta\lambda} = kN \tag{6.9}$$

and therefore, $\mathcal{R}$ is equal to the product of the order of diffraction and the number of interfering rays or rulings in the grating. To get an idea about the number of orders of magnitude involved, suppose, for instance, $\lambda = 5 \times 10^{-5}$ cm, and the grating has 1200 rulings per millimeter and length of 10 cm. Then $\Delta_{\min} \approx 0.04$ Å in the first order of diffraction and about 0.014 Å in the third order; the corresponding dispersion regions remain fairly wide, with 2500 Å and 1250 Å, respectively. Let us emphasize that the above calculation is the upper bound for resolution (it cannot be better than that!). It means that this is the limiting case resolution that would be achieved if the grating, lenses, and all other optical elements contained in the spectrometer were absolutely perfect.

To determine the dispersion region of a diffraction spectrometer, one starts from the precept that the different orders of diffracted radiation should not overlap with one another. Indeed, observe that if for some angle with value φ we observe radiation with wavelength λ in the first order, at the same angle we would be able to observe the second order of radiation with wavelength $\lambda/2$, the third order of radiation with wavelength $\lambda/3$, and so on. Since diffraction angles are the same for overlapping orders, it follows from Equation 6.1 that $\lambda_i k_i = \lambda_{(i+1)} k_{(i+1)}$, that is, $k\lambda = (k+1)(\lambda - \Delta\lambda)$, thus yielding the dispersion region:

$$\Delta\lambda = \frac{\lambda}{k+1} \tag{6.10}$$

As we had already mentioned in Chapter 5, the intensity in the first order of diffraction, which is the order most utilized in spectrometry, comprises only a small fraction of the input signal. It was for this reason that before profile diffraction gratings became available, diffraction spectrometers were regarded as significantly inferior to prism spectrometers in terms of light throughput. The situation changed dramatically after Raleigh came up with the idea of using blazed gratings, which was first realized by Wood (Figure 6.9). The figure shows that in the latter case one also has $\Delta = \Delta_1 + \Delta_2 = T(\sin\varphi + \sin\psi) = k\lambda$. In other words, diffraction angles will satisfy the same relation as they do for a non-blazed grating. However, the angular intensity distribution becomes different, due to an additional phase incursion that takes place. It can be shown that now the argument of the function $f_1(u)$, which stands for the

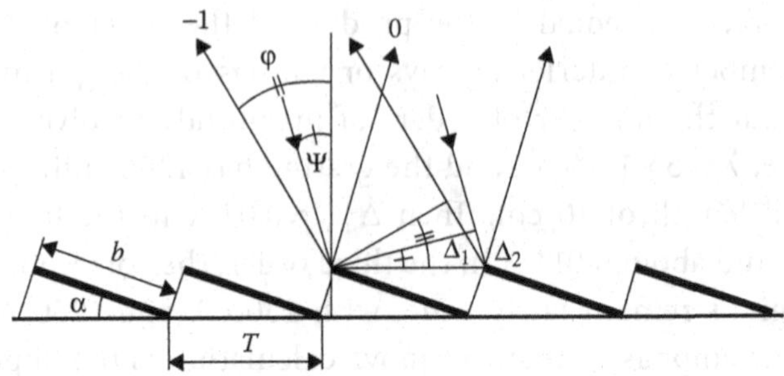

FIGURE 6.9 Diffraction on a blazed grating.

intensity distribution of harmonics in different orders of diffraction, can be expressed as follows:

$$u = \frac{\pi}{\lambda} b \cos(\psi + \alpha) \sin(\varphi + \psi + 2\alpha) \tag{6.11}$$

It can be seen that by changing the slope α of the facet of the grooves of the grating (α is usually called the blazing angle), one can actually achieve the concentration of radiation in the desired order of diffraction. The physical essence of this effect is illustrated in Figure 6.10. Suppose that the incident radiation is falling perpendicular to a groove facet and at the so-called Littrow angle ψ, corresponding to the first order of diffraction φ. Let us also choose the grating parameters (N, b, α) in such a way that in the case $\psi = \varphi$, the zero order would be directed along the groove facet

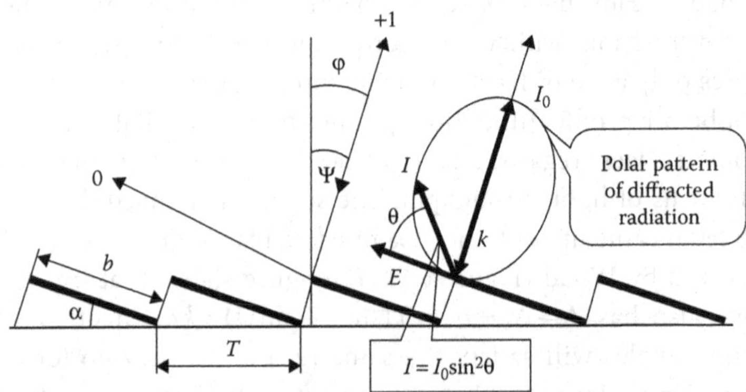

FIGURE 6.10 Concentration of diffraction radiation in the +1 order by a blazed grating.

or at least at some nearby angle, that is, $|\psi| \cong |\alpha|$. In this case, first, negative orders of diffraction simply do not occur, since radiation cannot propagate inside the grating. Second, since the wave vector $\boldsymbol{k}$ is perpendicular to the facet, the electromagnetic field vector $\boldsymbol{E}$ is parallel to the facet. The radiation intensity is then a maximum in the direction perpendicular to the facet and zero in the direction of $\boldsymbol{E}$ (see the dipole radiation diagram in Figure 6.10). The efficiency of such blazed gratings used, for example, in stretchers—devices designed to lengthen the time duration of femtosecond laser pulses—reaches up to between 90% and 95%. Basically, all of the incident radiation energy ends up being concentrated in the first order of diffraction.

Prisms were first used by Sir Isaac Newton as early as the seventeenth century (around 1672) for the purpose of expanding "white" light into its color spectrum. For a long time afterward, the prism played a key role as the core dispersive element in spectral devices, and prisms have not yet become completely anachronistic even in modern laboratory practice. It is worth noticing that the prism also once turned out to be very useful for measuring the coefficient of refraction and the dispersion of transparent substances.

The propagation of light rays in an isosceles prism with base T and vertex angle θ is shown in Figure 6.11. The figure shows that the angle between the wave vectors of radiation refracted by the prism is $\varphi = \alpha_1 - \beta_1 + \alpha_2 - \beta_2$,

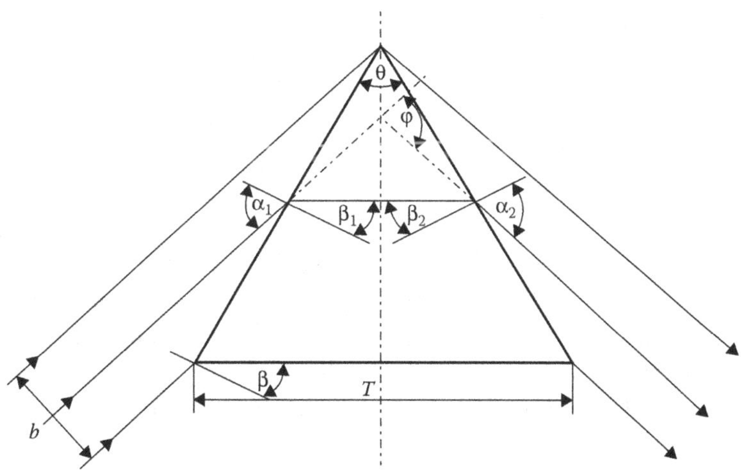

FIGURE 6.11 Light propagation through a symmetric prism.

the vertex angle being $\theta = \beta_1 + \beta_2$, so $\varphi = \alpha_1 + \alpha_2 - \theta$. The angles* α and β are related to each other via Snell's law:

$$\begin{aligned} \sin\alpha_1 &= n\sin\beta_1 \\ \sin\alpha_2 &= n\sin\beta_2 \end{aligned} \tag{6.12}$$

where:

n denotes the coefficient of refraction (refractive index) of the prism material

For a variety of reasons, the arrangement in which the light rays propagate symmetrically is by far the most preferable one and is shown in Figure 6.11. (This design, in particular, yields simultaneously the smallest possible deflection angle together with maximum spectral resolution.) We have $\alpha_1 = \alpha_2 = \alpha$, $\beta_1 = \beta_2 = \beta = \theta/2$, and†

$$\frac{d\varphi}{d\lambda} = \frac{2}{n}\tan\alpha\frac{dn}{d\lambda} = \frac{T}{b}\frac{dn}{d\lambda} \tag{6.13}$$

In accordance with the above minimal deflection principle, if one positions several prisms one after the other, the net dispersion of such a system equals the sum of the individual dispersions of each prism. Any other positioning of the prisms would cause the appearance of angular magnification, which would then have to be taken into account.

The prism's spread function $g(\varphi)$ is determined by diffraction on the limiting aperture (Figure 6.12)‡:

$$E(u) = E_0\frac{\sin^2 u}{u^2} \tag{6.14}$$

where:

$u = \pi b\sin\varphi/\lambda$

It is well known that the angular difference $\Delta\varphi$ between the main maximum ($u = 0$) and the first minimum ($u = \pi$) is equal to the ratio λ/b. Recall that two spectral lines are regarded as resolved in spectroscopy, provided that the angle γ between their wave vectors is more than

* These angles are taken with respect to the direction normal to the corresponding surface.

† If $\theta = 60°$, $(d\varphi/d\lambda) = (2/\sqrt{4-n^2})(dn/d\lambda)$.

‡ This function has FWHM $\beta \approx 0.9\pi$.

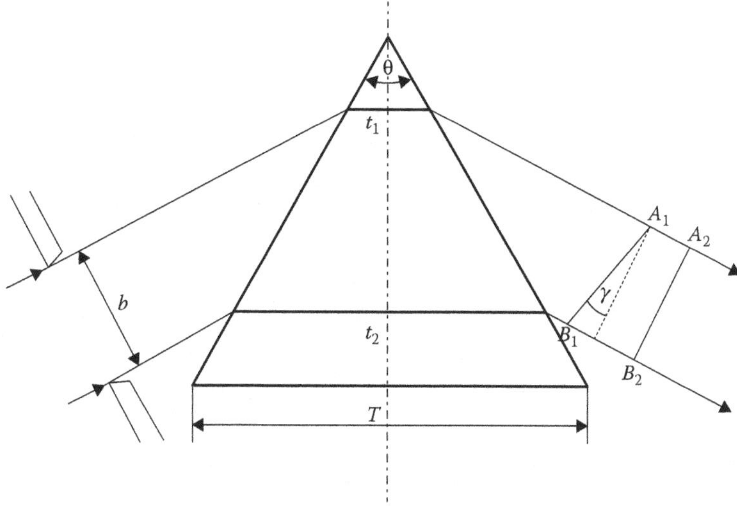

FIGURE 6.12 Wavefronts of light rays traveling through a prism for two nearby wavelengths λ and λ + Δλ.

or equal to Δφ. In Figure 6.12, we see the positions of two wave fronts A_1B_1 and A_2B_2 after the rays with two wavelengths λ_1 and λ_2 have been refracted inside the prism. The angle between the two wavefronts, that is, the angle between their normals given by the corresponding wave vectors, is*

$$\gamma = \frac{B_1B_2 - A_1A_2}{A_2B_2}$$

The fact that one wavefront is delayed with respect to the other, as well as their angular divergence, is caused by the difference in the corresponding refractive indices $n_1(\lambda_1)$ and $n_2(\lambda_2)$, as well as the geometric difference in the path lengths t_1 and t_2 in the upper and lower parts of the prism. Accordingly, we have

$$A_1A_2 = t_1(n_1 - n_2) = t_1\Delta n$$

$$B_1B_2 = t_2(n_1 - n_2) = t_2\Delta n$$

where:

$\Delta n = n_1 - n_2$

* Since the angle γ is small, we equate $\tan\gamma = \gamma$.

Since the width of the light beam $A_2B_2 = b$, it follows that

$$\gamma = \frac{t_2 - t_1}{b}\Delta n$$

Thus, the resolution condition $\gamma = \Delta\varphi = \lambda/b$ can be written as $\lambda = (t_2 - t_1)\Delta n$, whence

$$\frac{\lambda}{\mathrm{d}\lambda} = (t_2 - t_1)\frac{\mathrm{d}n}{\mathrm{d}\lambda} \tag{6.15}$$

For a consecutive array of prisms, we accordingly have

$$\frac{\lambda}{\mathrm{d}\lambda} = \sum_i (t_{2i} - t_{1i})\frac{\mathrm{d}n_i}{\mathrm{d}\lambda} \tag{6.16}$$

If incident radiation illuminates the whole surface of the prism ($t_1 = 0$, $t_2 = T$), then

$$\frac{\lambda}{\mathrm{d}\lambda} = T\frac{\mathrm{d}n}{\mathrm{d}\lambda} \tag{6.17}$$

Hence, it can be seen that the prism's spectral resolution depends on the dispersion of the refractive index, as well as the difference in the optical paths of the uppermost and lowermost rays traveling through the prism. Unfortunately, the derivative of refractive index $\mathrm{d}n/\mathrm{d}\lambda$ is a nonlinear function, which creates two unpleasant circumstances, which are described as follows: First, a nonlinearity of the λ scale in a spectrometer is definitely not a cause for enthusiasm as far as interpreting and processing the measured data is concerned. The second circumstance is the spectra end up being compressed in the red region and stretched in the blue region. This results in significant variation of spectral resolution. Note that the spectral resolution of the prism, although being a very useful characteristic for a variety of estimates, is not yet the spectral resolution of the whole spectral device. There is also at least a pair of lenses, not to mention miscellaneous issues, such as the homogeneity of the refractive index of the prism's material or the quality of its surface finish. A concrete measure of spectral resolution of a spectrometer is, of course, its spread function $g(\lambda)$, which for a variety of reasons is more reliable if measured experimentally than addressed theoretically.

There are two more problems with prism spectrometers: astigmatism and distortion of spectral lines. Astigmatism causes every point of the input slit be projected in the conjugate image plane as a short line segment perpendicular to the λ-axis. This clearly deteriorates the spatial resolution of the device in the direction of the slit. This alone would not cause the spectral resolution to deteriorate, were it not for the widening of the spectrum that occurs due to the phenomenon of spectral line distortion. The latter effect is illustrated in Figure 6.13. In the lower portion of the figure, one can see the light rays propagating in the X–Z plane. (The slit in the corresponding projection looks like a line segment of length h, commonly referred to as the *slit height*.) One can see that the light rays passing through the center ($x = 0$) of the slit meet the surface of the prism at a right angle. If $x \neq 0$, however, that is, for rays that do not pass exactly through the center of the slit, their angle of incidence onto the prism surface differs from a right angle by an amount that increases with the coordinate x. These noncentral rays will travel a longer distance through the body of the prism and will end up being deflected by a greater angle. As a result, the projected image of the off-center point will be shifted along the λ-axis toward the shorter wavelength direction. Consequently, spectral lines end up being bent, bulging out in the direction of longer wavelengths. Diffraction grating spectrographs are also not immune to this type of distortion, but they can be rendered negligibly small; they are dominated by the spread function FWHM, if one uses lens equipment with very large focal lengths (~1 m) and slits of relatively small heights (1–2 cm).

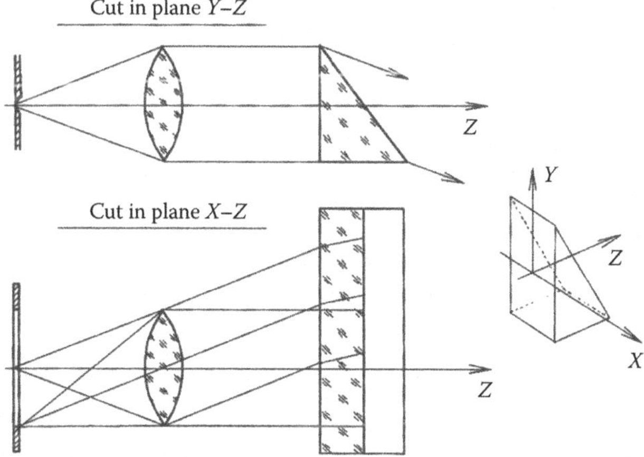

FIGURE 6.13 The principle of spectral line distortion by a prism.

6.3 HIGH-RESOLUTION SPECTRAL DEVICES

Spectrometers have spectral resolution $\mathcal{R} = \lambda/\Delta\lambda = kN = N\Delta_0/\lambda$ of the order of 10^4–10^5. Sometimes, however, there is a whole realm of measurements (superfine structure of spectral lines, isotopic shifts, etc.) that require a spectral resolution that is at least an order of magnitude greater to distinguish spectral components differing by between 10^{-2} and 10^{-4} Å. Devices capable of doing this are usually built on the basis of interferometer schemes and referred to as *high-resolution devices.* Strictly speaking, a diffraction grating also represents an interferometric device, because its high spectral resolution is the result of interference of a multitude of light beams. What is actually the difference? As we have seen, interference-based spectral devices have resolution $\mathcal{R} = \lambda/\Delta\lambda = kN = N\Delta_0/\lambda$, where Δ_0 is the difference in the optical paths of two neighboring beams and N is the total number of interfering beams. Diffraction grating spectrometers are characterized by a fairly small value of Δ_0, typically several times λ. Their high spectral resolution is due to the number of grooves N reaching into tens of thousands due to an increase in the size of the grating and/or the density of ruling. On the contrary, high-resolution devices have a fairly small number of interfering beams ($10 < N < 100$), but with $0.5 < \Delta_0 < 50$ cm. This easily allows for a spectral resolution of up to 10^{-4} Å, but the reverse side of the coin becomes a rather limited dispersion region. Indeed, the condition for having an interference maximum is multiplicity of the quantity Δ_0 at a whole number of wavelengths $\Delta_0 = k\lambda$ or $\Delta_0 = k'(\lambda + \Delta\lambda)$. The minimum value $\Delta\lambda$ corresponds to the difference $k - k' = 1$, from which it is easy to show that

$$\Delta\lambda \approx \frac{\lambda}{k} = \frac{\lambda^2}{\Delta_0} \tag{6.18}$$

This latter quantity is known as the device constant. Let us assume that $\Delta_0 = 5\,\text{cm}$ and $\lambda = 5 \times 10^{-5}$ cm. Then $\Delta\lambda = 5 \times 10^{-10}\,\text{cm} = 0.05\,\text{Å}$. Not much, as we see, and this is the price to pay for high resolution. For this very reason, high-resolution devices are often used in combination with additional monochromators or spectrometers. We will discuss such designs in more detail a little later in this section.

The above discussion suggests that as a basis for high-resolution devices, one needs an optical scheme that allows the interference of many beams with a large phase difference $2\pi\Delta_0/\lambda$ between each other. Historically, the well-known names here are the Michelson echelon, the

Lummer–Gehrcke plate, and the Fabry–Pérot etalon. The first two are, in fact, no longer used these days, so let us concentrate on the Fabry–Pérot etalon, bearing in mind that all three of the devices share the same operation principle.

The Fabry–Pérot interferometer, or etalon, consists of a pair of parallel glass or quartz plates facing each other, their inner faces being covered by reflective coatings. The exterior faces are coated or beveled to avoid additional unwanted interference. Light propagation inside the Fabry–Pérot interferometer is illustrated in Figure 6.14. The corresponding condition for an interference maximum can be stated as follows:

$$2T\cos\varphi = k\lambda \tag{6.19}$$

Let us find the dispersion of the device. Differentiating the above equation with respect to λ gives $\mathrm{d}\varphi/\mathrm{d}\lambda = -k/2T\sin\varphi$, and after eliminating k to yield $\mathrm{d}\varphi/\mathrm{d}\lambda = -1/\lambda\tan\varphi$, we find that for small angles φ,

$$\frac{\mathrm{d}\varphi}{\mathrm{d}\lambda} = -\frac{1}{\lambda\varphi} \tag{6.20}$$

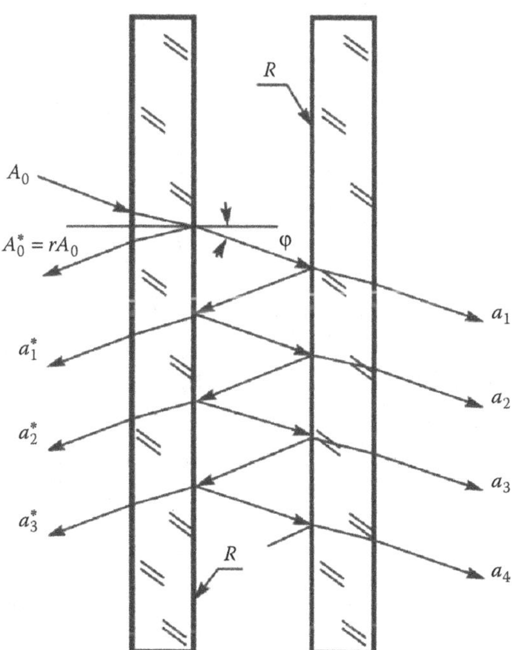

FIGURE 6.14 Light propagation in a Fabry–Pérot interferometer.

This shows that the dispersion of the etalon is independent of the distance T between the mirrors and it increases inversely with φ.

To calculate the net interference result, let us take advantage of the same method that we used earlier for the diffraction grating. The amplitudes of waves that have traveled through the interferometer form an infinite sequence of beams with vanishing intensities $a_1 = A_0(1-R)$, $a_2 = A_0R(1-R)e^{-i\delta}$, $a_3 = A_0R^2(1-R)e^{-i2\delta}$,..., which sum into a geometric series:

$$a_\Sigma = A_0(1-R)(1 + Re^{-i\delta} + R^2e^{-i2\delta} + R^3e^{-i3\delta} + \cdots) \tag{6.21}$$

where:

A_0 is the incident wave amplitude

R is the intensity reflection coefficient of the mirror coatings*

$\delta = 2\pi\Delta_0/\lambda = 4\pi T\cos\varphi/\lambda$ is the phase difference between a pair of adjacent rays (between the first one and the second one, the second one and the third one, and so on)

Summing the geometric series yields

$$a_\Sigma = A_0\frac{1-R}{1-Re^{-i\delta}}$$

and hence the intensity of radiation that has traveled through the interferometer

$$I = a_\Sigma a_\Sigma^* = I_0\frac{(1-R)^2}{(1-R)^2 + 4R\sin^2(\delta/2)} \tag{6.22}$$

The relationship indicated in the above equation is referred to as the Airy formula. (The above formulation does not take into account absorption occurring inside the mirror coatings. Although fairly small if one uses dielectric coatings, this should nonetheless be taken into consideration if the value of R is close to one.†) Equation 6.22 yields the spread function of the Fabry–Pérot etalon; it is plotted in Figure 6.15 for different values of R

* Recall that the intensity reflection coefficient $R = r^2$, where r is the amplitude reflection coefficient.

† Absorption can be taken into account by introducing into Equation 6.22 a corresponding coefficient based on the relation $T_r + A_b + R = 1$, combining the transmission T_r, the absorption A_b, and the reflection R of the mirrors. Note that $T_r = 1 - R$ if $A_b = 0$.

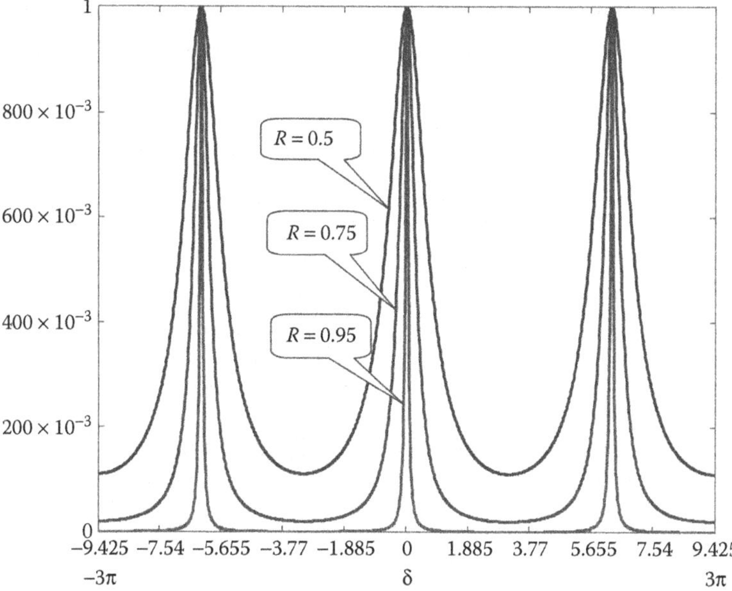

FIGURE 6.15 Far-region radiation intensity in Fabry–Pérot etalon for different values of the reflection coefficient of its mirrors.

and the function reaches its maxima for $\delta/2 = k\pi$. Then $\Delta_0 = k\lambda$ (the path between the mirrors equals an even number of half-waves) and $I = I_0$. As a criterion of resolution of the Fabry–Pérot etalon, one takes the FWHM β of the spread function. Two spectral lines will be regarded as separated, if their difference is at least β. From Equation 6.22, we find

$$\frac{I}{I_{\max}} = \frac{1}{1+[4R/(1-R)^2\sin^2(\delta/2)]} \tag{6.23}$$

from which we can get the angle (or the value of k) when $I = I_{0.5} = I_{\max}/2$. Let the spread function have its maximum at $\delta/2 = k\pi$ and $\delta/2 = (k+\Delta k)\pi$ corresponds to the value $I_{0.5}$. At the same time, $\sin(\delta/2) = |\sin(k+\Delta k)\pi| = |\sin\Delta k\pi| \approx |\Delta k\pi|$, provided that Δk is small, thus $\sin^2(\delta/2) = (\Delta k\pi)^2$. Substituting this value in Equation 6.23, we find the value of Δk, which realizes the value $I = I_{0.5} = I_{\max}/2$:

$$\Delta k = \frac{1-R}{2\pi\sqrt{R}}$$

It is to this value of Δk that the value $\beta/2$ should correspond, while the value β will correspond to two times the former quantity ($\delta k = 2\Delta k$):

$$\delta k = \frac{1-R}{\pi\sqrt{R}} \tag{6.24}$$

One can see from Equation 6.19 that when the incident light forms small angles with the etalon, one has $k = 2T/\lambda$. Differentiating this, we get $\lambda\delta k + k\delta\lambda = 0$, and hence

$$\mathcal{R} = \frac{\lambda}{\delta\lambda} = \left|\frac{k}{\delta k}\right| \tag{6.25}$$

Now substituting $k = 2T/\lambda$ and δk from Equation 6.24, we arrive at the resolution

$$\mathcal{R} = \frac{2\pi T}{\lambda}\frac{\sqrt{R}}{1-R} \tag{6.26}$$

and accordingly can identify the minimum size of a resolvable interval in terms of wavelength

$$\delta\lambda = \frac{\lambda^2(1-R)}{2\pi T\sqrt{R}} \tag{6.27}$$

To complete the formal aspect of the problem, let us comment on the meaning of the so-called effective number of interfering beams—the quantity N_{eff}. This is equal to the number of interfering beams of equal intensity that yields the same resolution as the infinite sequence with vanishing intensity in the Fabry–Pérot etalon:

$$\mathcal{R} = kN_{\text{eff}} = \frac{2T}{\lambda}N_{\text{eff}} \tag{6.28}$$

Combining Equations 6.28, 6.25, and 6.24, we obtain

$$N_{\text{eff}} = \frac{\pi\sqrt{R}}{1-R} \tag{6.29}$$

Having used only formal calculation methods, we have so far obtained a number of useful characteristics of the Fabry–Pérot interferometer. Indeed, paraphrasing A.N. Tikhonov, the advantage of mathematics is that it enables you to stop thinking; all you have to do is write down the original equations, perform the necessary formal operations on them, and arrive at the result. One cannot underestimate the role and power of mathematics for theoretical analysis of physical phenomena. However, solutions obtained formally are not enough to design experiments, nor are they sufficient to create new measuring devices, and in general, to lead to progress in physics. We must now ask ourselves about the physical meaning of the solutions obtained so far. Consider the arrangement when monochromatic radiation falls normally onto the mirror surfaces and the length T is a multiple of one-half the wavelength so that $T = k\lambda/2$. Then the phase difference between the interfering beams is $\delta = 4\pi T/\lambda = 2\pi k$, that is, $\delta/2 = \pi k$. According to Equation 6.22, the intensities of the radiation that is incident on and that has traveled through the interferometer turn out to be equal to one another: $I_0 = I$. How does it come about that a system consisting of two consecutive mirrors, each with transmission $(1-R)$ (which is sometimes only a few percent), has become transparent for radiation? The first answer that comes to mind is that the net output is a result of the cophasal summation of light beams, which previously have been reflected many times by parallel mirrors, leaving the interferometer. This is indeed correct. (In fact, cophasal light beams inside the interferometer are also summed.) Yet let us pay attention to the following situation, radiation inside the interferometer with intensity I_{inn} and outside it, where $I = I_0$ are separated by the exit mirror with transmission $(1-R)$. Clearly, this is possible only if I_{inn} exceeds I_0 by a factor of $1/(1-R)$. In other words, the radiation intensity and therefore the energy density inside the interferometer is $1/(1-R)$ times greater than that outside the interferometer. This phenomenon is well known to laser experts, who know by experience the importance of the optical strength of mirrors. For a similar reason, the radiation incident onto the interferometer is not reflected by the entrance mirror, which behaves as if it were transparent. Of course, there is reflection, but what happens (Figure 6.14) is that compensating beams a_1^*, a_2^*, a_3^*,... with opposite phase are traveling in the very same direction. Their superposition is the reason why it seems as if reflection does not occur.

Let us stress that a necessary condition for the above results to be valid is resonance, that is, when the length T equals a whole number of half-wavelengths $T = k\lambda/2$, that is, $\delta/2 = \pi k$. In addition, the situation has to be *stationary*. The fact of the matter is that two parallel mirrors represent an optical resonator or, if you prefer, an optical energy storage device.

The transition time Δt in the resonator will be precisely equal to the time it takes to accumulate the amount of energy inside the resonator, which corresponds to such a value I_{inn}, that would allow for radiation intensity $I = I_0$ at the output of the interferometer. Hence, we should point out that the relationships derived by Equations 6.22 through 6.29 will only work in the stationary situation, after a time interval Δt when the oscillations in the resonator have become stabilized. The estimate for Δt can be obtained from the quantity N_{eff}—in order for us to even approach the limiting spectral resolution, light must bounce back and forth inside the resonator for at least a number of times equal to N_{eff}. Naturally, the time it takes will be $\Delta t = N_{eff} 2T/c\cos\varphi$. How good is the above estimate? Unfortunately, in real life when the number of interfering rays is finite, there is no way to get a nice analytic expression like the Airy formula. The intensity of radiation that has traveled through the interferometer $I_{lim}(N, R, \delta)$ can be calculated for a finite number of rays, and by restricting the number of terms in the series (6.21) to $I_{lim}(N, R, \delta) = a_{lim} a^*_{lim}$, we are able to calculate the spectral resolution $\delta\pi$. The results of such calculations are shown in Figure 6.16. According to the figure, the measurement error in the case

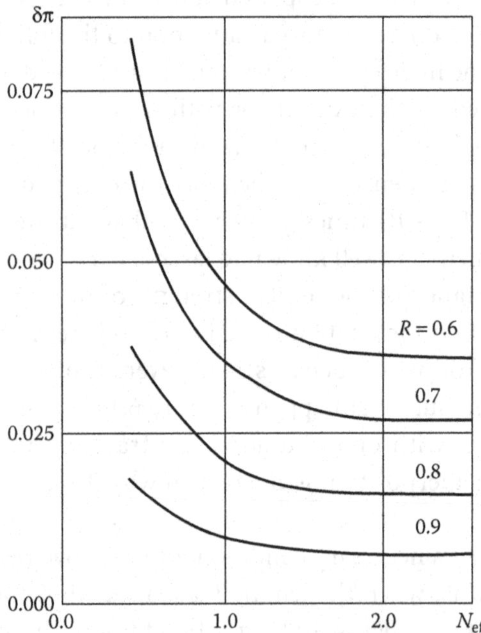

FIGURE 6.16 Measurement errors $\delta\pi$ versus the dimensionless transition time in a resonator N_{eff}.

$N_{\text{eff}} = 1$ is 1.3 times greater than the error for the stationary situation, but already in the case $N_{\text{eff}} = 2$, the error will differ from its limiting value only in the third digit after the decimal point. There is also another method for calculating the transition time of a resonator, which can be found from the dependence:

$$I_{\text{inn}}(t) = \frac{I_0}{1-R}\left\{1 - \exp\left[-\frac{c(1-R)}{2T}t\right]\right\} \tag{6.30}$$

If $N_{\text{eff}} = 2$, that is, with $\Delta t = N_{\text{eff}} 4T/c\cos\varphi$, the value $I_{\text{inn}}(t)$ will differ from the value $I_0/(1-R)$ also only in the third decimal place.

The quantity $\Delta t = N_{\text{eff}} 4T/c\cos\varphi$ is an estimate for the time resolution of the Fabry–Pérot interferometer, as it represents the minimum measurement duration that one needs to ensure that the measurement error δ be only 1.3 times greater than the minimum possible error. Naturally, if Δt_{ch} is the characteristic time of some physical phenomenon under investigation, then in order that the Fabry–Pérot interferometer provides reliable information about the process, one must have $\Delta t_{\text{ch}} \geq \Delta t$. Suppose that the distance between the mirrors is T = 10 cm and the reflection coefficient R = 0.95, then $\Delta t_{\text{ch}} \geq 4 \times 10^{-8}$ s $= 40$ ns ($\Delta t_{\text{ch}} \geq 0.1$ μs for R = 0.98 and T = 10 cm). Decreasing the distance T to 0.5 cm, we find $\Delta t_{\text{ch}} \geq 2 \times 10^{-9}$ s $= 2$ ns. In other words, in the latter case, the Fabry–Pérot etalon is still useless to conduct measurements in the picosecond range. If one continues decreasing the distance between the mirrors or the reflection coefficient, then one eventually loses the advantage in spectral resolution of the Fabry–Pérot etalon compared to common spectrometers.

Let us compare the resolution of the Fabry–Pérot interferometer with that of the diffraction grating which we analyzed earlier. Suppose that we are dealing with a wavelength $\lambda = 5 \times 10^{-5}$ cm, a distance between the mirrors T = 10 cm, and a reflection coefficient R = 0.95. The resolution of the Fabry–Pérot etalon $\delta\lambda_{\text{min}}$ will then be 2×10^{-4} Å, which is higher by two orders of magnitude than the resolution offered by the diffraction grating. Unfortunately, one cannot have everything; the dispersion region of the etalon will be catastrophically narrow (about 0.025 Å). It can be made proportionally wider by decreasing the parameter T, but this will cause the resolution to deteriorate in the same proportion. If the distance T is reduced to 2.5 mm, the etalon's dispersion region will increase to 1 Å, while the resolution will fall to 8×10^{-3} Å.

The latter dispersion region is still insufficient for analyzing spectra that are superpositions of variable wavelengths. In this case, one attempts to design a system that would combine a Fabry–Pérot device together with additional monochromators or spectrometers. As for the monochromator, one chooses the width of its exit slit in to ensure that signal frequencies outside the etalon's dispersion region are cut off. Then one can proceed to investigate the fine structure of a chosen spectral line or perhaps resolve several nearby ones. If used in combination with a spectrometer, the Fabry–Pérot device is positioned in front of the spectrometer and illuminated with a slightly divergent beam. If the beam was monochromatic, the interference pattern projected onto the entrance slit of the spectrometer would be a set of concentric circles, the difference in the radii of adjacent circles becoming smaller as the radii increase. This picture would clearly not develop if the width of the radiation spectrum exceeded the dispersion region. However, once one has accomplished wavelength decomposition, it becomes possible to observe clear multiple interference patterns, each one corresponding to a specific wavelength. Spectral lines will appear as if they have been cut through by lateral segments (the slit will cut out the central part of the interference pattern). Processing such measurement data, one can roughly estimate a particular wavelength from the position of a spectral line on the spectrogram. The interference pattern allows the wavelength to be determined with a high degree of accuracy.

6.4 SPECIAL METHODS OF SPECTRAL ANALYSIS

In this section, we briefly consider the physical principles underlying intracavity laser spectroscopy, as well as spectroscopy based on coherent anti-Stokes Raman scattering (CARS). These methods for spectral analysis are used mainly in molecular spectra research.

Intracavity laser absorption spectroscopy is a method for studying radiation absorption in matter, due to the special features of the latter's molecular structure. Before lasers were invented, these challenges were met by using common spectrophotometers, which actually did not guarantee high measurement accuracy in the case of strong absorption. The required frequency dependencies of the absorption coefficient $\mu(\omega)$ or the absorption cross section $\sigma_a(\omega)$ can be found from the following equation:

$$\frac{I(\omega)}{I_0(\omega)} = \exp\left[-\sigma_a(\omega)nL\right] = \exp\left[-\mu(\omega)L\right] \tag{6.31}$$

where:

$I_0(\omega)$ is the probe beam intensity

$I(\omega)$ is the intensity of the beam that has traveled through the object under investigation

n is the density of absorbing particles

L is the length of the absorption layer

The ratio $I(\omega)/I_0(\omega)$ can be measured experimentally either by frequency scanning, that is, by changing the narrow-band radiation frequency of the probing beam, or by using broadband probing radiation, where the bandwidth $\Delta\omega_p$ is much greater than the spectral width of the absorption line $\Delta\omega_a$. In the latter case, the radiation that has traveled through the sample is recorded by a spectrometer.

One can see from Equation 6.31 that for small values of $\mu(\omega)$, the ratio $I(\omega)/I_0(\omega)$ gets close to one and the measurement accuracy will be small. To improve this situation, one can sometimes succeed in effectively increasing the value of L by using multipass designs where the beam traverses the sample (say, a cell containing the gas under investigation) many times, being reflected back and forth by a system of mirrors. However, as the number of reflections increases, the intensity of the probe beam will decrease dramatically, owing to losses at the cells' windows and mirrors, and as the result of vignetting due to probe beam divergence, etc. (If, for instance, single-pass losses amount to 10%, then after 10 double return passes through the cell, the probe beam intensity will end up becoming smaller by a whole order of magnitude.) All these affects naturally lead to a loss of measurement accuracy. To compensate for these so-called *non-resonance* losses, the investigation sample itself, be it a transparent solid or a cell with liquid or gas, is placed inside the cavity of a laser. Then first, the presence of an active amplification medium in the cavity enables one to compensate for losses due to radiation being reflected from the mirrors, traveling through the cell windows, or any other optical elements inside the cavity. However, one achieves the same effect as in multipass systems; the probe beams traverse the sample many times after their multiple reflections from the mirrors.

If a cell filled with an absorbing substance is placed inside a laser cavity, the amplification factor per single pass between the mirrors, due to the presence of both amplifying and absorption media, can be written as

$$k(\alpha,\mu,L) = \exp\left[\alpha(\omega)L_1 - \mu(\omega)L_2\right] \tag{6.32}$$

where:

L_1 is the distance traveled in the active medium

L_2 is the length of the cell filled with the absorption medium

$\mu(\omega)$ is the absorption coefficient in the investigated medium

$\alpha(\omega) = \sigma(\omega)n$ is the amplification factor of the active medium, where $\sigma(\omega)$ and n denote, respectively, the cross section of the induced radiation and the density of atoms in the inverted state. (Note that in this case the generation threshold will be achieved at $r_{\text{eff}}^2 \exp\{2[\alpha(\omega)L_1 - \mu(\omega)L_2]\} = 1$ or $[\alpha(\omega)L_1 - \mu(\omega)L_2] = -\ln r_{\text{eff}}$, where r_{eff} is a coefficient that takes into account mirror losses as well as other *nonresonant losses.*) Applying the same summation algorithm for the beams inside the resonator that we have already used *apropos* of the Fabry–Pérot etalon (Section 6.3), we obtain*

$$I = I_0 \frac{[1 + rk(\alpha,\mu,L)]t}{[1 - rk(\alpha,\mu,L)]^2 + 4rk(\alpha,\mu,L)\sin^2(\delta/2)} \tag{6.33}$$

where:

I_0 is the radiation intensity in the active medium in the absence of the mirrors

I is the radiation intensity after the laser's exit mirror

r and t are, respectively, the reflection and transmission coefficients

$\delta = 2\pi\Delta_0/\lambda = 4\pi T\cos\varphi/\lambda$ is the phase difference between the adjacent beams (the first and the second, the second and the third, and so on) (as with the case of the Fabry–Pérot interferometer with optical path T between mirrors)

This measurement requires that the radiation bandwidth $\Delta\omega_p$ is much greater than the spectral width of the absorption line $\Delta\omega_a$. To ensure this, in practice one often uses dye lasers, that is, lasers that have active media with a wide luminescence region. Strictly speaking, in this probing method, the object under investigation is illuminated by radiation whose spectrum is not continuous, but has only the set of frequencies corresponding to the cavity's longitudinal modes. (In other words, the investigated object is probed discretely, with a wavelength step $\Delta\lambda$.) There is nothing wrong with this, since the spectral distance between the modes $\Delta\lambda = \lambda^2/\Delta_0$ is sufficiently small. For instance, if $\lambda = 10^{-4}$ cm and the cavity

* For more detail, see Landsberg, G.S., 2003. *Optics.* Moscow, Russia: Fizmatlit, Chapter 15.

has a length of 1 m, then $\Delta\lambda = 10^{-2}$ Å, which is much smaller than the spectral width of the absorption line. Assuming that the reflection coefficients of the cavity's mirrors are sufficiently large so that most of the radiation is contained in resonance modes ($\delta/2 = \pi k$), and that the small percentage of the radiation intensity in the regions between the modes is negligible, one can rewrite Equation 6.33 as follows:

$$I \cong I_0 \frac{[1 + rk(\alpha, \mu, L)]t}{[1 - rk(\alpha, \mu, L)]^2} \tag{6.34}$$

where:

$$k(\alpha, \mu, L) = \exp[\alpha(\omega)L_1 - \mu(\omega)L_2]$$

Solving Equation 6.34 for k and taking into account the fact that $2I/I_0 t >> 1$, we derive

$$k(\alpha,\mu, L) = \exp[\alpha(\omega)L_1 - \mu(\omega)L_2] \cong \frac{1 + \sqrt{2I_0 t/I}}{r} = A \tag{6.35}$$

and thus

$$\alpha(\omega)L_1 - \mu(\omega)L_2 = \ln A = \ln(1 + \sqrt{2I_0 t/I}) - \ln r \approx \sqrt{2I_0 t/I} - \ln r \tag{6.36}$$

Certainly, one can obtain the exact solution of Equation 6.33, although it appears rather cumbersome. However, we will demonstrate that this is unnecessary and that the main goal of the calculation involved is to show that the difference $[\alpha(\omega)L_1 - \mu(\omega)L_2]$ is equal to some quantity in the form $\ln A$, which is dependent on the resonator parameters (r, t) as well as the pumping power I_0. We emphasize that this method of spectroscopy does not require absolute measurements. The absorption $\mu(\omega)$ that is required can be found from the ratio of $\mu(\omega)$ to $\alpha(\omega) = \sigma(\omega)n$, since the cross section σ and the value of n are readily available. One needs to compare only the values of the signals measured in the spectrogram within the measured absorption line, where $\alpha(\omega_{in})L_1 - \mu(\omega_{in})L_2 = \ln A$ and those in its immediate vicinity, where $\mu(\omega_{out})L_1 = 0$ and $\alpha(\omega_{out})L_1 = \ln A$.

Although the accuracy of intracavity laser absorption spectroscopy is not very high (~10%), its gain in sensitivity is such that it gives an advantage of some six or seven orders of magnitude in comparison with spectrophotometry.

Coherent anti-Stokes Raman (light) scattering (CARS) represents an efficient method for studying the molecular structure of matter. It enables one to determine easily the normal frequencies of molecular oscillations and to draw conclusions about their symmetry and dynamics. The characteristics of Raman scattering spectra also enable one to analyze readily the composition of complex gas mixtures. Notwithstanding all this, in a sense, CARS represents just a recent step in the development of Raman spectroscopy. Therefore, before considering processes inherent in CARS, let us make a list of the physical phenomena involved in the usual Raman scattering or the Raman effect.

- As a result of the Raman effect, the light scattered by a medium* contains satellite frequencies in addition to the principal component with the probe radiation frequency ω_p. The satellite frequencies have values that are symmetric with respect to ω_p and separated from it by the quantity $|\omega_0|$, where ω_0 are the eigenfrequencies of oscillations that are characteristic of the medium under investigation and independent of the frequency ω_p of the probing pulse.
- The frequencies $\omega_r = \omega_p - \omega_0$ that are shifted in the "red direction" (i.e., in the direction of smaller values of wave frequency) are referred to as *Stokes components.* The frequencies $\omega_b = \omega_p + \omega_0$ that are shifted in the "blue direction" (i.e., in the direction of an increase in frequency) are known as *anti-Stokes components.* It is clear that in the former case a photon of the probing radiation loses a part of its energy to excite a molecule, whereas in the latter case it gains extra energy from an excited molecule. The number of molecules N_i in the excited state is considerably smaller than the number of molecules in the normal state N_k. Since both N_i and N_k result from random fluctuations due to random collisions, the corresponding intensities of the scattered radiation I_i and I_k within the solid angle 4π are proportional to the populations of the levels. Thus, $I_i \propto N_i$ and $I_k \propto N_k$ (Chapter 9), and therefore, the "red" satellites have much higher intensities than the "blue" ones.
- The eigenfrequency of molecular oscillations can be found from the relation $\hbar\omega_0 = \mathcal{E}_i - \mathcal{E}_k$, where $\mathcal{E}_i$ and $\mathcal{E}_k$ are the molecular discrete energy levels. According to Boltzmann statistics, the ratio of

* For more detail about light scattering processes, see Chapter 9.

the number of molecules in the excited state N_i to the number of unexcited molecules N_k is

$$\frac{N_i}{N_k} = \exp\left(-\frac{\mathcal{E}_i - \mathcal{E}_k}{kT}\right) = \exp\frac{\hbar|\omega_0|}{kT} \propto \frac{I_i}{I_k}$$

from which one can see, in particular, that the intensity of a blue satellite should increase with temperature. This is indeed the case as confirmed by the experiment. For a variety of reasons, it is anti-Stokes component that is fruitful from the point of view of measurement and subsequent data analysis and interpretation. However, as we have seen, it is the anti-Stokes component that lacks intensity under usual Raman scattering conditions, which obviously does not help the measurements.

For coherent anti-Stokes Raman scattering, this problem is solved in two ways: by increasing the intensity of the scattered anti-Stokes component and by drastically improving the collection efficiency of scattered photons. The first task is clear: One has to generate a sufficiently large number of excited molecules in the investigated medium. This is not a novelty in experimental physics, since resonance luminescence is based on the same method. However, one encounters considerable obstacles against exciting molecules this way, as it requires a tunable source of radiation in the far-infrared region that does not exist even today. However, the CARS method uses a different approach to the resonant excitation of molecules. It utilizes two laser beams in the visible or near-infrared region of the spectrum with frequencies ω_1 and ω_2, which are focused simultaneously into the sample volume. The frequencies ω_1 and ω_2 are chosen in such a way that their difference is equal to ω_0—the eigenfrequency of the molecular oscillator to be excited. If the beams are powerful enough, then due to nonlinearity, they give rise to a beat wave inside the medium, whose frequency equals $(\omega_1 - \omega_2) = \omega_0$ and which serves to excite the molecular oscillators. Simultaneously, or after a time sufficient for the excited molecules to reach a sufficient concentration, one shines a probe beam through the sample. (In fact, by using femtosecond lasers with their broadband radiation, one often arrives at the situation when several, rather than one, molecular oscillators can be excited.) Yet this is not the main thing about the CARS method. Its principal feature is that the forced oscillations of the molecules produced by the same electromagnetic waves turn out to be coherent. This remarkable fact has two key nontrivial consequences.

First, as a result of the probe beam scattering on coherently oscillating molecules, the amplitudes of the scattered waves also add up coherently, so that their overall intensity turns out to be proportional to the square of the number of excited molecules, rather than to its first power as in the usual Raman effect. Second, the scattered photons propagate just within a narrow spatial angle, rather than the whole range of 4π, as would be the case in the usual Raman effect. The point is that if ω_p is the probing radiation frequency, ω_0 is the eigenfrequency of the molecular oscillator, and ω_s is the scattered radiation frequency, they must satisfy the following relationships:

$$\hbar\omega_s = \hbar\omega_p + \hbar\omega_0$$

as well as

$$\boldsymbol{k}_s = \boldsymbol{k}_p + \boldsymbol{k}_0$$

The first equation is just the conservation of energy and the second one is responsible for the conservation of momentum. Since $\omega_0 << \omega_p \approx \omega_s$, the angle θ between the wave vectors $\boldsymbol{k}_p$ and $\boldsymbol{k}_s$ is fairly small, although often sufficient to have the probe and scattered beams diverge from one another. The scattered radiation propagates, therefore, in the direction of the vector $\boldsymbol{k}_s$, and its divergence is determined by the beam's size in the lateral section, that is, by diffraction.

These two circumstances, namely, the proportionality of the scattered radiation intensity to the square of the number of excited molecules and the fact that the scattered radiation is generated only within a narrow solid angle, give CARS spectroscopy an advantage in terms of sensitivity of some five or six orders of magnitude compared to the usual Raman spectroscopy.

REFERENCES

1. Sivukhin, D.V. 2002. *Optics.* Moscow, Russia: Fizmatlit.
2. Zaidel, A.N., Ostrovskaya, G.V., and Ostrovskii, Yu.I. 1972. *Technique and Practice of Spectroscopy.* Moscow, Russia: Nauka.

CHAPTER 7

Interferometry and Shadowgraphy Methods

7.1 INTERFEROMETRY

Interferometry is one of the most accurate measurement methods. There are practically no other devices that allow length measurements on the scale of one-hundredth or even up to one-thousandth of an optical wavelength. Today's optical industry would not exist without interferometric devices. They provide an unconditional guarantee of quality of optical devices. Interferometry is used successfully in measuring airflow around the models of planes and rockets in wind tunnels, simulating spacecraft reentry in dense atmospheric layers and processes of generation of shock and detonation waves, obtaining information about the density distribution in plasma formations: This is far from a complete list of fundamentally important applications of interferometry.

Interferometry does not allow direct measurement of time intervals. However, it does enable one to measure optical path lengths and thus the time that it takes light to travel along the path. The classical series of experiments conducted by Michelson and Morley provides a textbook example of such measurements. The so-called Michelson–Morley experiment paved the way toward relativity theory and is regarded as one of the most important experiments in the history of science. It was *the time difference of the propagation of two light beams*, which had simultaneously emerged from the same origin in two different directions and returned after having been reflected by mirrors, which was measured in these experiments. The measurement accuracy was equal to 3×10^{-17} s. At the time, it was an unprecedented world record!

The output of interferometric measurements is an interferogram, or an interference pattern, representing the result of interference of two light waves: one wave that has traveled through or been reflected from a studied object, called the *subject wave*, and the other, a coherent *reference wave*. The latter usually has a plane wavefront. Thus, in order to obtain an interferogram, one should superimpose the two waves in some plane. Devices affecting this superimposition are called interferometers. Today there are dozens of various types of these devices. Rather than considering the multitude of their designs, let us focus on the physical aspects of interferometry, its principal methods, and the ways of processing interferograms.

7.2 LIGHT INTENSITY DISTRIBUTION IN AN INTERFEROMETRIC PATTERN

Suppose we are dealing with a Mach–Zehnder interferometer (Figure 7.1) and the source of the radiation is a single-frequency laser generating a plane light wave $U(t) = U_0 e^{i\omega t}$ that is propagating along the Z-axis. In the above equation, U_0 is the amplitude and ωt is the phase of the wave. Consider the wave amplitude distribution on a screen,* which has been placed behind the interferometer in some plane oriented parallel to the X–Y plane. On the screen, the reference wave $U_1(t)$, which has not passed

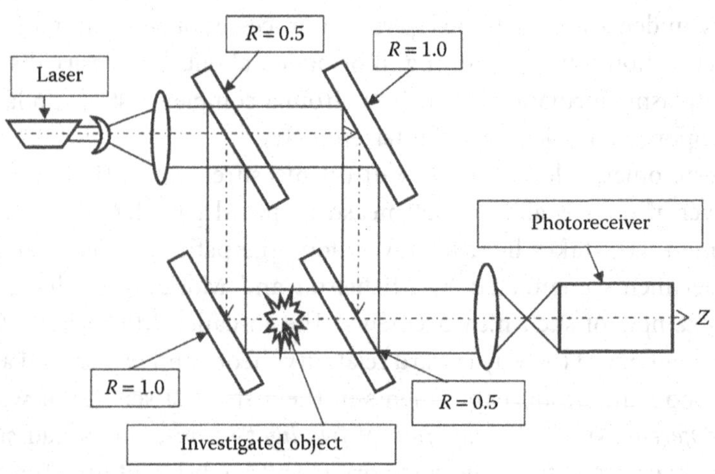

FIGURE 7.1 Mach–Zehnder interferometer.

* It should be clear that in the setting of our mental experiment, the role of the "screen" will be played by a photographic film, a CCD matrix, a cathode of a photomultiplier, or a similar device.

through the object under investigation, will have accumulated some phase delay relative to the initial wave. Thus, we have

$$U_1(t) = U_{01}\exp[i\omega(t - Z_1/c)] \tag{7.1}$$

where:

U_{01} is the amplitude of the reference wave

$\omega = 2\pi\nu$ is its angular frequency (where ν is the frequency)

t denotes the time

Z_1 is the distance from the observation point, namely, the screen, to the source at the origin

The wave $U_2(t)$, which has traveled through the subject under investigation, will of course also be delayed with respect to the emitted wave as follows:

$$U_2(t) = U_{02}\exp\left\{i\omega\left[t - \frac{(n-1)\Delta Z}{c} - \frac{Z_2}{c}\right]\right\} \tag{7.2}$$

where:

U_{02} is the amplitude of the subject wave

$(n - 1)\Delta Z$ is the increase in length of the optical path due to the fact that the subject wave has been propagating through the subject that is being studied, which is characterized by a refractive index n

Let $\theta_1 = \omega(t - Z_1/c)$ and $\theta_2 = \omega\{t - [(n-1)\Delta Z/c] - Z_2/c\}$ denote, respectively, the phases of the reference and subject waves. The sum of these two waves at any point in space where they are superimposed will then be

$$U_\Sigma(t) = U_{01}\exp(i\theta_1) + U_{02}\exp(i\theta_2) = U_\Sigma\exp(i\theta_\Sigma) \tag{7.3}$$

where:

U_Σ is the amplitude of the resulting wave

θ_Σ is its phase

This situation is illustrated in Figure 7.2. A photolayer, a photocathode, a charge-coupled device (CCD) matrix, or almost any other photodetector would react upon the luminosity or intensity $I_\Sigma = U_\Sigma U_\Sigma^*$ of the light

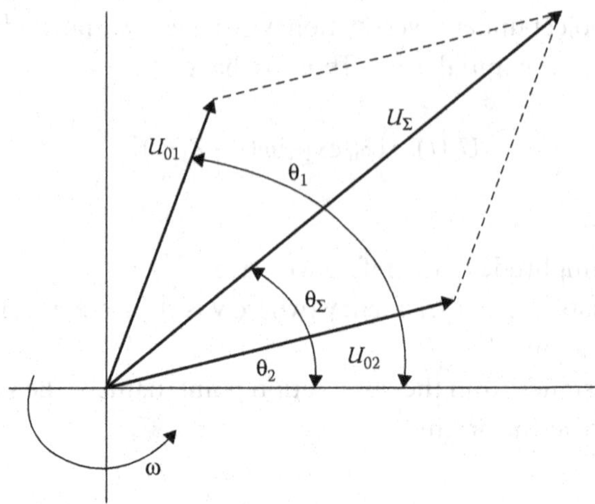

FIGURE 7.2 Sum of two waves at any point in space where they are superimposed.

incident onto it, with U_Σ^* denoting the complex conjugate of the resulting wave. The net intensity is as follows:

$$\begin{aligned} I_\Sigma &= U_\Sigma \exp(\mathrm{i}\theta_\Sigma) U_\Sigma^* \exp(-\mathrm{i}\theta_\Sigma) \\ &= [U_{01}\exp(\mathrm{i}\theta_1) + U_{02}\exp(\mathrm{i}\theta_2)][U_{01}\exp(-\mathrm{i}\theta_1) + U_{02}\exp(-\mathrm{i}\theta_2)] \\ &= U_{01}^2 + U_{02}^2 + U_{01}U_{02}\{\exp[\mathrm{i}(\theta_1 - \theta_2)] + \exp[-\mathrm{i}(\theta_1 - \theta_2)]\} \end{aligned} \tag{7.4}$$

Let us do some algebra using the well-known Euler identity. We have

$$I_\Sigma = U_{01}^2 + U_{02}^2 + 2U_{01}U_{02}\cos(\theta_1 - \theta_2) \tag{7.5}$$

Substituting the values of θ_1 and θ_2 into the identity (7.5), we get

$$I_\Sigma = U_{01}^2 + U_{02}^2 + 2U_{01}U_{02}\cos\left[\frac{\omega t - \omega Z_1}{c} - \frac{\omega t + \omega(n-1)\Delta Z}{c} + \frac{\omega Z_2}{c}\right]$$

and replacing ω/c by $2\pi/\lambda$, we find that (assuming that the arms of the interferometer have equal length $Z_1 = Z_2$)

$$I_\Sigma = U_{01}^2 + U_{02}^2 + 2U_{01}U_{02}\cos[2\pi(n-1)\Delta Z/\lambda] \tag{7.6}$$

This equation determines the light intensity at any point of the interference pattern in the plane of the screen as a function of the distribution of the refractive index $n(x,y,z)$ inside the studied subject. The latter quantity determines the effective optical path length:

$$L(x,y) = \int_{Z_1}^{Z_2} n(x,y,z)\mathrm{d}z = \bar{n}(x,y)\Delta Z(x,y)$$

where:

The quantities Z_1 and Z_2 determine the boundaries of the studied object

$$\Delta Z(x,y) = Z_2(x,y) - Z_1(x,y)$$

Clearly, at the end of the day, the distribution of light in the interference pattern is determined by the curvature of the subject wavefront, whose appearance has been caused by the various phase shifts that the subject wave will have acquired while traveling through the investigated object: $2\pi\Delta Z(x,y)\bar{n}(x,y)/\lambda = \Phi(x,y)$. Having accepted the above notation for these phase shifts, we can rewrite Equation 7.6 as follows:

$$I_\Sigma = U_{01}^2(x,y) + U_{02}^2(x,y) + 2U_{01}(x,y)U_{02}(x,y)\cos[\Phi(x,y)] \tag{7.7}$$

In order to take measurements using interferometry, the interferometer is usually adjusted in such a way that the reference wave arrives at the screen at some angle to the subject wave. We will further show that this is useful, for it enables one to determine the sign of the subject wave curvature. Indeed, suppose the subject wave arrives normal to the plane of the screen and forms some angle β with the reference wave. This angle β is exactly the angle between the wave vectors $\boldsymbol{k}_{\text{sub}}$ and $\boldsymbol{k}_{\text{ref}}$ of the subject and reference waves with the vector $\boldsymbol{k}_{\text{ref}}$ having being rotated in the X–Z plane around the Y-axis. Then, in the absence of any object of investigation, assuming once again that the arms of the interferometer have equal length, the difference in the optical path lengths of the reference and subject waves, will cause their phases to be shifted with respect to one another by the amount $2\pi x\sin\beta/\lambda$, which for small β can be replaced by $2\pi\beta x/\lambda$. (Which wave has a phase advance or a phase delay depends on the signs of x and β.) Figure 7.3 illustrates this argument, with $l = x\sin\beta$ denoting the difference in the optical paths of the interfering waves, so that $l/\lambda = (x\sin\beta)/\lambda$ yields the delay normalized by one period of the light wave and $2\pi x\sin\beta/\lambda$ gives the phase delay in radians. The spatial intensity distribution in the interference pattern can be found, taking into account the fact that in the plane of the screen the phase of the reference wave depends linearly on x

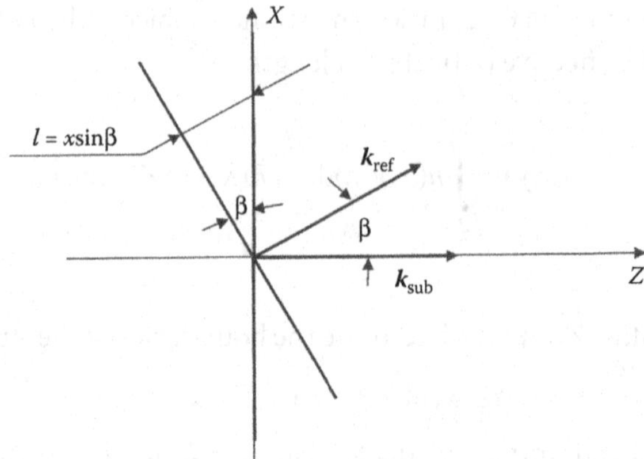

FIGURE 7.3 Phase shift during interference of two waves with plane wavefronts.

and is equal to $2\pi\beta x/\lambda$, whereas the phase of the subject wave is a function of both x and y, which has been denoted as $\Phi(x,y)$, that is,

$$I_{\Sigma}(x,y) = U_{01}^{2}(x,y) + U_{02}^{2}(x,y) + 2U_{01}(x,y)U_{02}(x,y)\cos[\Phi(x,y) + 2\pi\beta x/\lambda] \tag{7.8}$$

If the mirrors of the interferometer are adjusted to ensure that $\beta = 0$, then the intensity distribution in the interference pattern will be described by Equation 7.7. As one can see from Equation 7.7, it is impossible to distinguish between $\Phi(x,y)$ and $-\Phi(x,y)$ from the interference pattern, namely, to distinguish phase fronts with positive and negative curvatures from one another. The cosine is an even function. But if we slant the reference ray by some nonzero angle β with respect to the subject wave, then within the cosine, together with $\Phi(x,y)$, we will acquire an extra term $2\pi\beta x/\lambda$, whose sign does depend on whether β is positive or negative. Now if, for instance, the derivatives $\partial\Phi(x)/\partial x$ and $\partial(2\pi\beta x/\lambda)/\partial x$ have different signs (Equation 7.8), then the interference lines will be stretched out along the X-axis; conversely, they will get closer together if the signs are the same. Figure 7.4 illustrates this scenario in the case when the subject wave has a spherical wavefront. In the example shown in Figure 7.4a, $\beta = 0$ and the interference pattern in the X–Y plane takes the form of concentric circles, which are the same, no matter whether the subject wave is converging or diverging, provided that, of course, the wavefront curvature has the same absolute value in both cases. For negative β and divergent subject wave (Figure 7.4b), depending on whether the derivatives $\partial\Phi(x)/\partial x$ and $\partial[\Phi(2\pi\beta x/\lambda)]/\partial x$ have

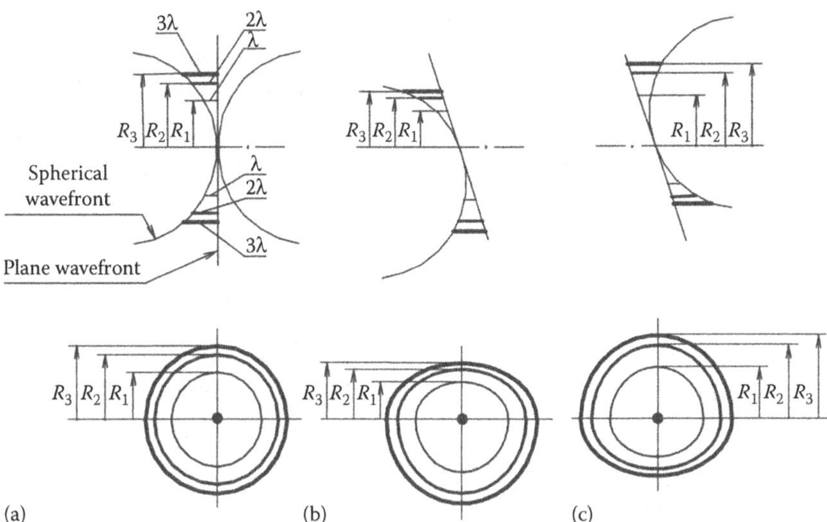

FIGURE 7.4 Interference of waves with plane and spherical wavefronts (negative and positive curvatures); (a) $\beta = 0$ where the interference pattern takes the form of concentric circles for both converging and diverging subject waves; (b) the situation with negative β and divergent subject wave; (c) β is negative and the subject wave is convergent.

the same (the upper part of the diagram) or different signs (the lower part), the interference lines are, respectively, compressed or stretched. Therefore, in Figure 7.4b, the circles have become ellipses, the lines in the upper part of the pattern are compressed, and the lines in the lower part are stretched out. The situation with a convergent subject wave is the same, up to the inversion of the geometric effect (Figure 7.4c); now the lines in the upper part of the pattern are stretched out and in the lower part the lines are compressed.

Unfortunately, the model considered above is an idealization and cannot possibly be put fully into practice for the following reasons: First, no matter how hard one tries, it is impossible to ensure homogeneity of interfering beams in the plane of the screen; neither the initial intensity distribution nor the reflection coefficient distribution for the semitransparent mirrors involved in the interferometer design are constant functions, namely, $U(x,y) \neq \text{const}$, $R(x,y) \neq \text{const}$. Second, it is impossible to manufacture mirrors with ideally flat faces. Let us recall that we are aspiring to measure the quantity $\Phi(x,y)$ with a precision of the order of one-tenth, or perhaps even one-hundredth of a wavelength, which means that ideally the mirror surfaces should have been manufactured with a precision in excess of $\lambda/100$. This would be difficult to say the least, if not impossible, so we have to somehow come to terms with this inevitable evil. Is there a way out? Whether we want it or not, this aspect of the matter has to be given attention. This can be done formally by introducing an extra term $\varphi(x, y)$ within the cosine, whose role will be to account for the above systematic parasitic phase shifts. In view of this, let us rewrite Equation 7.8 as follows:

$$I(x,y) = U_{01}^2(x,y) + U_{02}^2(x,y) \\ + 2U_{01}(x,y)U_{02}(x,y)\cos[\Phi(x,y) + 2\pi\beta x/\lambda + \varphi(x,y)] \tag{7.9}$$

Now we have a correct equation for the interferogram intensity $I(x,y)$, which would not be of so much interest, had it not contained implicitly the quantity $\Phi(x,y)$, in which we are interested, since it in principle encodes information about the parameters of the object under investigation. All is well, but we have so far only one equation for five unknowns: $I_{01}(x,y)$, $I_{02}(x,y)$, $\Phi(x,y)$, $2\pi\beta x/\lambda$, and $\varphi(x,y)$. What shall we do? The solution is to use the same equation several times. First, by alternately blocking the reference and subject beams in the interferometer, we can identify the distributions $I_{01}(x,y)$ and $I_{02}(x,y)$. Furthermore, by recording a so-called unperturbed interferogram, that is, what happens in the absence of the object under investigation, we can find the sum of the quantities $2\pi\beta x/\lambda$ and $\varphi(x,y)$, and only then calculate $\Phi(x,y)$. Having recorded a pair of interferograms, the "unperturbed" one as well as the one taken in the presence of the object under investigation, the dependence $\Phi(x,y)$ can now be found graphically. This procedure is illustrated in Figure 7.5 for the model

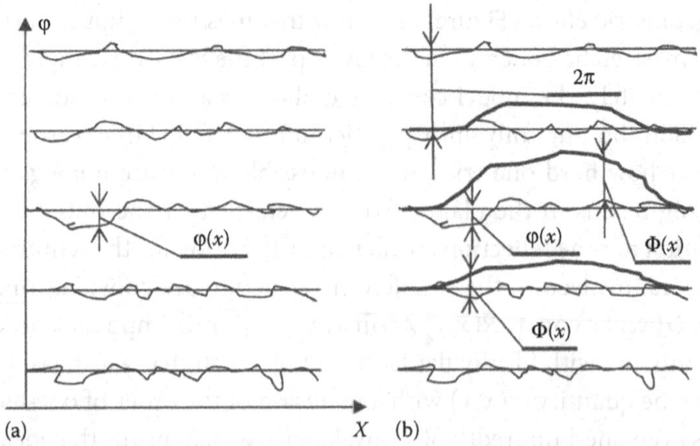

FIGURE 7.5 Graphical interferogram processing: (a) so-called unperturbed interferogram, that is, $\varphi(x)$; (b) $\Phi(x) + \varphi(x)$.

case of a one-dimensional (1D) object, when the phase shift in question is a function of only one spatial variable, so the notation $\Phi(x,y)$ is replaced by $\Phi(x)$.

7.3 PROCESSING INTERFEROGRAMS

Using the procedure described earlier, it becomes feasible to find the desired quantity $\Phi(x,y)$, which describes the distribution of the light intensity of the interference pattern, from Equation 7.9. For flat objects with uniform thickness $\Delta Z = \text{const}$ (for instance if you test the quality of optical glass plates) this enables one to find the distribution of the refractive index $n(x,y)$ averaged in the Z-direction. Alternatively the product $n(x,y)\Delta z(x,y)$ can be found if $\Delta z(x,y) \neq \text{const}$. It is possible because

$$\frac{2\pi}{\lambda}\int_{Z_1}^{Z_2} n(x,y,z)\mathrm{d}z = \frac{2\pi}{\lambda}\bar{n}(x,y)\Delta Z = \Phi(x,y) \tag{7.10}$$

However, a three-dimensional (3D) distribution $n(x,y,z)$ cannot, in principle, be recovered by methods of traditional interferometry since a function of three variables (3D) cannot be restored from its two-dimensional (2D) "projection" $\Phi(x,y)$.

Nevertheless, in some cases, it turns out that with some *a priori* information at hand, one can determine the distribution of the refractive index for 3D objects as well. This primarily concerns objects characterized by axial symmetry, such as cylinders or spheres, where for each cross section the function $n(r)$ is, in fact, 1D. [Contour lines $n(r) = \text{const}$ are shown in Figure 7.6.] Let us shine light through such an object perpendicular to the X–Y plane and along its cross-sectional chords in the Z-direction (Figure 7.6). Then for a fixed value of y, the net phase shift along each chord crossing the cross section of the cylinder or sphere will be only a function of the x-coordinate, which is as follows:

$$\Phi(x) = \frac{2\pi}{\lambda}\int_{-Z}^{Z} n(r)\mathrm{d}z \tag{7.11}$$

where:

$$|Z| = \sqrt{R^2 - x^2}$$

After changing the variable ($z = \sqrt{r^2 - x^2}$, whence $\mathrm{d}z = r\mathrm{d}r/\sqrt{r^2 - x^2}$) and introducing $U(x) = \Phi(x)\lambda/2\pi$, we obtain

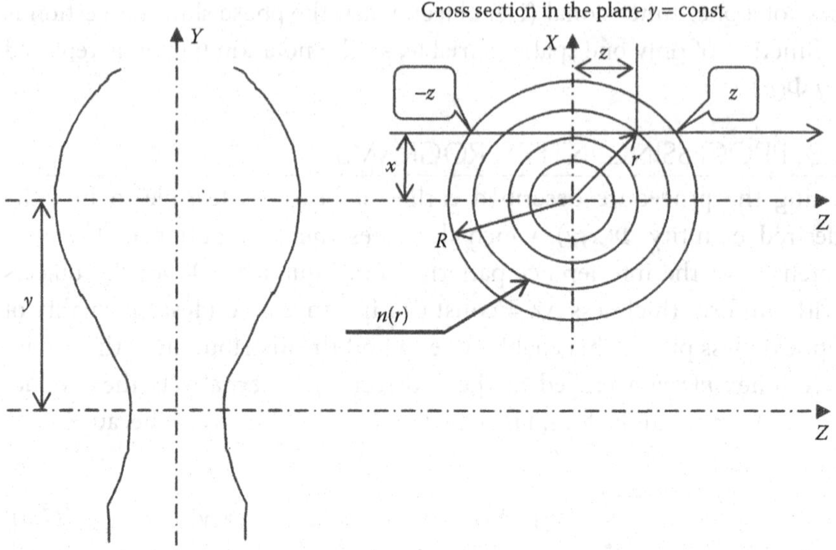

FIGURE 7.6 Processing of an object with axial symmetry (Abel equation).

$$U(x) = 2\int_{x}^{\infty} \frac{n(r) r \mathrm{d}r}{\sqrt{r^2 - x^2}} \tag{7.12}$$

This is an Abel equation, whose solution is well known:

$$n(r) = \frac{1}{\pi}\int_{r}^{R} \frac{U'(x)\mathrm{d}x}{\sqrt{x^2 - r^2}} \tag{7.13}$$

To conclude this section, let us note that throughout the discussion of interferometric measurements it has been assumed that the object under investigation is illuminated by practically parallel rays propagating along the Z-axis. This assumption is valid only when the gradients of the refractive index $\partial n/\partial x$ and $\partial n/\partial y$, which are responsible for the bending of the beams propagating in the investigated medium, are sufficiently small. If this is not the case, different rays that have traveled different optical distances and therefore have different phase shifts could in principle arrive at to the same point on the screen. This complication would make the whole processing procedure for interferograms into such a major headache that no man in his right mind would ever take on such an enterprise.

7.4 SHADOWGRAPHY AND SCHLIEREN TECHNIQUES

There is a whole variety of situations in which we are interested not in the spatial distribution of the refractive index itself but only in its gradient $\nabla n(x, y, z)$. (A typical situation arises in the study of shock waves.) As a matter of fact, before the full-scale advent of interferometry into optical technology, shadowgraphy methods, and above all the so-called *Tepler* or *Schlieren method*, had provided the main techniques for measuring the homogeneity of glass and the quality of optical equipment in general, even including large-scale astronomical devices. The limitation concerning gradients in the refractive index addressed at the end of Section 7.3 not only does not result in a negative effect on shadowgraph methods, but moreover, the greater the gradients $\partial n/\partial x$ and $\partial n/\partial y$, the pattern that is recorded by applying the latter techniques becomes clearer and has higher contrast. As far as the shadowgraph pattern per se is concerned, any one of us has seen it many times when looking on a hot sunny day along a road where upward moving air currents become well discernible with a naked eye. One can be tempted to take a photograph of this pattern, which regrettably does not have very high contrast. The poor contrast is caused by two main circumstances: first the low diffraction efficiency of the subject of the snapshot, and second, and most importantly, the extremely high intensity of the zero-order diffraction that is involved. In addition, when filming a shadowgraph, the contrast of the image will be proportional (here for simplicity we confine the analysis to 1D only) to the second derivative $\mathrm{d}^2n/\mathrm{d}x^2$. This certainly does not cause much enthusiasm as far as processing the image is concerned, even in the case when the final goal is to determine just the first derivative $\mathrm{d}n/\mathrm{d}x$, and one is not concerned with reconstructing the distribution of the quantity $n(x)$ itself.

How can one improve the image contrast? The recipe is simple—just block out the zero order of diffraction. An optical device, which implements this idea, is shown in Figure 7.7. This technique is called the dark-field, or Schlieren (or Tepler) method. The figure clearly indicates that we are dealing here with just the ordinary filtration in the Fourier plane: A movable screen (a so-called Foucault knife-edge), whose edge is parallel to the Y-axis, is blocking the zero order of diffraction by cutting off all the rays that have not been refracted by the object under investigation. Choosing the right ratio of focal lengths for a pair of confocal lenses L_1 and L_2 enables one to transfer the desired scale of the image to a photographic device.

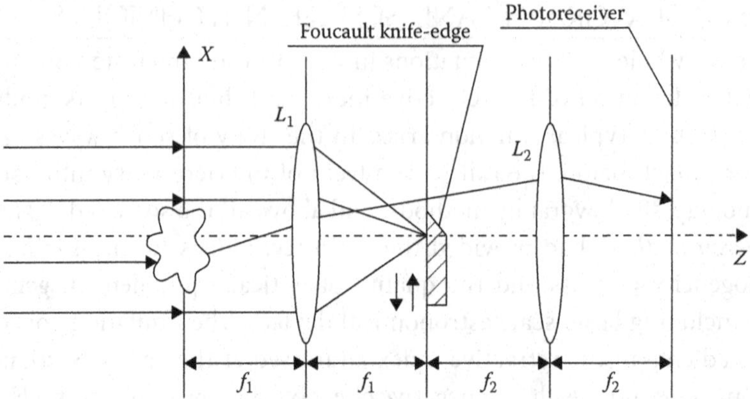

FIGURE 7.7 Tepler or Schlieren method.

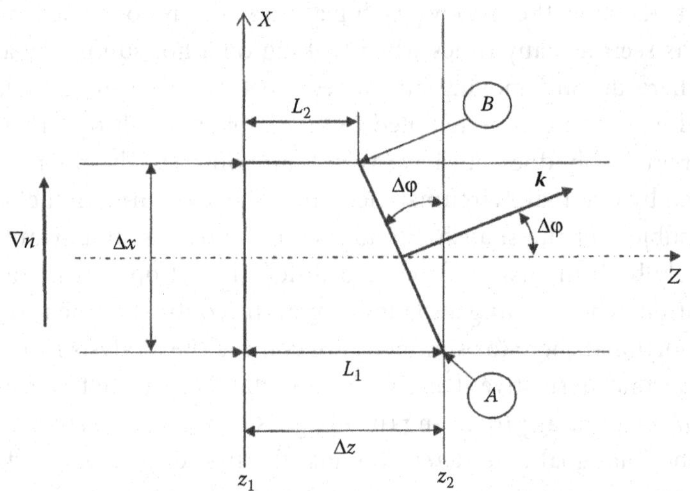

FIGURE 7.8 Rays refracted by the studied object.

The angle by which the rays have been refracted by the studied object can be found as follows: Suppose that we initially had two rays separated from one another by a distance Δx, both propagating along the Z-axis, and the refractive index n of a thin layer of the medium in the Z-direction can be regarded as being constant over small intervals Δz (Figure 7.8). Let $\partial n/\partial y = 0$ and $\partial n/\partial x = \mathrm{d}n/\mathrm{d}x$. Choose a time interval Δt such that the lower ray, which propagates through a part of the medium with a refractive index n, will travel the distance $\Delta z = L_1 = c\Delta t/n$. The path of the upper ray is characterized by a higher refractive index, and therefore, over the same time Δt it will travel a shorter distance

$L_2 = c\Delta t/(n+\Delta n)$. As a result, the radiation wavefront (the line *AB*) will rotate around the *Y*-axis by the angle $\Delta\varphi \approx \tan\varphi = (L_1 - L_2)/\Delta x$. Accordingly, the angle between the wave vector $\boldsymbol{k}$ normal to the wavefront and the *Z*-axis will also be $\Delta\varphi$. The difference in optical lengths is $(L_1 - L_2) = c\Delta t/n - c\Delta t/(n+\Delta n) = c\Delta t\Delta n/n(n+\Delta n) \approx \Delta n\Delta z/n$, so in view of $c\Delta t = n\Delta z$, as $\Delta x \to 0$ we have $(n+\Delta n) \to n$, and therefore, $n(n+\Delta n) \to n^2$. Hence, for small Δx, we have $\Delta\varphi = (L_1 - L_2)/\Delta x = \Delta n\Delta z/\Delta x n$. Passing to the limit, we arrive at $\mathrm{d}\varphi = (\mathrm{d}n/\mathrm{d}x)n^{-1}\mathrm{d}z$, whence

$$\varphi(x) = \int_{Z_1}^{Z_2} \frac{\mathrm{d}n}{\mathrm{d}x} n^{-1}(z)\mathrm{d}z \tag{7.14}$$

where:

Z_1 and Z_2 are the quantities that mark the boundaries of the object that is being illuminated

x is the so-called impact parameter, which is the coordinate of the transmission beam before it encounters the object

The integral (7.14) enables one therefore to calculate the refraction angle as a function of x. Naturally, in order that the beam does not get cut off by the Foucault knife-edge blocking the zero order of diffraction, the value of φ must be greater than θ—the divergence angle of the transmitted radiation. Conversely, the minimum refraction angle that can be measured this way is $\varphi_{\mathrm{min}} \geq \theta$. The Foucault knife-edge is equipped with a micrometer adjustment mechanism that enables one to shift it along the *X*-axis, cutting off progressively greater angles, which will then disappear from the resulting image.

For large gradients $\mathrm{d}n/\mathrm{d}x$, one has to deal with a path integral along the curvature path followed by the ray:

$$\varphi(x) = \int_{L} \frac{\mathrm{d}n}{\mathrm{d}x} n^{-1}(x,z)\mathrm{d}l \tag{7.15}$$

where:

$$\mathrm{d}l = \sqrt{(\mathrm{d}x)^2 + (\mathrm{d}z)^2}$$

Let us come back to Figure 7.7. We can see that the section of the object lying above the *Z*-axis will be cut off by the Foucault knife-edge. In the special case of objects with axial symmetry, this will only entail some worsening of the image intensity, but in general the upper section of the object will be lost once and for all. What can possibly be done about it? A common approach, instead of using the Foucault knife-edge, is to stretch a wire parallel to the Y-axis in the Fourier plane so that it crosses the Z-axis. The diameter of the wire is chosen so that $\varnothing = f\theta$, where f is the focal length of the lens L_1. However, there is another way that appears to be much more elegant: one obtains a negative image of the zero order of diffraction by placing a film in the Fourier plane and making an exposure without the investigated object. The photographic negative thus obtained is then placed on the Z-axis in the Fourier plane in order to annihilate the zero order of diffraction when the picture is taken.

CHAPTER 8

Holographic Methods of Research

THE ESSENTIAL FEATURE OF holography is obtaining a special kind of diffracting object (a hologram), such that after being illuminated by a plane wave, it will restore the subject wave, that is, the wave that has been reflected by or traveled through the object under investigation. The difference between a hologram and a normal photograph is that the latter retains only information about the spatial amplitude distribution of the subject wave, whereas the former also contains information about the wave's phase distribution or its phase wavefront. This is achieved due to the fact that in holography, as in interferometry, there are two waves that arrive at the photoreceiver. The subject wave carrying information about the object under investigation is allowed to interfere with a plane reference wave. The result of this interference enables one to identify the phase distribution of the subject wave. Typical optical devices used to produce holograms are shown in Figure 8.1a and b. One can see that in both diagrams a highly coherent plane wave is split into two parts. The first part, which usually has higher intensity, is directed onto the object under investigation and is reflected from it (Figure 8.1a) or passes through it (Figure 8.1b). Thereupon, it arrives at the photoreceiver (typically a photographic layer), where it meets the second part of the coherent wave, which has merely been reflected by a plane mirror. We have a subject wave in the form:

$$U_{\text{sub}}(x, y) = a_{\text{sub}}(x, y)e^{i\Phi(x,y)}$$

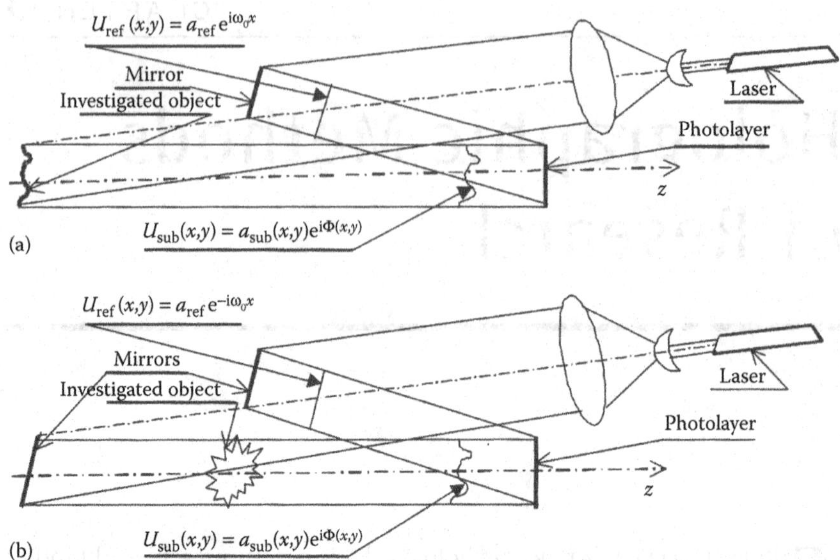

FIGURE 8.1 Typical optical devices used in holography. (a) Reflective object; (b) transparent object.

which arrives at the photoreceiver—the future hologram. In the above equation, $\Phi(x, y)$ represents the phase distribution of the subject wave and $a_{sub}(x, y) = |U_{sub}(x, y)|$ is the amplitude distribution. The reference wave is written as follows:

$$U_{ref}(x, y) = a_{ref}e^{-i\omega_0 x}$$

The superposition of the two waves on the photographic layer then becomes

$$\begin{aligned} U_\Sigma(x, y) &= U_{sub}(x, y) + U_{ref}(x, y) \\ &= a_{sub}(x, y)e^{i\Phi(x,y)} + a_{ref}e^{-i\omega_0 x} \end{aligned}$$

so the resulting intensity

$$\begin{aligned} I(x, y) &= U_\Sigma(x, y)U_\Sigma^*(x, y) \\ &= a_{ref}^2 + a_{sub}^2(x, y) + a_{ref}a_{sub}(x, y)\left\{e^{i[\Phi(x,y)+\omega_0 x]} + e^{-i[\Phi(x,y)+\omega_0 x]}\right\} \qquad (8.1) \\ &= a_{ref}^2 + a_{sub}^2(x, y) + 2a_{ref}a_{sub}(x, y)\cos[\Phi(x, y) + \omega_0 x] \end{aligned}$$

After the photographic layer has been developed (creating a hologram), it will reveal a periodic structure, which is essentially an interferogram that has recorded information about both the subject wave's phase $\Phi(x,y)$ and amplitude $a(x,y)$ distributions. Let us now consider the basic principles of the holographic method together with some technical considerations and requirements that enable us to make a clone copy of the subject wave.

8.1 PHOTOGRAPHIC LAYERS IN HOLOGRAPHY

A photographic layer is usually utilized as the receiver in holography for the simple reason that for quantities $\Phi(x,y)$ with large full width at half maximum (FWHM), the spatial frequencies of the interference lines on a hologram can be as high as 10^4 cm^{-1}. Unfortunately, today there are no other photoreceivers capable of resolving images with such high spatial frequencies. To consider some issues of holography, for reasons that will become clear shortly, one usually deals with the quantity T, which is the transmission of the photographic layer rather than its density $D=\lg(I_0/I)$. The intensity transmission of a photographic layer is defined by definition (like D which is defined by definition also) via the relation $I=I_0T$. In both cases, I_0 denotes the intensity of the light wave incident on the developed photographic layer, I being the intensity of the wave that has traversed it. In other words, $T=I/I_0$. Since the intensity of a light wave equals the square of its amplitude, the amplitude transmission of the developed photolayer will be defined as $t=\sqrt{T}$. Let us see what happens to the light traveling through the photographic layer as the number of developed grains increases. For instance, we are interested in the dependence $I(n)$, where n is the areal density of the developed grains, that is, the number of grains of the photographic layer per unit area. If σ denotes the ensemble average of the effective cross section of a single grain, the product σn will yield the fraction of the photographic layer's surface area through which no light will traverse. Therefore, $\mathrm{d}I=-I\sigma\mathrm{d}n$, and after integration, we find $\ln I=-\sigma n+C$. The constant of integration C can be determined from the condition $I=I_0$ for $n=0$. Hence, $\ln I=-\sigma n+\ln I_0$, $\ln(I/I_0)=\ln T=-\sigma n$, and

$$\lg(I/I_0)=\lg T=-D=-\sigma n\lg e \tag{8.2}$$

However, as we have already found out in Chapter 4 that, in the "linear" (on the logarithmic scale) section of the characteristic curve, the density D of the photographic layer depends on exposure $\mathcal{E}$ as follows:

$$D=\gamma(\lg\mathcal{E}-\lg\mathcal{E}_0) \tag{8.3}$$

In the above equation, $\mathcal{E} = I^* \Delta t$ is the exposure, that is, the energy that has been deposited on the unit surface area of a photographic layer during the exposure time Δt by a light flux with intensity I^*, $\gamma = D'(\lg\mathcal{E})$ is the contrast, and $\lg\mathcal{E}_0$ is the exposure value at the point of intersection of the extrapolation of the linear section of the characteristic curve with the horizontal coordinate axis (Figure 4.2). Clearly, the higher the photosensitivity S_λ of the photographic layer, the smaller will be the value of $\lg\mathcal{E}_0$. Photosensitivity is defined as the reciprocal of the exposure necessary to achieve $D = 1$.* Let us take advantage of Equation 8.3 to calculate the value of $\lg\mathcal{E}_0$. Since $\mathcal{E}_\lambda = 1/S_\lambda$ for $D = 1$, then by definition we have $\gamma[\lg(1/S_\lambda) - \lg\mathcal{E}_0] = 1 = \lg 10$. After a little algebra we find $\mathcal{E}_0 = 1/10^{1/\gamma} S_\lambda$. Then we can rewrite Equation 8.3 as follows:

$$D = \gamma \lg k\mathcal{E} \tag{8.4}$$

thereby introducing a special coefficient, characteristic of a particular photographic layer $k = 10^{1/\gamma} S_\lambda$. Since from Equation 8.2 we have $\lg T = -D$, $\lg T = -\gamma \lg k\mathcal{E}$, we also get

$$T = (k\mathcal{E})^{-\gamma} \tag{8.5}$$

The above equation essentially makes a statement, which is not completely unexpected, that the increase of exposure and photosensitivity of a photographic layer reduces the transmission of the developed negative image. However, if a photographic layer is developed with a reversal process or a positive image is printed, the situation reverses and transmission increases as the result of greater exposure or increased photosensitivity. Formally, this is reflected by the fact that $\gamma > 0$ for a negative image (Figure 4.22) and $\gamma < 0$ for a positive image. The processing of a photographic layer is regarded as optimal if the latter quantity reaches the value $\gamma_{\text{opt}} = -2$. In this case, $T = (k\mathcal{E})^{-\gamma} = (k\mathcal{E})^2$, and the amplitude transmission of the photographic layer becomes

$$t(x,y) = \sqrt{T(x,y)} = \sqrt{[k\mathcal{E}(x,y)]^2} = k\mathcal{E}(x,y) = kI^*(x,y)\Delta t = \mathcal{K}I^*(x,y) \tag{8.6}$$

where:

$\mathcal{K} = k\Delta t$

It follows that the spatial distribution of the amplitude transmission $t(x,y)$ of a photographic layer is equal, up to a constant multiplier $\mathcal{K}$, to the

* By the definition of photosensitivity, the normalizing density $D = D_0 + 1$. However, low-sensitivity holographic layers have $0.01 < D_0 < 0.02$, thus differing from by some 1% or 2%. Therefore, the condition $D = 1$ describes the situation with a reasonable accuracy.

spatial distribution of the light flux intensity $I^*(x,y)$, which has affected the exposure of the photographic layer.

8.2 FORMATION OF HOLOGRAMS AND RECONSTRUCTION OF WAVEFRONTS

Let the two plane electromagnetic waves $U_1(x) = a_1$ and $U_2(x) = a_2(x)\exp(i\omega_0 x)$ (where ω_0 is a spatial frequency describing the phase change in the x-direction) be incident on a photographic layer. The first wave, referred henceforth to as the reference wave, travels in the direction normal to the X–Y plane where the photographic layer is situated. The second wave, henceforth called the subject wave, makes an angle β with the normal to the plane, so β becomes the angle between the two corresponding wave vectors. This results in a phase difference

$$\Delta\Phi(x) = \frac{2\pi\beta x}{\lambda} = \omega_0 x$$

depending on x in such a way that at a distance $x = 1/\omega_0$ the phases of the two waves differ from one another by one radian. The resulting amplitude in the X–Y plane will be

$$U_\Sigma(x) = a_1 + a_2(x)\exp(i\omega_0 x) \tag{8.7}$$

and the overall intensity in the plane of the photographic layer is as follows:

$$\begin{aligned} I(x) &= U_\Sigma(x)U_\Sigma^*(x) \\ &= a_1^2 + a_2^2(x) + a_1 a_2(x)\exp(i\omega_0 x) + a_1 a_2(x)\exp(-i\omega_0 x) \\ &= a_1^2 + a_2^2(x) + 2a_1 a_2(x)\cos(\omega_0 x) \end{aligned} \tag{8.8}$$

where:

$U_\Sigma^*(x)$ denotes the complex conjugate of the wave $U_\Sigma(x)$

Suppose that all the values $I(x)\Delta t$ lie within the so-called *photographic width*, that is, on the linear section of the photographic layer's characteristic curve. Then, according to Equation 8.6, the amplitude transmission of the photographic layer, developed with reversal to the value $\gamma_{opt} = -2$, can be written as follows:

$$\begin{aligned} t(x) &= \mathcal{K}I(x) = \mathcal{K}[a_1^2 + a_2^2(x) + a_1 a_2(x)\exp(i\omega_0 x) + a_1 a_2(x)\exp(-i\omega_0 x)] \\ &= \mathcal{K}[a_1^2 + a_2^2(x) + 2a_1 a_2(x)\cos(\omega_0 x)] \end{aligned} \tag{8.9}$$

Note that the quantities $t(x)$, $\mathcal{K}I(x)$, and $\mathcal{K}[a_1^2 + a_2^2(x) + 2a_1a_2(x)\cos(\omega_0 x)]$ are dimensionless.

Thus, the equation that is obtained for the transmission of the photographic layer $t(x)$ comes as a result of interference of the two plane waves that are incident on it—the reference wave $U_1(x) = a_1$ and the subject wave $U_2(x) = a_2(x)\exp(i\omega_0 x)$. In this particular case, as shown by Equation 8.9, the quantity $t(x)$ represents the sum of a constant component $\mathcal{K}[a_1^2 + a_2^2(x)]$ and an oscillatory component $\mathcal{K}2a_1a_2(x)\cos(\omega_0 x)$, whose spatial periodicity is $1/2\pi\omega_0$. The diffraction efficiency of such a sinusoidal transparency depends on the depth of the spatial modulation* of its transmission, that is, it is determined by the structure of the developed photographic layer $t(x)$. [The structure of the quantity $t(x)$ has in turn been defined by the amplitudes of the interfering waves and the angle between their wave vectors.] If such a transparency is illuminated by a spatially uniform wave with a flat wavefront, then due to the diffraction phenomenon in the transparency, the wave a, falling perpendicularly on the transparency, will be split into three waves traveling separately thereafter. The first wave a_0, the zero order, will travel in the direction of the original incident wave. The second wave $a_{+1}\exp(i\omega_0 x)$, namely, the first positive order of diffraction, and the third wave $a_{-1}\exp(-i\omega_0 x)$, corresponding to the first negative order, will travel at angles, respectively, $\pm\beta$ to the left and right of the zero order. It is clear that the maximum intensity in the side orders will be observed in the case when the spatial modulation depth is 100%. In this special case, $a_0 = a/2$ and $a_{+1} = a_{-1} = a/4$. In the general case, as it was shown earlier in Chapter 5, a should be multiplied by $t(x)$ in order to yield the final outcome of diffraction.

To do so, let us use Equation 8.9:

$$\begin{aligned} U_{\text{rec}}(x) &= at(x) = a\mathcal{K}I(x) \\ &= a\mathcal{K}[a_1^2 + a_2^2(x) + a_1a_2(x)\exp(i\omega_0 x) \\ &\quad + a_1a_2(x)\exp(-i\omega_0 x)] \\ &= a\mathcal{K}[a_1^2 + a_2^2(x)] + aa_1\mathcal{K}[a_2(x)\exp(i\omega_0 x)] \\ &\quad + aa_1\mathcal{K}[a_2(x)\exp(-i\omega_0 x)] \end{aligned} \tag{8.10}$$

* Recall that the spatial modulation depth is formally defined by the modulation index $m = (t_{\max} - t_{\min})/t_{\max}$, where t is the amplitude transmission coefficient of the transparency (Chapter 5).

First of all, we would like to draw the reader's attention to the fact that we have obtained, apart from a constant multiplier, a copy $aa_1\mathcal{K}[a_2(x)\exp(\mathrm{i}\omega_0 x)]$ of the subject wave and a wave $aa_1\mathcal{K}[a_2(x)\exp(-\mathrm{i}\omega_0 x)]$, which is the complex conjugate of the above copy. Indeed, the subject wave $U_2(x) = a_2(x)\exp(\mathrm{i}\omega_0 x)$ was traveling in the same direction, its amplitude distribution $a_2(x)$ being directly proportional to the amplitude of the restored copy. It should be clear that any wave $U(x, y) = a(x, y)\exp[\Phi(x, y)]$ [here $a(x, y)$ denotes the wave's amplitude and $\Phi(x, y)$ denotes its phase distribution] can be expanded as a superposition of plane waves, and therefore, the above result applies in the general case. In essence, the above constitutes the basis of holography; the intensity of the interference field that arises from adding together the subject and reference waves results in a special diffracting structure of the developed photographic layer. This structure is such that in the process of reconstruction of the hologram, when it is illuminated by a plane wave, there arises, due to diffraction, a copy of the subject wave and its complex conjugate.

Note that the wave vector of the reconstruction wave is not necessarily perpendicular to the hologram. Suppose, for example, that the hologram (8.9) is reconstructed using a wave $a\mathrm{e}^{-\mathrm{i}\omega_0 x}$, which meets the hologram at an angle $-\beta$ to it ($\beta = \omega_0\lambda/2\pi$). Then

$$\begin{aligned} U_{\text{rec}}(x) &= a\mathrm{e}^{-\mathrm{i}\omega_0 x}t(x) \\ &= a\mathrm{e}^{-\mathrm{i}\omega_0 x}\mathcal{K}[a_1^2 + a_2^2(x) + a_1 a_2(x)\exp(\mathrm{i}\omega_0 x) \\ &\quad + a_1 a_2(x)\exp(-\mathrm{i}\omega_0 x)] \\ &= a\mathcal{K}[a_1^2 + a_2^2(x)]\mathrm{e}^{-\mathrm{i}\omega_0 x} + aa_1\mathcal{K}[a_2(x)] + aa_1\mathcal{K}[a_2(x)]\mathrm{e}^{-2\mathrm{i}\omega_0 x} \end{aligned} \tag{8.11}$$

This shows that the first order of the restored wave propagates in the direction normal to the hologram, the zero order propagates at an angle $-\beta$ to it (i.e., in the direction of the reconstruction wave), and the first negative order makes with it an angle -2β.

Observe that in the case of overexposure of the photographic layer, where, for example, we end up having a rectangular rather than sinusoidal structure, we will encounter a sequence of diffraction orders with vanishing amplitudes. The angle, which the first order wave will make with the zero order, will be proportional to $\omega_0 x$; furthermore, it will be proportional to $2\omega_0 x$ in the second order and so on.

The above procedure enables us to describe the process of recording and reconstructing holograms in its most general form. Let us return to Equation 8.1, which represents the intensity of two interfering waves incident on a photographic layer, namely, the subject wave $U_{sub}(x,y)=a_{sub}(x,y)e^{i\Phi(x,y)}$ and the reference wave $U_{ref}(x,y)=a_{ref}e^{-i\omega_0 x}$:

$$\begin{aligned}I(x,y) &= U_\Sigma(x,y)U_\Sigma^*(x,y)\\ &= a_{ref}^2 + a_{sub}^2(x,y) + a_{ref}a_{sub}(x,y)\left\{e^{i[\Phi(x,y)+\omega_0 x]} + e^{-i[\Phi(x,y)+\omega_0 x]}\right\}\\ &= a_{ref}^2 + a_{sub}^2(x,y) + 2a_{ref}a_{sub}(x,y)\cos\left[\Phi(x,y)+\omega_0 x\right]\end{aligned}$$

Assume once again that all the values $I(x,y)\Delta t$ are within the photographic dynamic range, namely, that they lie on the linear section of the photographic layer's characteristic curve. Suppose that the layer has been developed with reversal up to the value $\gamma_{opt}=-2$. In this case, according to Equation 8.6, we can write down the equation for the amplitude transmission of the photographic layer as follows:

$$\begin{aligned}t(x,y) &= \mathcal{K}I(x,y)\\ &= \mathcal{K}\Big(a_{ref}^2 + a_{sub}^2(x,y)\\ &\quad + a_{ref}a_{sub}(x,y)\left\{e^{i[\Phi(x,y)+\omega_0 x]} + e^{-i[\Phi(x,y)+\omega_0 x]}\right\}\Big)\\ &= \mathcal{K}\left\{a_{ref}^2 + a_{sub}^2(x,y) + 2a_{ref}a_{sub}(x,y)\cos[\Phi(x,y)+\omega_0 x]\right\}\end{aligned} \tag{8.12}$$

To restore the hologram (8.12), let us use the wave $ae^{-i\omega_0 x}$, which is incident on the hologram at an angle $-\beta(\beta=\omega_0\lambda/2\pi)$. Then

$$\begin{aligned}U_{rec}(x) &= ae^{-i\omega_0 x}t(x,y)\\ &= ae^{-i\omega_0 x}\mathcal{K}\Big(a_{ref}^2 + a_{sub}^2(x,y)\\ &\quad + a_{ref}a_{sub}(x,y)\left\{e^{i[\Phi(x,y)+\omega_0 x]} + e^{-i[\Phi(x,y)+\omega_0 x]}\right\}\Big)\\ &= a\mathcal{K}[a_{ref}^2 + a_{sub}^2(x,y)]e^{-i\omega_0 x} + a\mathcal{K}\left[a_{ref}a_{sub}(x,y)e^{i\Phi(x,y)}\right]\\ &\quad + a\mathcal{K}\left\{a_{ref}a_{sub}(x,y)e^{-i[\Phi(x,y)+2\omega_0 x]}\right\}\end{aligned} \tag{8.13}$$

This equation shows that the output of the reconstruction process is a copy $aa_{ref}\mathcal{K}\{a_{sub}(x,y)\exp[i\Phi(x,y)]\}$ (with a constant multiplier) of the subject wave and the complex conjugate of the copy, namely, the wave $aa_{ref}\mathcal{K}(a_{sub}(x,y)\exp\{-i[\Phi(x,y)+2\omega_0 x]\})$. Indeed, the subject wave $U_{sub}(x,y)=a_{sub}(x,y)e^{i\Phi(x,y)}$ has the same phase front $\Phi(x,y)$, its amplitude distribution $a_{sub}(x,y)$ being proportional to the amplitude distribution of the restored copy. The first term in Equation 8.13 represents a zero-order wave, which travels in the direction of the reconstruction wave. A subject wave clone, which is the second term in Equation 8.13, is traveling in the direction normal to the hologram, and the third term, the clone's complex conjugate, makes an angle $2\beta=\omega_0\lambda/\pi$ with the first-order wave.

Naturally, if one takes a snapshot of the wave $aa_{ref}\mathcal{K}\{a_{sub}(x,y)\exp[i\Phi(x,y)]\}$, by letting it enter the lens of a camera; then with a proper choice of exposure one would end up having the same distribution of film density as if a standard photograph of the object had been taken. By using various reconstruction schemes, the same wave can also be used to create shadow images and interferograms.

8.3 RECONSTRUCTION TECHNIQUES

A reconstruction scheme designed, among other purposes, to obtain holographic interferograms is shown in Figure 8.2. Mirrors M_1, M_2, M_3, and M_4 form a device, splitting the reconstruction beam into two parts and then directing both of the resulting beams onto a hologram Γ, so that they arrive at a small angle to each other. The intensities of each of the two beams can be varied by inserting attenuating filters or changing the reflection coefficients of the corresponding mirrors. The Fourier filtration is realized by couple of confocal lenses L_1 and L_2. A mask diaphragm provides filtration in the Fourier plane of lens L_1. Another lens L_3 transmits the reconstructed image to a film or some other photoreceiver, such as a charge-coupled device (CCD) camera.* Diffraction on the hologram

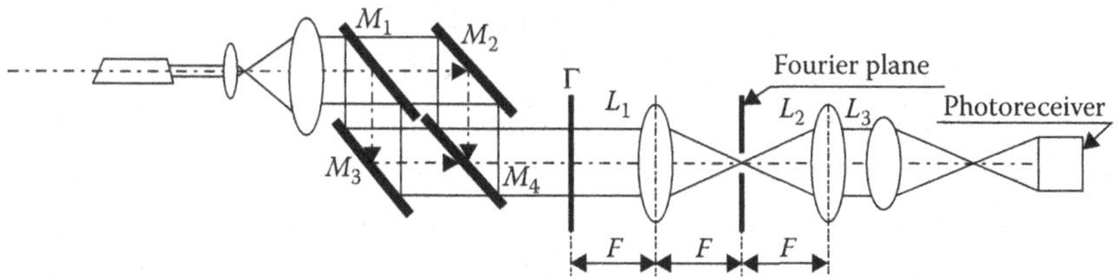

FIGURE 8.2 Schematic of hologram reconstruction techniques.

* We would like to point out that in the reconstruction process of holograms that have been recorded under conditions when the quantity $\Phi(x,y)$ varied over a wide range, one needs to take advantage of high-resolution wide-aperture optics. Otherwise information about high-gradient regions, which cause strong refraction of the subject beam, will be lost.

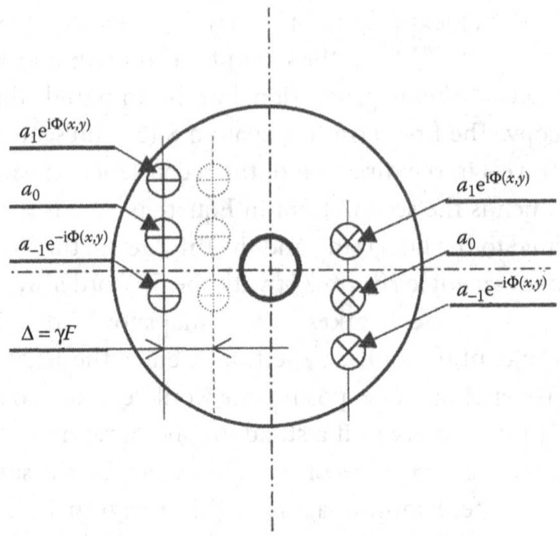

FIGURE 8.3 Typical situation in the Fourier plane.

causes each of the two waves that illuminate it to split into three, so that in the focal plane of the first lens there appear two copies of the Fourier spectra, corresponding to the two illuminating waves. Figure 8.3 shows what happens in the Fourier plane. The points in the Fourier plane, onto which the diffracted waves have been "focused," correspond to a more or less random positioning of the mirrors. As the incidence angle of either of the two beams illuminating the hologram changes, it causes the corresponding Fourier spectrum in the focal plane of the lens L_1 to move. Thus, for example, if we turn the mirror M_3 (Figure 8.2) by an angle $\gamma/2$, one of the Fourier spectra, as shown in Figure 8.3, will shift by the distance $\Delta = \gamma F$, where F is the focal length of the lens L_1. Therefore, by tilting the mirrors M_2 and M_3 and moving the Fourier spectra in the focal plane of the lens L_1, we are able to align selected orders of diffraction with the aperture opening, thereby affecting filtration of the required orders of diffraction. As a result, if six waves have arrived at the lens L_1, only two waves will remain after filtration. This fact is widely used in producing holographic interferograms. For example, if we filter through the aperture the zero order of one copy of the Fourier spectrum and the first order of the other, we will obtain exactly the same interferogram as if we were dealing with a standard interferometer design.

Let us demonstrate this by considering a scenario in which most of the information about an object under investigation is contained not in the amplitude, but rather in the phase distribution $\Phi(x, y)$ of the subject wave. This is the case in a whole variety of applications when one deals with transparent objects, such as gas media, plasma physics objects, optical elements of laser devices. Another application is for an object that is completely reflecting. This situation occurs when dealing with mirrors and dense plasmas, etc. As shown in Figure 8.4a, let us align the subject wave $a_1 e^{i\Phi(x,y)}$ with a diaphragm aperture. As a reference wave, let us use the zero-order diffraction wave, which we will direct on to the lens L_1 at an angle $\beta^* = \lambda\overset{*}{\omega}_0/2\pi$,* that is, the reference wave will be given as $a_0 e^{-i\overset{*}{\omega}_0 x}$. In this arrangement, the center of the zero-order diffraction spot will not exactly coincide with the center of the aperture, but will rather be

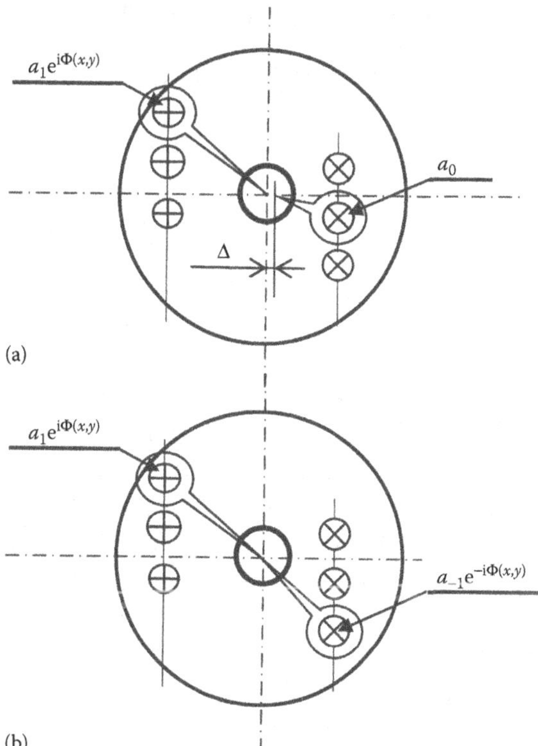

FIGURE 8.4 Different scenarios of the filtration in the Fourier plane. (a) The subject wave and zero order of the reference wave are aligned with a diaphragm aperture. (b) The same order, but opposite signs of the subject waves are aligned with a diaphragm aperture.

* Note that as a rule $\beta^* \ll \beta$—the angle at which the reference beam was incident on the hologram when it was recorded.

positioned away from it by some distance $\Delta = \beta^* F$, where F represents the focal length of the lens L_1. Since, as we agreed, the amplitudes of these two waves do not depend on the coordinates, namely, $a_0(x,y) = \text{const}$ and $a_1(x,y) = \text{const}$, the amplitude of their superposition in the photoreceiver plane, located behind the lens L_3, will be equal to

$$U_\Sigma(x,y) = a_1 e^{i\Phi(x,y)} + a_0 e^{-i\omega_0^* x}$$

and have the intensity

$$\begin{aligned} I(x,y) &= U_\Sigma(x,y)U_\Sigma^*(x,y) \\ &= a_0^2 + a_1^2 + a_0 a_1 \exp\{i[\omega_0^* x + \Phi(x,y)]\} \\ &\quad + a_0 a_1 \exp\{-i[\omega_0^* x + \Phi(x,y)]\} \\ &= a_0^2 + a_1^2 + 2a_0 a_1 \cos\left[\frac{2\pi\beta^* x}{\lambda} + \Phi(x,y)\right] \end{aligned} \tag{8.14}$$

where:
 β^* is the angle between the subject and reference waves

The equation obtained clearly indicates that the luminosity distribution in the interference pattern produced by the restoration technique corresponds exactly to the luminosity distribution in a usual interferogram (Chapter 7).

Now let us match the aperture with the components of the Fourier spectrum have the same order, but opposite signs. For instance, consider the interference of the orders plus one and minus one of the subject wave. This is shown in Figure 8.4b. The resulting complex amplitude in the image plane can be written as

$$U_\Sigma(x,y) = a_1 e^{i\Phi(x,y)} + a_{-1} e^{-i\Phi(x,y)}$$

so the luminosity distribution in the interference pattern will become

$$\begin{aligned} E(x,y) &= U_\Sigma(x,y)U_\Sigma^*(x,y) \\ &= a_1^2 + a_{-1}^2 + a_1 a_{-1} \exp[i2\Phi(x,y)] + a_1 a_{-1} \exp[-i2\Phi(x,y)] \\ &= a_1^2 + a_{-1}^2 + 2a_1 a_{-1} \cos[2\Phi(x,y)] \end{aligned} \tag{8.15}$$

One can see that in this case we have gained a twofold increase in sensitivity. The sensitivity can be increased further by filtering in still higher order harmonics. This way, modifying the above calculation in an obvious way to embrace the second orders of diffraction, we arrive at the intensity distribution in the interference pattern, which yields a fourfold increase in sensitivity:

$$I(x,y) = a_2^2 + a_{-2}^2 + 2a_2a_{-2}\cos[4\Phi(x,y)] \tag{8.16}$$

However, using even higher orders has a very negative effect on the signal-to-noise ratio. Indeed, the intensity of higher order diffraction beams, or the diffraction efficiency, decreases in higher orders, while noise caused by phantom scattering of the restoration wave on the hologram becomes considerably greater. As numerous measurements indicate, phantom scattering caused by the noise due to the photographic layer on which the hologram is being recorded is independent of angle, which means that the spectral distribution of the spatial frequencies of the scattered component can be regarded as white noise. In view of this, the intensity of the noise reaching the radiation detector turns out to be proportional to the area of the aperture. However, one can see from Equations 8.14 through 8.16 that the diffraction angles that have the highest information content *apropos* of the phase distribution $\Phi(x,y)$ grow in proportion with the order of diffraction. Therefore, one should increase the diameter of the aperture in direct proportion to the order of diffraction, that is,

$$\frac{S_1}{S_n} = \left(\frac{1}{n}\right)^2 \tag{8.17}$$

where:

S_n is the area of the aperture for the nth diffraction order

Simultaneously, the noise component will increase by the factor of n^2.

Let us now discuss one of the most important features of holographic interferometry, which is its insensitivity to constant phase distortions introduced by the optical system. As earlier, we denote their aggregate by $\varphi(x,y)$. In order to take the latter quantity into account, one uses the same photographic layer to produce two holograms: a preliminary one, recorded without any object under investigation, and a definitive one, recorded with the object present. (In the interval between the two exposures, one changes the angle which the reference wave makes with the photographic layer.) The first hologram obviously records information about the phase distortions due to the optical system $\varphi(x, y)$. The second hologram records the sum of phase differences due to the object under investigation and the phantom distortions of the optical system, namely $[\Phi(x, y) + \varphi(x, y)]$. Clearly, after illuminating such a hologram with two beams, we will end up, as a result of diffraction, with twelve, rather than six waves as earlier. Now we can use our restoration technique to filter out the distortions. Indeed, let us align the aperture in such a way that the only waves to enter it are $U(x, y) = a_1 e^{i\varphi(x,y)}$ and $U(x, y) = a_1 e^{i[\Phi(x,y)+\varphi(x,y)]}$, in which case the net complex amplitude in the image plane becomes

$$U_\Sigma(x, y) = a_1 e^{i[\Phi(x,y)+\varphi(x,y)]} + a_1 e^{i\varphi(x,y)}$$

yielding the luminosity distribution

$$\begin{aligned} E(x, y) &= U_\Sigma(x, y) U_\Sigma^*(x, y) \\ &= a_1^2 + a_{-1}^2 + 2a_1 a_{-1} \cos[\Phi(x, y)] \end{aligned} \tag{8.18}$$

Evidently, the phase distortions $\varphi(x, y)$ introduced by the optical system have been completely compensated in the final interferogram.

The key difference between a standard interferogram and a hologram* consists of the fact that an interferogram enables one only to reconstruct the shape of a wavefront, that is, to calculate the quantity $\Phi(x, y)$. A hologram, however, allows for complete *a posteriori* (i.e., possibly done after the experiment in a laboratory environment) restoration of the subject wave $U_{sub}(x, y) = a_{sub}(x, y) e^{i\Phi(x,y)}$. It is precisely this feature of holograms which enables one to implement the procedures described earlier for increasing sensitivity and compensation for phantom-phase shifts $\varphi(x, y)$. This, however, does not exhaust all the numerous benefits of holography by a long way. For example, within the same reconstruction scheme, one can also obtain Tepler images by blocking the central near-axis region of

* Essentially, the difference between a standard interferogram and a hologram consists merely in the value of the angle β between the wave vectors of the reference and subject waves. Of the two summands inside the cosine, in the former case one has $\Phi(x, y) \approx (2\pi x \sin\beta)/\lambda$, whereas in the latter case $\Phi(x, y) << (2\pi x \sin\beta)/\lambda$. This enables one to achieve spatial separation and filtering of different orders of diffraction through the process of hologram restoration.

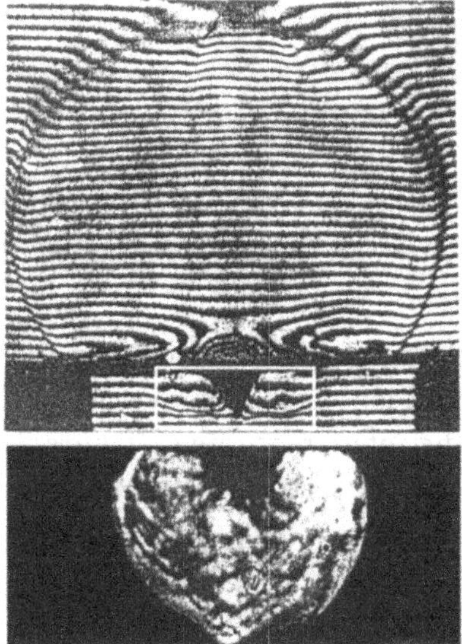

FIGURE 8.5 Shock wave formation in a residual gas after a thin metal foil has been exposed to a laser pulse. The interferogram (top) and Teplerogram (bottom, showing the section marked by a rectangle on the upper image) have been obtained from the same hologram by implementing the restoration method that is described in the text.

the aperture by a nontransparent screen, thereby filtering out the high spatial frequencies. To illustrate this, Figure 8.5 shows a pair of interferogram/Teplerogram images that have been obtained from the same hologram. The figure shows the results of the impact of high-intensity laser radiation on a thin (6 μm) aluminum foil. A focused laser pulse, with energy ~100 J and duration 2 ns, creates a luminosity ~10^{14} W cm^{-2} on the foil's surface within a focal spot with diameter ~250 μm producing a pressure of the order of several megabars. As a result, a small section of the foil that has been exposed to the laser is heated to a temperature of some 10–15 eV and is accelerated in the direction of the laser beam. A spherical shock wave is generated in the residual gas (10 torr) and propagates in the direction opposite to the beam. It is worth mentioning that all the inhomogeneities of the heated and accelerated matter become very clearly identifiable on a Tepler image.

Let us mention one more possible application of the restoration technique which we are considering. By changing the angle β^*, at which the

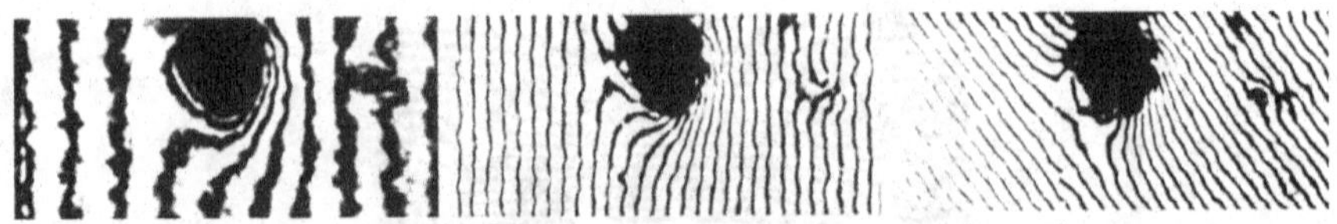

FIGURE 8.6 Interferograms obtained as a result of restoration of the same hologram for different values of the incidence angle direction of the zero-order diffraction beams.

diffracted wave of the zero order is incident on the lens L_1, we are able to alter the line separation on the interferogram, as well as to change the orientation of the interference lines (Figure 8.6). In a number of cases, this turns out to be extremely useful for further processing of interferograms.

CHAPTER 9

Electromagnetic Probing

NOWADAYS, ELECTROMAGNETIC PROBING OF various objects represents a widely used and powerful method of experimental physics [1–3]. We have already encountered some of its applications in interferometry and holography in Chapters 7 and 8, which were based on measuring the phase of an electromagnetic wave as it was traveling through or reflected from some object. In this chapter, we will mainly focus on the scattering* of electromagnetic waves [4], after having first considered some aspects of light propagation in transparent isotropic media. Scattering occupies a special niche among measuring methods, as, in contrast to the majority of other optical methods, it actually allows the determination of local parameters of the object under investigation, rather than integrals either along a signal's path through the object or over a range of directions, in which the object's emissions are being observed. We stress, however, that we will not consider here nonlinear phenomena, and also the analysis in this chapter will be confined to the visible and infrared region of the spectrum. The X-ray region will be addressed in Chapter 10.

9.1 ELECTROMAGNETIC WAVE PROPAGATION IN MEDIA

When physicists say a "transparent medium," they mean a solid-state matter, liquids, gases, or plasma, where the so-called gray losses, namely, light absorption along its path through the medium, are negligible. Thus, the main result of the interaction of radiation with matter consists of altering the way the radiation propagates through the medium. The factor,

* Scattering of a light wave is a process arising out of weak scattered waves, whose frequencies and directions are different from the frequency and the propagation direction of the main wave.

which basically shows how many times the speed of light in the medium is smaller than it is in a vacuum, is called, as a tribute to geometric optics, the *index of refraction* or the *refractive index*. One has the equation for the refractive index $n = \sqrt{\varepsilon\mu}$, where ε denotes the relative permittivity or dielectric constant and μ is the magnetic permeability of the medium. (All transparent media have, in fact, μ approximately equal to one. This excludes plasma with a magnetic field "frozen into" it, where the value of μ can reach up to five or six orders of magnitude.) The dielectric constant ε is the coefficient showing how the total electric field in the medium is changed when the polarization phenomenon takes place:

$$\varepsilon E = E + 4\pi P = E + 4\pi n_e e x_0 \quad (9.1)$$

where:

- E is the amplitude of the electric field of the electromagnetic wave in vacuum
- P is the electric field caused by polarization
- e is the charge of the electron
- n_e is the density of electrons
- x_0 is the amplitude of the electron oscillation in the field of the electromagnetic wave

The quantity x_0 can be found from the following standard ordinary differential equation describing forced oscillations of the electrons where E is the electric field of the wave, t is time, m is the electron mass, ω is frequency of the electromagnetic wave and ω_0 is eigenfrequency of atoms or ions (resonant frequency):

$$\ddot{x} - \gamma\dot{x} + \omega_0^2 x = \frac{e}{m} E e^{-i\omega t} \quad (9.2)$$

whose well-known solution is

$$x = x_0 e^{-i\omega t} \quad (9.3)$$

with

$$x_0 = \frac{eE}{m(\omega_0^2 - \omega^2 + i\omega\gamma)} \approx \frac{eE}{m(\omega_0^2 - \omega^2)} \quad (9.4)$$

(Away from the resonance $x_0 \approx eE/m(\omega_0^2 - \omega^2)$, since $|\omega_0^2 - \omega^2| >> |i\omega\gamma|$). Hence,

$$P = n_e e x_0 = \frac{n_e e^2 E}{m(\omega_0^2 - \omega^2 + i\omega\gamma)} \approx \frac{n_e e^2 E}{m(\omega_0^2 - \omega^2)} \quad (9.5)$$

$$\varepsilon = n^2 = 1 + \frac{4\pi n_e e^2}{m(\omega_0^2 - \omega^2 + i\omega\gamma)} \approx 1 + \frac{4\pi n_e e^2}{m(\omega_0^2 - \omega^2)} \tag{9.6}$$

and*

$$n \approx 1 + \frac{4\pi n_e e^2}{2m(\omega_0^2 - \omega^2)} \tag{9.7}$$

The latter equation is valid first of all away from the resonance (where $\omega_0 \approx \omega$) and it has also been derived under the assumption that there was a single oscillator in the medium. However, the truth is that in a real medium, there is a whole variety of oscillators, with the set of resonant frequencies ω_k. That is why in a gas, for instance, one actually has to take a summation over all the molecular or atomic energy levels of the mixed gas, using the Boltzmann distribution†:

$$n \approx 1 + \frac{4\pi e^2}{2m} \sum_{k=1}^{N} \frac{n_k f_k}{\omega_k^2 - \omega^2} \tag{9.8}$$

The quantity f_k in the above equation denotes the so-called *oscillator strength*, or the effectiveness of oscillators with a given frequency ω_k. It is a coefficient describing the contribution of the oscillators with this resonant frequency into the dispersion. (The reader should not be intimidated by the apparent complexity of Equation 9.8 since it remains accurate enough after merely confining the summation to just a few values of ω_k, which are close to ω.)

The complex value of the refractive index can be written as $\sqrt{\varepsilon} = n(1 - i\chi)$, and one can use Equation 9.6 to find its real and imaginary parts in the two equations that follow. The former corresponds to phase velocity, and the latter to damping of the wave.

$$n^2(1 - \chi^2) = 1 + \frac{4\pi n_e e^2(\omega_0^2 - \omega^2)}{m\left[(\omega_0^2 - \omega^2)^2 + \omega^2\gamma^2\right]} \tag{9.9}$$

* Note that Equation 9.7, as well as Equation 9.8, enables us to calculate the quantity n with good accuracy only if the second term on the right-hand side of these two identities is much smaller than one.

† Equation 9.8 is useful for a variety of estimates. However, more precise values of the refractive index (as well as the values of f_k), which we can find in the reference literature, have been determined, of course, by experiment.

and

$$2\chi n^2 = \frac{4\pi n_e e^2 \omega\gamma}{m[(\omega_0^2 - \omega^2)^2 + \omega^2\gamma^2]} \tag{9.10}$$

Near the resonance, we have $\omega_0 \approx \omega$, and therefore, $(\omega_0^2 - \omega^2) = (\omega_0 + \omega)(\omega_0 - \omega) \approx 2\omega(\omega_0 - \omega)$, which enables us to simplify the above equation to

$$n^2(1 - \chi^2) = 1 + \frac{8\pi n_e e^2 (\omega_0 - \omega)}{m\omega_0[4(\omega_0 - \omega)^2 + \gamma^2]} \tag{9.11}$$

and

$$2\chi n^2 = \frac{4\pi n_e e^2 \gamma}{m\omega_0[4(\omega_0 - \omega)^2 + \gamma^2]} \tag{9.12}$$

Taking into account the fact that $\chi^2 << 1$, let us find the near-resonance equation for $n(\omega)$:

$$n(\omega) \approx 1 + \frac{4\pi n_e e^2 (\omega_0 - \omega)}{m\omega_0\left[4(\omega_0 - \omega)^2 + \gamma^2\right]} \tag{9.13}$$

This function is shown in Figure 9.1. We should pay attention to the fact that near to the resonance, the quantity $n(\omega)$ is a decreasing function of ω. It is due to this particular feature that the latter dependence $n(\omega)$ is called the *abnormal dispersion.*

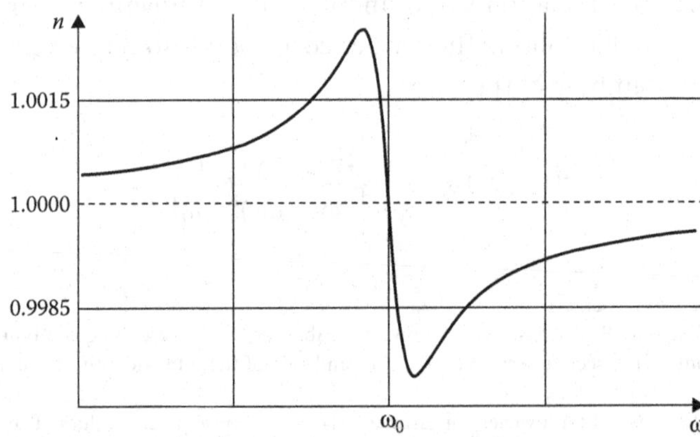

FIGURE 9.1 The dependence of $n(\omega)$ near a resonance.

In plasma, the equation for free electron oscillations ($\omega_0 = 0$) can be written as

$$m\ddot{x} = eE\exp(-\mathrm{i}\omega t) \tag{9.14}$$

Hence, $x_0 = -\dfrac{eE}{m\omega^2}$, $P = -\dfrac{n_e e^2}{m\omega^2}E$, and

$$\varepsilon = 1 - \frac{4\pi n_e e^2}{m\omega^2} = 1 - \frac{\omega_p^2}{\omega^2} \tag{9.15}$$

In the above equation, $\omega_p = \sqrt{4\pi n_e e^2/m} \approx 10^4\sqrt{n_e}$ is the plasma frequency, and finally

$$n = 1 - \frac{4\pi n_e e^2}{2m\omega^2} = 1 - \frac{\omega_p^2}{2\omega^2} \tag{9.16}$$

It turns out that the refractive index of plasma is less than one. This means that if we could manufacture, as a mental experiment, a double convex lens made of plasma, it would operate as a diverging lens.

The equation for the dispersion curve in plasma follows from Equation 9.15: $k(\omega) = \sqrt{\omega^2 - \omega_p^2/c^2}$. Therefore, for the phase and group velocities, we have, respectively, the equations $u_p = (\omega/k) = c\sqrt{\omega^2/\omega^2 - \omega_p^2}$ and $u_g = (\mathrm{d}\omega/\mathrm{d}k) = (c^2/u_p)$. Note that in plasma $u_p u_g = c^2$.

9.2 SCATTERING BY DENSITY FLUCTUATIONS

Huygens–Fresnel formalism does not allow the light beams traveling along straight lines in homogeneous continuous media to deviate by angles exceeding the diffraction angles. Such a deviation would indeed be an impossibility if the density of the propagation medium was perfectly uniform space-wise. Indeed, suppose a light beam with flat wavefront and diameter D is traveling through a homogeneous medium. Consider some plane perpendicular to the propagation direction, that is, normal to the wave vector $\boldsymbol{k}_0$ (Figure 9.2). Assume that the wave field in this plane is $A(t) = A_0 e^{-\mathrm{i}\omega\omega}$. Take a small volume ΔV of the medium in the plane of the wavefront: It contains $N = \Delta V n$ particles, where n denotes here medium density. Clearly, the electromagnetic field caused by the radiation from the electrons (free or bound) located in the wavefront plane inside the volume ΔV that find themselves in the oscillatory field of the primary light wave will be proportional to

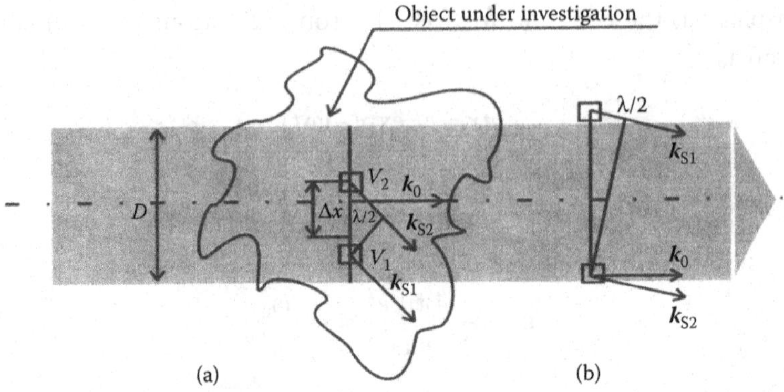

FIGURE 9.2 Electromagnetic wave scattering from density fluctuations. (a) $\Delta x < D$; (b) $\Delta x = D$.

$N = \Delta V n$: $A_e(t) \propto A_0 \Delta V n e^{-i\omega t}$. At some observation point, the scattered field will then be $A_S(t) = kA_0 \Delta V n \exp\{-i\omega[t-(L/c)]\}$, where L is the distance from the receiver to the source of radiation and k is a coefficient showing how many times the scattered field amplitude is smaller than the primary wave amplitude A_0. (The nature and magnitude of scattered radiation will be discussed in more detail later.) It follows from Figure 9.2 that if the scattering angles are not too small, we can always find another volume with the same value ΔV in the plane of the wavefront at some distance Δx from the former volume, such that the second volume will be located at a distance $L + \lambda/2$ from the radiation source. The sufficient condition to make it possible is that $\Delta x \sin\theta = \lambda/2$, where θ is the angle between the two wave vectors $\boldsymbol{k}_0$ and $\boldsymbol{k}_S$, that is, between the directions of the probing and the scattered radiation. Naturally, the beams $\boldsymbol{k}_{S1}$ and $\boldsymbol{k}_{S2}$ will interfere and annihilate each other if they have opposite phases. Will there be a match to every volume ΔV? Certainly not. As the angle θ decreases, Δx grows, and when it reaches $\Delta x = D$ (Figure 9.2b), the matching volume will find itself outside the limits of the propagating beam. But this corresponds to the case $2\theta = \lambda/D$, that is, θ is just the diffraction angle.

Therefore, if the probed object had a strictly constant spatial density or concentration, scattering beyond the diffraction angle would be impossible in principle. However, temperature fluctuations cause the density to fluctuate, and as a result, complete annihilation does not occur. It is clear that in order to determine the values of the scattered light field and its intensity in the direction $\boldsymbol{k}_S$, one should take the sum of contributions from each volume ΔV, lying in the plane of the wavefront:

$$A_\Sigma(t) = kA_0\Delta V\sum_j n_j \exp\left[-i\omega(t-\varphi_j)\right],\ I(t) = A_\Sigma(t)\cdot A_\Sigma^*(t) \quad (9.17)$$

where:

$\varphi_j = L_j/c$

Let us represent n_j in the form $n_j = \bar{n} + \delta n_j$, where $\bar{n}$ is the mean density and δn_j is a random fluctuation. It follows that

$$\begin{aligned} A_\Sigma(t) &= kA_0\Delta V\sum_j \left(\bar{n} + \delta n_j\right)\exp\left[-i\omega\left(t-\varphi_j\right)\right] \\ &= kA_0\Delta \dot{V}\sum_j \delta n_j \exp\left[-i\omega\left(t-\varphi_j\right)\right] \end{aligned} \quad (9.18)$$

as we have seen $\sum_j \bar{n}\exp[-i\omega(t-\varphi_j)] = 0$.

Then

$$I_\Sigma \propto A_0^2\sum_j \left(\delta n_j\right)^2 = A_0^2\sum_j \left(\overline{\delta n}\right)^2 = A_0^2\sum \bar{n} = A_0^2 N_\Sigma \quad (9.19)$$

because $\overline{\delta n} = \sqrt{\bar{n}}$. Note that N_Σ is the total number of radiating electrons inside the volume $V = \sum_j \Delta V$. Thus, the intensity of the scattered wave turns out to be proportional to the total number of electrons in the probing field.

9.3 BRILLOUIN SCATTERING

Density inhomogeneities can be caused not only by temperature fluctuations but often by various waves propagating in a medium. A typical example of the type of radiation scattering that they produce is Brillouin scattering—which is the case where an electromagnetic wave is scattered from a standing sound wave—which has been excited in a solid, a liquid, a gas, or a plasma by a corresponding oscillator. Figure 9.3 illustrates the principle of Brillouin scattering for a plane electromagnetic wave with wave vector $\boldsymbol{k}_0$ traveling in the Z-axis direction. The wave is scattered from density fluctuations caused by a standing sound wave that has been created by two sound waves with wave vectors $\boldsymbol{q}$ and $-\boldsymbol{q}$, traveling in two opposite directions. The angle between the direction of $\boldsymbol{q}$ and the X-axis

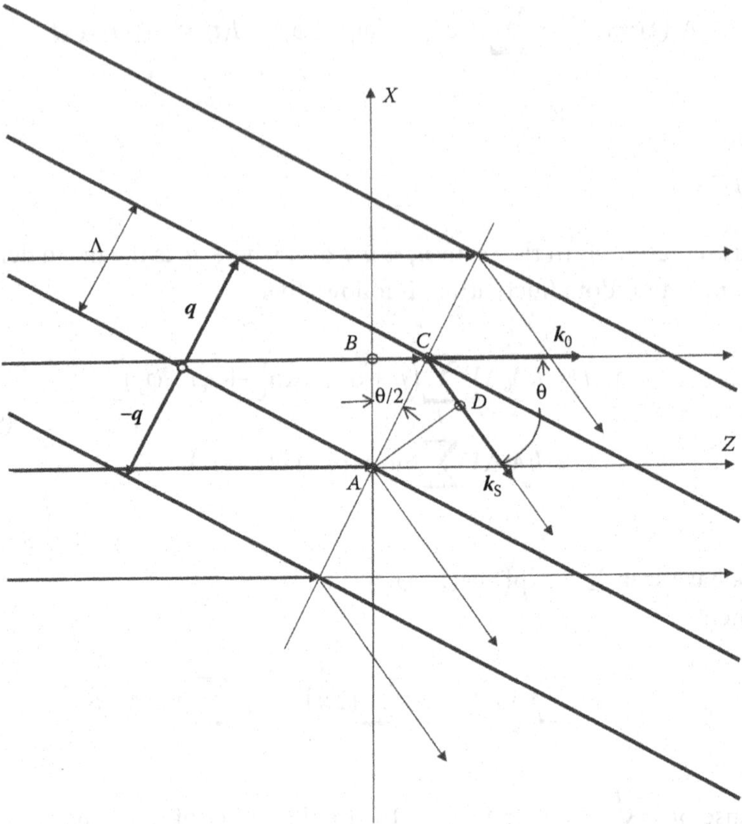

FIGURE 9.3 Brillouin scattering.

in the figure is $\theta/2$. If we follow the approach of L.I. Mandelstam and view Brillouin scattering as a diffraction phenomenon on the spatially periodic structure produced by the standing sound wave, it is easy to realize that the angle between $\boldsymbol{k}_0$ and the wave vector $\boldsymbol{k}_S$ of the scattered or diffracted wave will be equal to θ (the angle of *incidence* equals the angle of *reflection*). Let us verify that this is indeed the case. We draw a line perpendicular to the direction of $\boldsymbol{k}_S$ through the origin. If the line AD coincides with the plane of the wavefront of the scattered wave, its phase at the points A and D should be the same. It should also be the same at the points A and B, since both of these points lie in the plane of the wavefront of the incident wave, which is perpendicular to $\boldsymbol{k}_0$. Therefore, to ensure that the waves at the points A and D are in phase, it is necessary and sufficient that the path length B–C–D be equal to an integral number of wavelengths $n\lambda$. One can clearly see in Figure 9.3 that the segment lengths BC and CD are both equal to $\Lambda\sin(\theta/2)$, where Λ is the wavelength of the sound wave in the media. Hence, the result is as follows:

$$2\Lambda\sin\left(\frac{\theta}{2}\right) = n\lambda \tag{9.20}$$

which is nothing but the well-known Bragg condition. We will soon see that this result can also be obtained in a much easier and purely formal way, without the lengthy argument to which we have resorted in order to elucidate all the phenomenological details of what is going on.

The harmonic behavior of the spatial density of the medium due to the standing sound wave accounts for the effect that light scattered from it will be modulated (Figure 9.4). This accounts for the appearance of two spectral components in the scattered radiation:

$$E(t) \propto E_0\cos(\omega t)\cdot\cos(\Omega t) = \frac{1}{2}E_0\left[\cos(\omega_0 + \Omega)t + \cos(\omega_0 - \Omega)t\right] \tag{9.21}$$

where:

Ω denotes the angular frequency of the sound wave

Brillouin scattering can be described adequately using the quasi-quantum approximation. With this in mind, let us represent the standing sound wave as a superposition of two waves with identical amplitudes, traveling in opposite directions. Let the sound wave vectors be $\boldsymbol{q}$ and $-\boldsymbol{q}$. Due to energy and momentum conservation, we have

$$\hbar\omega_0 - \hbar\omega_S = \pm\hbar\Omega \tag{9.22}$$

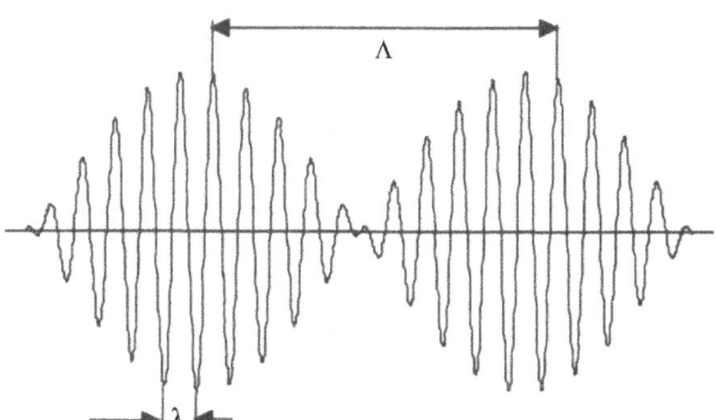

FIGURE 9.4 Modulation of scattered radiation.

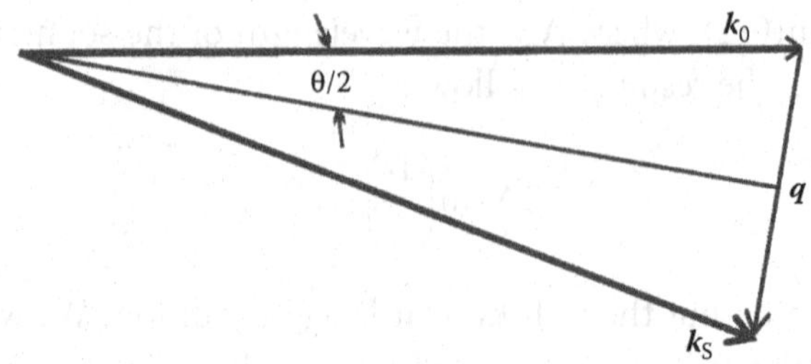

FIGURE 9.5 Brillouin scattering—Vector diagram.

and

$$\hbar \boldsymbol{k}_0 - \hbar \boldsymbol{k}_S = \pm \hbar \boldsymbol{q} \tag{9.23}$$

Equation 9.22 demonstrates that the physical reason for the spectral components to appear is the result of photon scattering on phonons. A photon either gains or loses the amount of energy equal to the energy of the phonon involved in the scattering. Equation 9.23, whose graphical representation is shown in Figure 9.5, allows the derivation of a number of equations that will prove rather useful later.

First, Figure 9.5 clearly shows that if $\boldsymbol{k}_0 \approx \boldsymbol{k}_S = \boldsymbol{k}$

$$|\boldsymbol{q}| = 2|\boldsymbol{k}|\sin\left(\frac{\theta}{2}\right) \tag{9.24}$$

We now substitute in the above equation alternately $|\boldsymbol{q}| = 2\pi/\Lambda$ or $|\boldsymbol{q}| = \Omega/v_S$, where v_S denotes the speed of sound and $|\boldsymbol{k}| = 2\pi/\lambda$ or $|\boldsymbol{k}| = \omega/c$, respectively. Since the absolute values of the two wave vectors $\boldsymbol{k}_0$ and $\boldsymbol{k}_S$ can be regarded as approximately equal, namely, $|\boldsymbol{k}_0| \approx |\boldsymbol{k}_S|$, we first obtain that

$$\lambda = 2\Lambda \sin\left(\frac{\theta}{2}\right) \tag{9.25}$$

which is an analog of the Bragg condition and also that

$$\Omega = 2\frac{v_S}{c}\omega \sin\left(\frac{\theta}{2}\right) \tag{9.26}$$

or

$$\frac{\Delta\omega}{\omega} = 2\frac{v_S}{c}\sin\left(\frac{\theta}{2}\right) \tag{9.27}$$

Thus, we can see that the spectral shift of each scattered component is indeed caused by the Doppler effect. Finally, we find that the difference

$\boldsymbol{k}_0 - \boldsymbol{k}_S$, which is called the difference vector and is usually denoted by $\bar{\boldsymbol{k}}$, will have an absolute value equal to

$$|\bar{\boldsymbol{k}}| = \frac{4\pi \sin(\theta/2)}{\lambda_0} \tag{9.28}$$

where:

$$\lambda_0 = \frac{2\pi c}{\omega_0}$$

9.4 THOMSON, RAYLEIGH, AND RESONANCE SCATTERING

The theory of radiation gives the following equation for the power of radiation scattered by an oscillating electron:

$$P_S = \frac{2}{3}\frac{e^2 \overline{\ddot{x}}^2}{c^3}$$

where:

- c is the speed of light
- $\overline{\ddot{x}}$ is the mean square acceleration of the electron
- e is its charge

Differentiating Equation 9.3 twice, we find $\ddot{x} = -\omega^2 x_0 e^{-i\omega t}$, and hence

$$\overline{\ddot{x}}^2 = \omega^4 x_0^2 = \frac{e^2 E_0^2 \omega^4}{m^2(\omega_0^2 - \omega^2 + i\omega\gamma)^2} \tag{9.29}$$

Now, in view of the fact that $I_0 = cE^2/4\pi$, where I_0 is the intensity of the probe radiation (i.e., $E_0^2 = 4\pi I_0/c$), and using the equation $r_0 = e^2/mc^2$ for the so-called classical electron radius, we obtain

$$P_S = I_0\left[\frac{8\pi}{3} r_0^2 \frac{\omega^4}{(\omega_0^2 - \omega^2 + i\omega\gamma)^2}\right] \approx I_0\left[\frac{8\pi}{3} r_0^2 \frac{\omega^4}{(\omega_0^2 - \omega^2)^2}\right] \tag{9.30}$$

Let us denote the quantity in the square brackets as σ. It has the dimension of area and plays the role of the scattering cross section. Indeed, if we multiply both sides of the above equation by the duration Δt of the probe pulse and the electron density n, the resulting quantity $P_S \Delta t n = I_0 \sigma \Delta t n$ will have the meaning of the energy scattered by n electrons per unit volume.

TABLE 9.1 Thomson, Rayleigh, and resonance scattering

	Frequency Conditions	Scattering Cross-section	Type of Scattering
1	$\omega_0 << \omega$	$\sigma = (8\pi/3)r_0^2$	Thomson scattering
2	$\omega_0 >> \omega$	$\sigma = (8\pi/3)r_0^2(\omega/\omega_0)^4$	Rayleigh scattering
3	$\omega_0 = \omega$	$\sigma = (8\pi/3)r_0^2(\omega^2/\gamma^2)$	Resonance scattering

It is clear that $P_S\Delta tn/h\nu$ represents the number of photons scattered from a unit volume into the solid angle 4π.

The results of three different possible cases, namely, when $\omega_0 << \omega$, $\omega_0 >> \omega$, and $\omega_0 = \omega$, are shown in Table 9.1.

This table indicates that Thomson scattering has a cross section of only $\sigma \approx 10^{-25}\,\text{cm}^2$; for Rayleigh scattering, it increases as $1/\lambda^4$, and since we always have $\gamma << \omega$, the resonance scattering cross section is larger by many orders of magnitude than either of the other two scattering cross sections.

Let us emphasize that the quantity σ characterizes the full effective scattering cross section from a single electron into the solid angle 4π. The radiation patterns of polarized and nonpolarized scattering radiation are shown in Figure 9.6. The differential scattering cross section, namely, scattering into the differential solid angle $d\Omega$ in the direction making an angle φ to the direction of the vector $\boldsymbol{E}$ of the probing wave, can be expressed as follows:

$$d\sigma = r_0^2 \sin^2 \varphi d\Omega \tag{9.31}$$

It should be clear that $\int_{4\pi} d\sigma = (8\pi/3)r_0^2$.

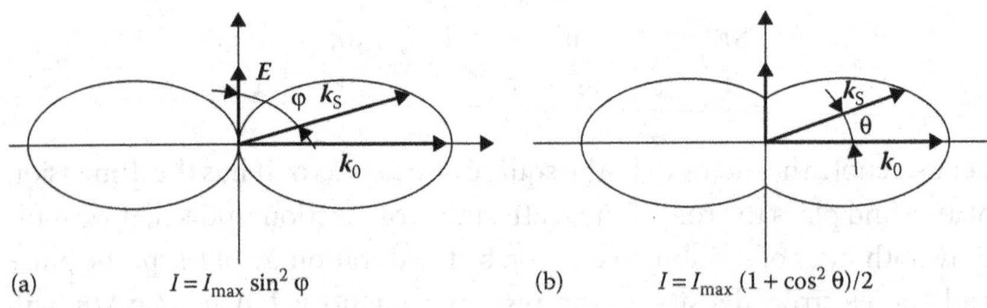

FIGURE 9.6 (a) Polarized and (b) Nonpolarized scattering radiation pattern.

9.5 SCATTERING FROM MOVING ELECTRONS

Let an electron move with velocity $\boldsymbol{v}_e$ and the vectors $\boldsymbol{k}_0$, $\boldsymbol{k}_S$, and $\boldsymbol{v}_e$ are directed as shown in Figure 9.7. It is clear that due to the Doppler shift, the oscillations of the electron will have the frequency

$$\omega_e = \omega_0 - \frac{\boldsymbol{v}_e \cdot \boldsymbol{k}_0}{c|\boldsymbol{k}|}\omega_0 = \omega_0\left(1 - \frac{\boldsymbol{v}_e \cdot \boldsymbol{k}_0}{c|\boldsymbol{k}|}\right) \tag{9.32}$$

Since the electron is moving, the radiation measured by a fixed receiver will in turn to be shifted in frequency with respect to that of the oscillating electron:

$$\omega_S = \omega_0\left(1 - \frac{\boldsymbol{v}_e \cdot \boldsymbol{k}_0}{c|\boldsymbol{k}|}\right)\left(1 + \frac{\boldsymbol{v}_e \cdot \boldsymbol{k}_S}{c|\boldsymbol{k}|}\right) \tag{9.33}$$

Disregarding the small term v_e^2/c^2, we arrive at the equation

$$\omega_S = \omega_0\left(1 + \frac{\boldsymbol{v}_e \cdot \bar{\boldsymbol{k}}}{c|\boldsymbol{k}|}\right) \tag{9.34}$$

with the difference vector $\bar{\boldsymbol{k}} = \boldsymbol{k}_0 - \boldsymbol{k}_S$. Hence

$$\Delta\omega = \omega_S - \omega_0 = \omega_0 \frac{\boldsymbol{v}_e \cdot \bar{\boldsymbol{k}}}{c|\boldsymbol{k}|} = \boldsymbol{v}_e \cdot \bar{\boldsymbol{k}}$$

as indeed $|\boldsymbol{k}| = (\omega_0/c)$.

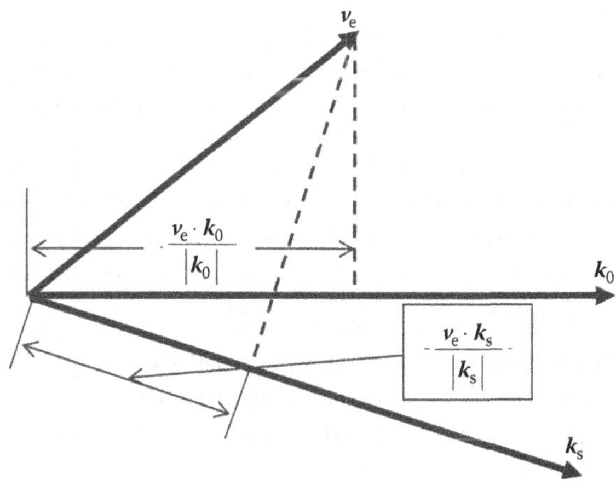

FIGURE 9.7 Scattering from moving electrons—Vector diagram.

Consider now the spectrum of the scattered radiation from electrons that have a Maxwellian distribution with a temperature T_e. For an arbitrary direction, the distribution function is

$$dn_e = A\exp\left(-\frac{mv^2}{2kT}\right)dv \tag{9.35}$$

where:

factor A is determined by system of units

Let us regard ω_0 as the frequency origin. Then, $\omega = \bar{k}\cdot v_e$, $dv = d\omega/\bar{k}\cos\varphi$, and, as previously, $|\bar{k}| = 2(\omega_0/c)\sin(\theta/2)$. Furthermore,

$$dn_e = B\exp\left[-\frac{m\omega^2c^2}{4\omega_0^2 2kT\sin(\theta/2)}\right]d\omega \tag{9.36}$$

where:

factor B is determined by system of units

Therefore, we can easily calculate the value of the FWHM of radiation scattered at an angle θ:

$$\Delta\lambda_{1/2} = 4\lambda_0\sin\frac{\theta}{2}\sqrt{\frac{kT_e}{mc^2}\ln 2} \tag{9.37}$$

The above equation shows that the temperature T_e can be determined via relative measurements and there is no need to measure the absolute magnitude of the scattered radiation to calculate T_e.

9.6 COLLECTIVE SCATTERING IN PLASMA

The physics of collective scattering is demonstrated in Figure 9.8. The figure compares two different scenarios—when the probing wave has a wavelength, which is either much smaller (Figure 9.8a) or much greater (Figure 9.8b) than the Debye radius r_D. In the former case, the electrons inside the Debye sphere are moving in essentially random directions in the electric field without generating any space charge, and therefore, one deals with scattering from free electrons. In the latter case, all the electrons inside the Debye sphere are displaced in the same direction; such a displacement creates a space charge and therefore a restoring force. Clearly, this generates plasma oscillations with frequency ω_p. This physically clear explanation, constrained by the rather stringent condition that $\lambda << r_D$, cannot account for the situation shown in Figure 9.9, where

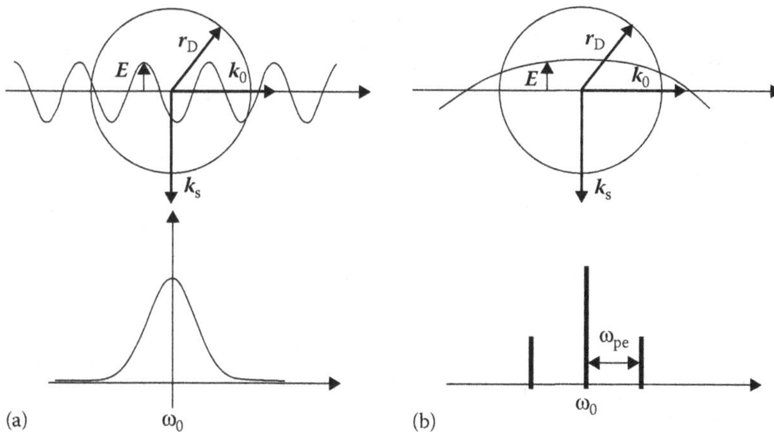

FIGURE 9.8 The two limiting cases of light scattering: (a) incoherent scattering when the wavelength of the probing radiation is smaller than the Debye radius; (b) coherent scattering when the wavelength of the probing radiation is larger than the Debye radius.

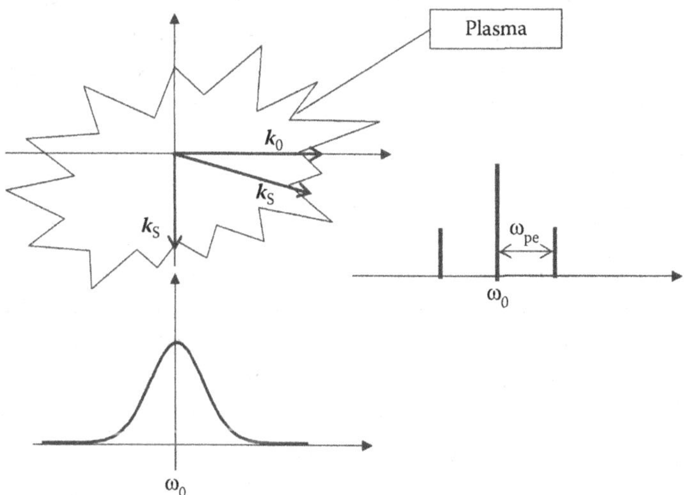

FIGURE 9.9 Coherent scattering when the probing wavelength and Debye radius are approximately equal.

one has Thomson scattering in the direction perpendicular to the probe beam and collective scattering at small angles to the direction of the vector $\boldsymbol{k}_0$. Another point of view proves to be more fruitful here; photons are being scattered not only from fluctuations of free electrons but also from plasma oscillations, or plasmons, similar to the scattering from phonons considered in Section 9.3. The only difference is that the spatial distribution of the wave vectors of plasma oscillations is chaotic, and therefore, in principle, one can have scattering at any angle. Similar to the conservation law Equations 9.22 and 9.23, we can state that

$$\omega_0 - \omega_S = \omega_p \tag{9.38}$$

and

$$\boldsymbol{k}_0 - \boldsymbol{k}_S = \boldsymbol{k}_p \tag{9.39}$$

where:

ω_p and $\boldsymbol{k}_p$ denote, respectively, the frequency and wave vector of the plasma oscillations

From Figures 9.5 and 9.9, we can see that in order to have scattering at large angles, the absolute value of the vector $\boldsymbol{k}_p$ should be of the order of magnitude as the absolute values of $\boldsymbol{k}_0$ and $\boldsymbol{k}_s$. This cannot hold in a series of cases, since quite often oscillations with $|\boldsymbol{k}_p| \approx |\boldsymbol{k}_0|$ cannot exist in plasma due to the phenomenon of Landau damping. (This is indeed the case in Figure 9.9.) The possibility of collective scattering in plasma is described in terms of the so-called Salpeter criterion:

$$\alpha = \frac{1}{kr_D} = \frac{\lambda_0}{r_D 4\pi \sin(\theta/2)} \tag{9.40}$$

Collective scattering takes place if $\alpha > 1$. Note that collective scattering enables one to find the electron density in plasma from merely relative measurements of the displacement of spectral components, since $\omega_p^2 = 4\pi ne^2/m$. It is also worth mentioning that in a whole variety of cases when collective scattering takes place, one encounters the phenomenon of splitting of its central maximum.* The latter is caused by light scattering from waves of ion plasma frequency ω_{pi}. Thus, one can also use the values of the spectral components $(\omega_0 \pm \omega_{pi})$ to find the ion density in plasma and once again from relative measurements only. It has been demonstrated in Tokamak facilities using a powerful microwave source (a gyrotron) and will be installed as a diagnostics on the International Thermonuclear Experimental Reactor (ITER) facility. However, we must acknowledge the fact that this technique, although so promising in principle, has so far found application only in a few rather specific scenarios.

* It is necessary to use powerful microwave radiation to study this phenomenon.

9.7 INSTRUMENTAL REALIZATION

There are three main measurement methods for scattered radiation. The first method collects radiation scattered by the plasma into a small solid angle and therefore enables us to obtain information about plasma temperature and density in a small volume centered on the region where the direction of observation intersects the beam of the probing laser (Figure 9.10a). The time of the measurement is determined by the time of the laser pulse. Radiation scattered by the plasma is directed to a spectrometer or a polychromator.* Having passed through the latter device, the radiation from different spectral regions is usually measured by an assembly of photomultipliers. Alternatively, instead of a spectrometer or a polychromator, one can use a series of narrow band-pass filters transmitting in different spectral intervals. It is not possible to find the velocity distribution of the electrons this way, but at least one can measure their temperature if we assume the plasma to be in thermodynamic equilibrium where the scattered radiation has a Gaussian distribution (Section 9.5), whose parameters can be determined by only a few measurements. This simple realization has only one flaw: Each measurement accounts for temperature only at a single spatial point in the object under investigation, as well as at a single time instant. Let us make a technical note that along the direction of the probe beam one usually installs a series of light traps or "baffles." Their role is to reduce the stray light background arising from the probe beam scattering on the input and output windows of the vacuum chamber.

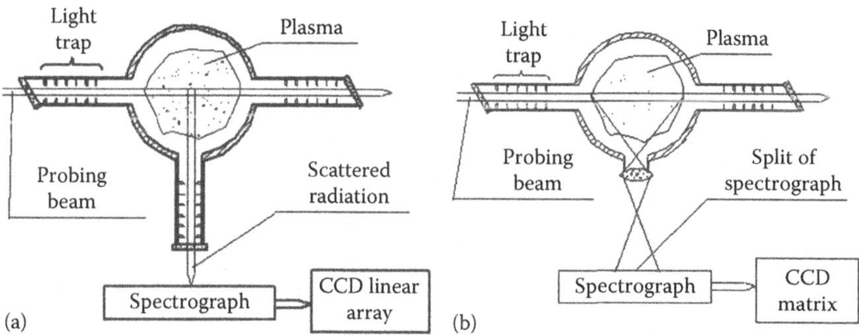

FIGURE 9.10 Scattered radiation measurements: (a) scattered radiation coming from single volume of plasma, (b) scattered radiation coming from each point along the path of the beam.

* These spectral devices have been described in detail in Chapter 6.

In the second scheme, the line through the plasma that is illuminated by the probing beam is imaged onto the input slit of a spectrometer (Figure 9.10b) in such a way that the direction of propagation of the laser beam is optically conjugate to the spectrograph slit. This enables us to measure plasma temperature along the whole path of the probing beam, since we will have the spectra of scattered radiation coming from each point along the path of the beam. Naturally, one needs to use a two-dimensional measurement device here such as an electron-optical framing camera or a high-sensitivity charge-coupled device (CCD) camera with a cooled matrix.

The third approach is the so-called light detecting and ranging (LIDAR)* scheme (Figure 9.11). In this device, plasma is being probed by a very short laser pulse (its duration τ determines the spatial resolution $\Delta Z = c\tau/2$) and the backward scattered radiation is directed to an electron-optical framing camera, recording the spectrum as a function of time $I = I(t,\lambda)$. Clearly, in this case there is a one-to-one correspondence between the time and space position of the laser pulse. This enables us to measure the temperature and density distributions along the path of the probing beam, since, as in the previous case, the scattered radiation spectra

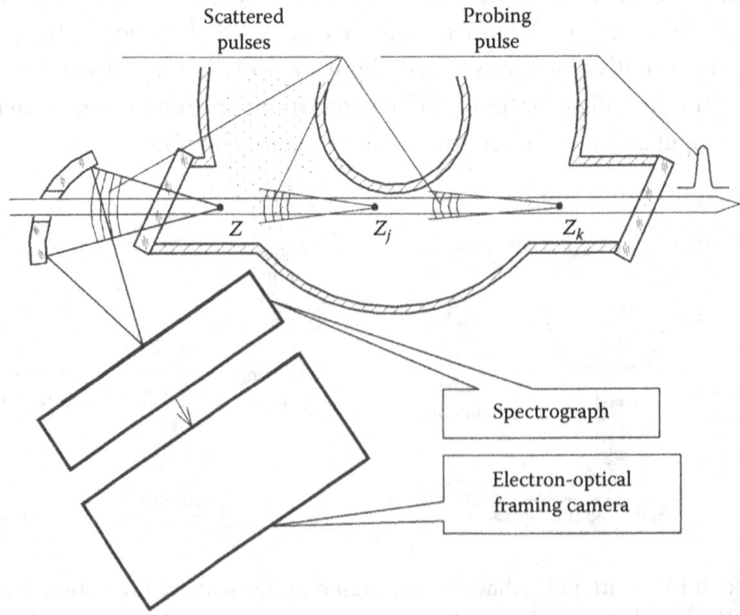

FIGURE 9.11 The principle of operation of LIDAR Thompson scattering.

* LIDAR has been called this way by analogy with radio detecting and ranging (RADAR).

from essentially each point along the path of the probing beam become available. All the three methods described above are widely used today.

How powerful should the probing beam be? Let us estimate the necessary energy. It should be clear from Equation 9.30 that the number of photons scattered during a time interval Δt into the solid angle 4π is $P_S \Delta t\eta/h\nu = I_0 \sigma \Delta t\eta/h\nu = \mathcal{E}\sigma\eta/h\nu$, where $\mathcal{E} = I_0 \Delta t$ is the energy density of the probing laser beam. The quantity η denotes the fraction of the scattered radiation that is collected (usually this is in the range of 2×10^{-3}–2×10^{-5} depending on the distance from the scattering volume to the receiving lens). If, γ describes the losses inside the optical channel (typically 0.2–0.5), μ is the quantum efficiency of the photoreceiver (~0.3) and finally, if D is the diameter of the probing beam, we have the equation $N_e = (\pi D^2 \mathcal{E}\eta\gamma\mu/4h\nu)n\sigma$ for the number of electrons emitted from the photoreceiver cathode or recorded in the CCD matrix. Suppose we consider Thomson scattering ($\sigma \approx 10^{-25}\,\text{cm}^2$). The typical electron density of large-scale fusion devices of the Tokomak type is $n \approx 10^{14}\,\text{cm}^{-3}$, the probe radiation wavelength is 1.054 μm, and the laser impulse energy $\pi D^2 \mathcal{E}/4$ is about 1 J.* Then, if one wishes to have a spatial resolution of about 1 cm, namely, to be able to measure scattered radiation coming from one cubic centimeter of the medium, the number of scattered photons reaching the aperture of the measurement device will be about 10^4. Furthermore, the number of photoelectrons will be about 600, with the statistical error $1/\sqrt{N_e}$ of some 4% for density measurements and two to three times more for measurements of temperature. These estimates turn out to be in good agreement with experiment reality.

In conclusion, we present the results of laser probing of plasma in the Joint European Torus (JET) device, at time moments before and after the injection of a deuterium pellet.† The laser used in this experiment operated with a repetition frequency of 1.2 Hz. An electron-optical framing camera was used as the radiation receiver. We can see from Figure 9.12 that after the deuterium pellet injection, the plasma density in the core of the plasma increased by more than an order of magnitude and was accompanied by an order of magnitude drop of the plasma temperature.

* Let us note that continuous-wave (CW) lasers could have been used for Thomson scattering diagnostics if you deal with low temperature plasmas.

† www.jet.efda.org.

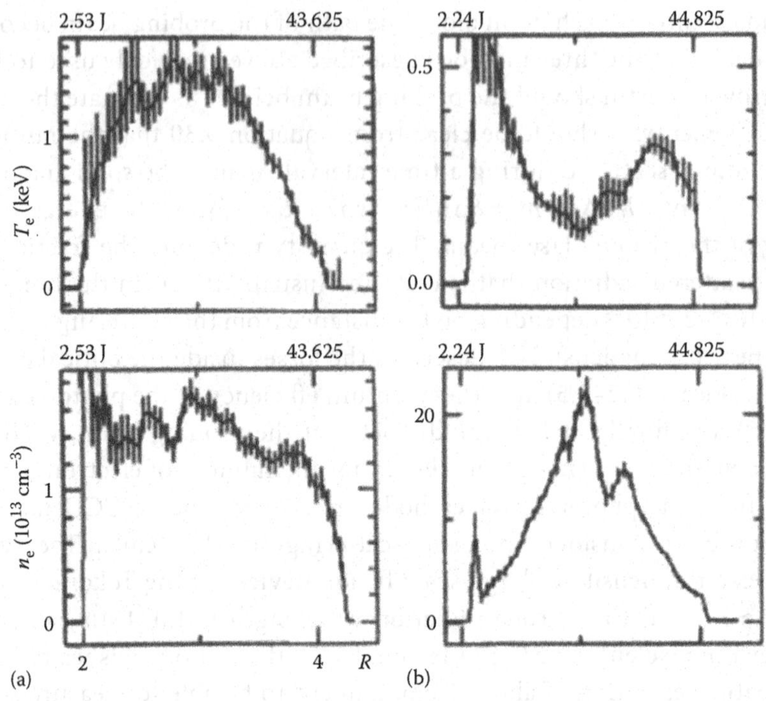

FIGURE 9.12 Typical results of temperature and density profile measurements with LIDAR Thompson scattering in the JET experiment: profiles before (a) and after (b) injection of a pellet of solid deuterium. (From EFDA JET Culham Science Centre, Abingdon, United Kingdom)

REFERENCES

1. DeSilva, A.W. 2000. The evolution of light scattering as a plasma diagnostic. *Contributions to Plasma Physics* 40: 23.
2. Pergament, M.I. 2000. *Plasma Diagnostics by Radiation Scattering.* Encyclopedia of Low Temperature Plasma, vol. 2, pp. 569–572. Moscow, Russia: Nauka.
3. Sheffield, J. 1975. *Plasma Scattering of Electromagnetic Radiation.* London: Academic Press.
4. Born, M. and Wolf, E. 2003. *Principles of Optics.* Cambridge: Cambridge University Press.

CHAPTER 10

Measurements in the X-Ray Region of the Spectrum

X-RAYS NOWADAYS HAVE APPLICATIONS in a wide range of investigations and measurements, from biology and medicine to crystallography and metallurgy, each area calling for its own detailed consideration of the techniques that are used. Naturally, one cannot possibly cover such a vast scope in a single discussion, and an attempt to do so would be counterproductive. We will therefore concentrate our attention only on those X-ray measurement methods that are used in experimental physics and mainly confine the spectral interval considered to the so-called far (or vacuum) ultraviolet (VUV) and soft X-ray (SXR) regions of the spectrum.

10.1 SPECTRAL DOMAIN AND SOURCES OF X-RAY RADIATION

More precisely, we are going to consider the wavelength range from 200 to about 0.25 nm (corresponding to photon energies ranging from some 6 to 5000 eV).* It is no accident that these wavelengths are so common in experimental physics applications. First, it is in this wavelength range that electromagnetic radiation actively interacts with atomic structures, thus

* If wavelengths are given in nanometers and photon energy in electron volts, the transition between the two systems of units can be made using the relationship $\hbar\omega\lambda = 1239.842$ eV × nm.

becoming a tool for their exploration and analysis.* Second, the radiation emitted by very high-temperature objects is strongest in this range, whether it is from high-temperature plasma in a laboratory or from *astrophysical objects* in the universe. Third, the fact that these wavelengths are comparable with intra-atomic distances in a solid-state matter opens up a vast opportunity for various types of measurement.

The wavelength range in question is arbitrarily divided into the VUV (200–10 nm) and the SXR (10–0.25 nm) regions. The boundary between the two regions is defined only for convenience. The overall long and short wavelength boundaries are determined by the fact that outside this range there is no need to use vacuum devices, but the choice of an intermediate wavelength of 10 nm to separate the VUV from the SXR is entirely arbitrary. The whole VUV/SXR range is characterized by a high absorption rate in air at an atmospheric pressure. This creates a variety of additional obstacles for measurement design, such as forcing us to use vacuum devices, preventing the use of refraction-based optical instruments, and restricting the possible range of reflection angles.

Electromagnetic radiation in the spectral region under discussion has a number of special features and can, according to its physics, be divided into radiation related to free–free transitions, free–bound transitions, and bound–bound transitions. The first type of radiation is commonly referred to as *bremsstrahlung* or *braking radiation*. As we have established earlier in Chapter 9, the power radiated by a single moving electron is proportional to the square of its acceleration $\ddot{x}^2$ along its trajectory. In plasma, for instance, an electron radiates as it moves along a curved trajectory when accelerated in the field of an ion. The other scenario for radiation of exactly the same nature is synchrotron radiation,† namely, electromagnetic radiation emitted by electrons moving in a magnetic field, which causes their trajectories to acquire curvature. To create such curved trajectories, one uses a so-called undulator or wiggler. This is a magnet that generates a strong alternating transverse magnetic field, forcing the electron trajectories to attain a wiggly shape, reminiscent of the sine curve. An electron beam moving this way loses energy that transforms into a stream of photons emitted inside a narrow cone around the

* It is worth mentioning that infrared radiation, which in a similar way interacts with molecular objects, has played a decisive role in research toward the identification of molecular structures and still remains today a dependable tool for molecular analysis.

† Sometimes, a hot, magnetically confined fusion plasma is also a source of synchrotron radiation.

axis of the electron beams. In both cases, the spectrum of the radiation that is generated is a continuum.

Free–bound transitions pertain to so-called recombination radiation, where free electrons are trapped in vacant bound states in ionized atoms, releasing their excess energy as photons. In plasma, processes of this kind intensify when plasma cools, because the decrease in electron velocity naturally increases the recombination cross section. Interestingly enough, the recombination process results in the phenomenon of population inversion, because at first an electron is more likely to occupy a higher energy bound state, with a subsequent transition to a ground state by releasing a photon carrying away the excess energy. We note in passing that this mechanism is being seriously explored in research toward the development of X-ray lasers.

Finally, let us describe bound–bound transitions, namely, electron transitions within the atomic structure. Let us subdivide the photon emission processes into emission from the upper versus intermediate atomic shells. In the former case, a photon or electron exciting an atom shares with it a fraction of its energy, causing the transition of a bound electron in the level with the same principal quantum number to a higher energy state. When the excited state decays (with a typical lifetime ~10^{-8} s except for metastable levels), the relaxation process takes place, whereby the atom returns to its initial state, and a photon whose energy equals the energy difference between the two energy states is emitted. Such a mechanism is largely responsible for the generation of radiation in the visible spectrum range. However, the radiation from highly ionized atoms, for example, hydrogen-like ions of carbon or oxygen, which are commonly observed in high-temperature plasma, falls into the X-ray range. (These are the same Balmer, Lyman, and Paschen series, only shifted into the X-ray region!)

A sufficiently energetic electron or photon may be capable of exciting an atom by knocking an electron out from one of its inner shells. As relaxation occurs, the freed vacancy in the energy level, say, K, can be filled by an electron from one of the higher levels L, M, N,…, where K, L, M, N,… are energy levels characterized by principal quantum numbers 1, 2, 3, 4,…, respectively. As in the previous case, the energy of the emitted photon will be equal to the difference in the energy states of the atom before and after relaxation. Note that if the vacancy, say, on the level K, is filled by an electron from (say) level L, the latter level in turn becomes a vacancy, which will then be filled by electrons from still higher levels. Radiation generated by an atom undergoing such a cascade of transitions is known as the

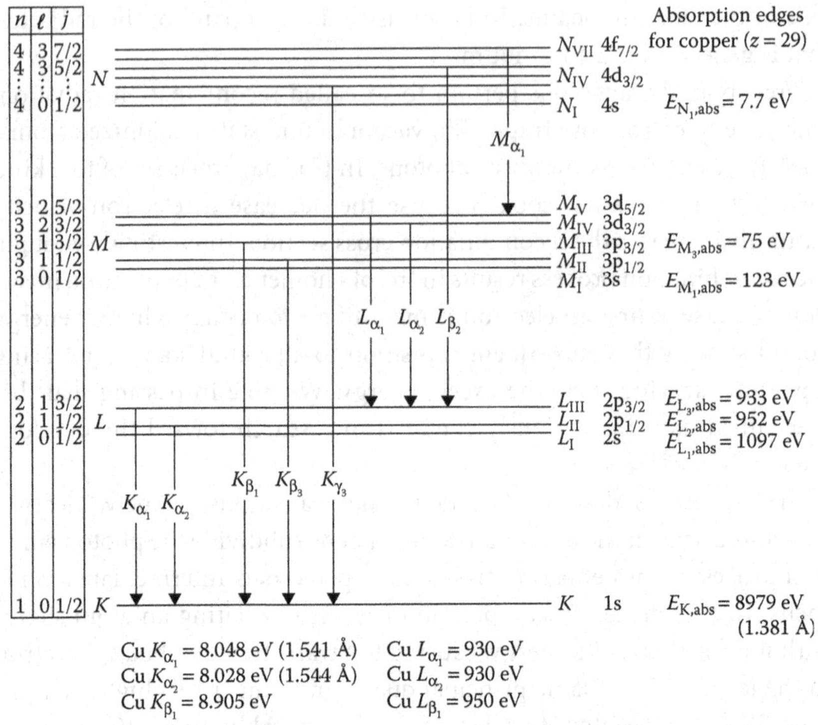

FIGURE 10.1 The energy structure of the copper atom. (From Attwood, D. T., *Soft X-Rays and Extreme Ultraviolet Radiation—Principles and Applications*. Cambridge University Press, Berkeley, CA, 2007. With permission.)

characteristic radiation. Its discrete spectrum consists of a relatively small number of spectral lines, making it a very useful tool for identification of the chemical composition of an object under investigation. This is illustrated in Figure 10.1, showing an example of the transitions responsible for generation of the *K* and *L* series spectra in a copper atom.

The physical mechanisms listed above form the basis of artificial sources of X-ray radiation. We will mainly consider those sources, which are used by experimental physicists. The oldest but still a very common device (although it is used mostly in medicine) is the Roentgen, or X-ray tube, depicted in Figure 10.2, which shows the principal elements of this device: a heated cathode, an electron emitter, a cooled anode, and power supplies. The upper limit for the generated X-ray spectrum is determined by the applied voltage, and the lower bound is determined by the transmission through the walls of the tube. This is a simple, reliable, and reasonably compact device. Its efficiency in transforming electrical energy

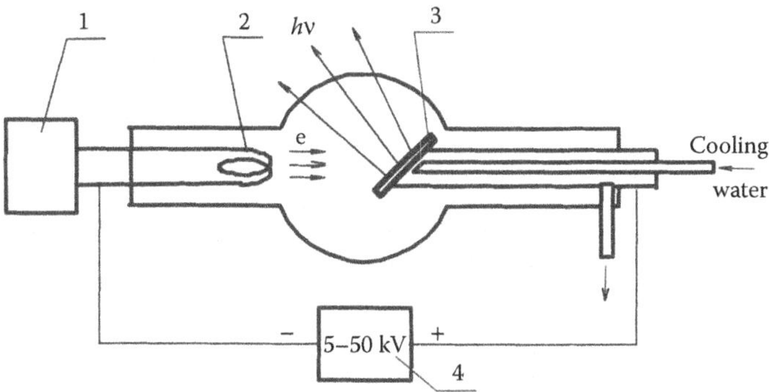

FIGURE 10.2 X-ray tube: 1, power source; 2, heated cathode; 3, anode.

into X-ray radiation is only 1%, but this does not constitute its main flaw. Unfortunately, the lower bound of the generated spectrum lies outside the VUV/SXR range, which has a decisive importance for experimental physics. Second, the X-ray tube represents an extended source of X-ray emission, whereas the majority of X-ray investigation techniques would prefer a pinpoint light source of radiation. Third, the continuous radiation spectrum can also be an inconvenience, as monochromatic radiation is usually required, and this means that in addition to the X-ray tube, one needs to set up a monochromator, which also has to be a vacuum type.

A variant of such an X-ray device with a monochromator was designed for an X-ray spectrophotometer in the Leningrad (now St. Petersburg) State University in the 1960s. The spectrophotometer does not have an X-ray tube *per se*. Instead, an emitter containing a source of electrons and a complex cathode are placed in a single vacuum container together with the other elements of the device. The emitter's anode takes the form of a multifaceted prism, whose faces are covered with thin plates made of different materials with different values of Z (Figure 10.3). The prism can be rotated, without compromising the vacuum conditions, around its vertical axis to allow the electron beam to impinge on the selected face. The electron beam is formed by a focusing projector, and the accelerating voltage is carefully controlled to ensure that the energy of the electrons corresponds to the ionization energy of the level that is being excited. Looking at the design of this machine, one can tell that it was intended for the generation of characteristic radiation. Its discrete and very sparse spectral composition facilitates the production of monochromatic radiation, in particular, by allowing us to use simple reflection–absorption filters (Section 10.3).

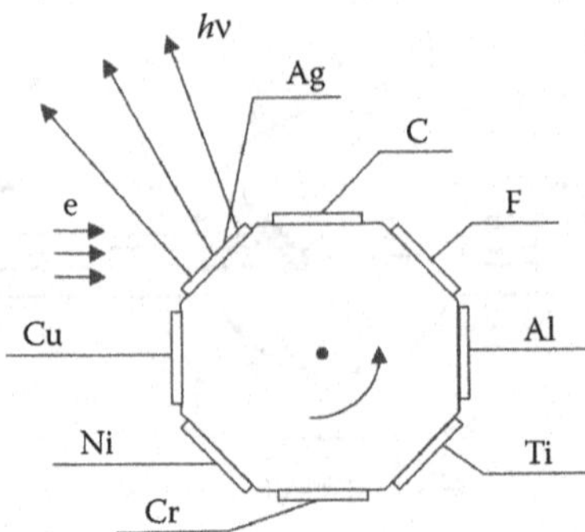

FIGURE 10.3 Anode of an X-ray radiator designed to generate characteristic radiation.

In recent years, synchrotron radiation has become widely used. Unfortunately, a synchrotron, mainly due to its large size, cannot provide a universal portable source of radiation for different large-scale experiments. One has to transport the object of investigation to the synchrotron and not the other way around. There are several synchrotrons that are used as X-ray radiation sources for collective experimental purposes in the so-called general-use centers equipped with workstations, where physicists, biologists, chemists, and potentially anyone who requires an X-ray source for their measurements can run their experiments. However, additional equipment is needed to make use of the X-ray source: Monochromators, detectors, and any other additional devices for a particular experiment are usually not included into the center's package and are the responsibility of the investigators using the facilities.

Sometimes, capillary discharges are used to generate VUV. The divergence of the radiation is then defined by geometry alone, which is basically the ratio of the diameter of the capillary to its length. In some sense, this is a point source, although its size, namely, the diameter of the output face of the capillary, is not really so small. Genuinely, "point" sources of X-ray radiation are built on different principles. First, the source of radiation is a plasma, which has been created by a focused laser beam. Measurements show that if a copper target is hit by a laser beam with a focal spot intensity of some 10^{14} W cm^{-2}, about 50% of the beam's energy will be converted into X-rays, with the photon energy exceeding 100 eV. The size of the radiating region is approximately equal to the size of the focal spot (30–100 μm). Moreover, the time duration of the X-ray output approximately equals that of the laser pulse that generates it. This allows experimentalists to

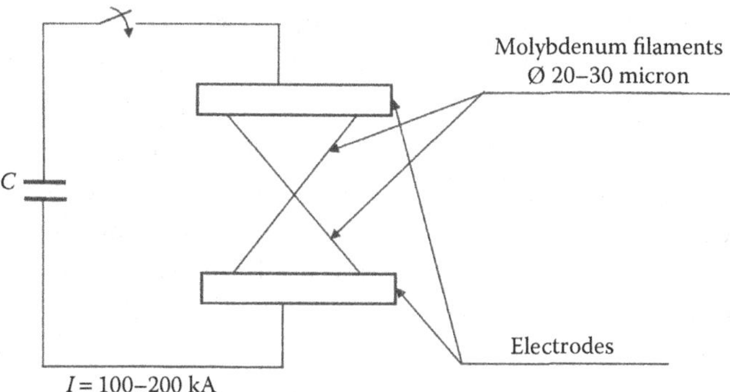

FIGURE 10.4 Point source of X-ray radiation (X-pinch).

take X-ray "snapshots" with nanosecond exposure times. Another point source of X-ray radiation that is used today is the so-called X-pinch, which is in essence a microscopic-sized Z-pinch with a localized neck region. A schematic of such a device is shown in Figure 10.4. The conductor configuration with metal filaments of a few tens of microns in diameter causes a dense and hot plasma formation to appear locally near to the point where the conductors intersect. The size of this region, when photographed through a filter that blocks photons with energies under 1 keV, is less than several microns.

A more detailed description of X-ray radiation sources and the underlying physical phenomena can be found in the substantial monograph by David Attwood [1].

10.2 X-RAY DETECTORS

The oldest, yet in fact still a very reliable, X-ray detector is a photographic emulsion (see Section 4.4). The main obstacle arising in this context is the strong absorption of SXR and VUV radiation in the uppermost layer of the photoemulsion. This causes a dramatic reduction of photographic width and restricts the surface density of activated grains and thus their total number. This entails a sharp reduction of the signal-to-noise ratio. There are two ways of trying to deal with these flaws: (1) to cover the emulsion with a thin layer of a luminescent substance, thus converting the X-ray radiation into visible light, and (2) to use the recently developed so-called nongelatin emulsions. These emulsions are transparent to SXRs, and therefore, the grains of such an emulsion layer are activated throughout its whole depth.

As a matter of fact, scintillators—although widely used for measurements of hard X-rays and gamma and corpuscular radiation—are rarely used in the VUV/SXR range. This is not so surprising because scintillators used in the former measurements play an intermediary role for delivering radiation propagating through the air to the photocathodes of vacuum photoelectron devices (photomultipliers, electron-optical cameras). They convert radiation, which by itself would not be able to penetrate through the window of the corresponding vacuum device, to the visible spectral range, where the penetration is no longer an issue. This is not a problem for VUV/SXR radiation, although this is partially for a different reason. Radiation in this wavelength range has to be transported in vacuum in any case, while the photoeffect quantum yield for such relatively energetic quanta is high enough even without any special kind of photocathodes. Therefore, in this spectral region, one can effectively use open-ended photoelectron multipliers (PEMs), such as the device shown schematically in Figure 10.5, where radiation falls directly onto the first dynode, which plays the role of a photocathode. One can also use microchannel multipliers (MCMs) or their assemblies, which were described earlier in

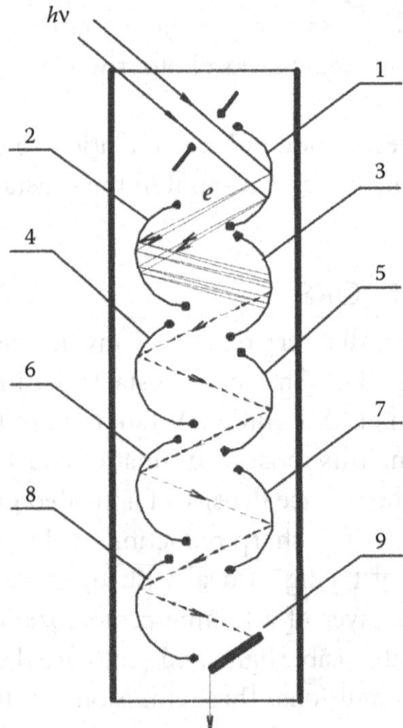

FIGURE 10.5 Open photoelectron multiplier: 1–8, dynodes; 9, collecting electrode.

Chapter 4 (see Section 4.2). In this case, the SXR beam illuminates the facets of an MCM. In electron-optical converters (EOCs), a photocathode for SXR radiation ($h\nu > 100$ eV) is made of gold foil with a thickness of several hundred nanometers. When the front face of such a photocathode is illuminated, electrons are emitted from the reverse face of the foil. The time resolution of electron-optical cameras operating in this wavelength region is typically in the range of 10–15 ns.

Photoelectron multipliers are first and foremost used to record weak radiation beams, because their gain (namely, the electron multiplication coefficient) can be as high as 10^6–10^7. However, their time resolution is not particularly good: The half-width of the transfer function, even in high-quality, specially designed versions, still exceeds several tens of nanoseconds. Open photodiodes can be used to measure sufficiently intense beams. Figure 10.6 shows a typical arrangement. The operation principle of such a device should be clear from the figure: A photocathode, which is maintained at a negative potential, is illuminated through a grid anode that is transparent to the radiation. The photocathode is connected to a cable via a low-inductance capacitor. In the design of photodiodes with high time resolution, special attention is paid matching the photodiode

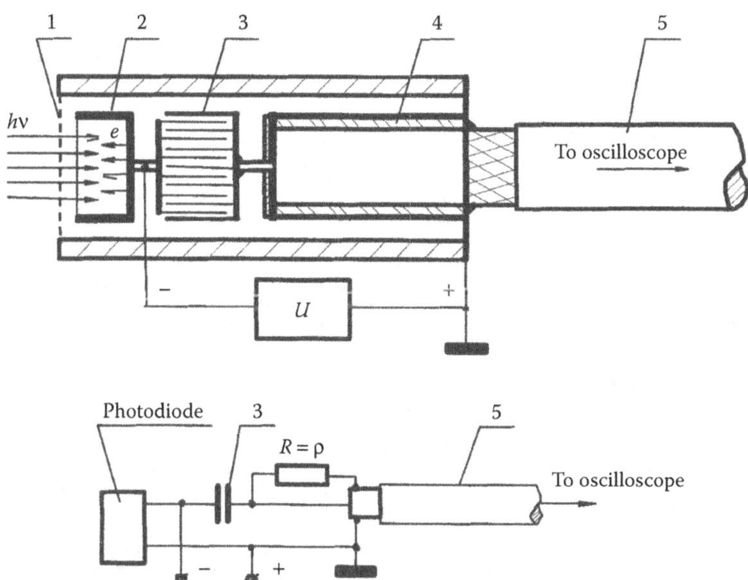

FIGURE 10.6 Construction and equivalent circuit of an open photodiode: 1, grid anode; 2, photocathode; 3, low-inductance capacitor; 4, matching resistor; 5, coaxial cable.

with the cable transmitting the signals. This illustrates exactly to the kind of design problems that have been discussed in Chapter 3. With careful matching of the two elements, it is possible to achieve a typical time resolution of 0.2–0.3 ns.

Calorimeters, from the viewpoint of the physics of rapid processes, have no time resolution whatsoever, but they represent the only device that allows reliable absolute measurements within the spectral range under consideration. A schematic of the construction and integration of a thermocouple calorimeter designed for VUV/SXR measurements by the Russian Research Institute for Physics, Technology and Radio Engineering (the Russian acronym is VNIFTRI) is shown in Figure 10.7. The calorimeter is a copper–constantan thermopile, with some 1000 thermal junctions supported on an aluminum plate. A dielectric coating of some 50 μm thickness and a coiled equivalence resistor are positioned on the top of the "hot" thermal junctions. The resistor is made of 20-μm-thick manganin foil and serves to calibrate the detector. The calibration process consists of applying rectangular electric current pulses of known duration and amplitude to the equivalence resistor. This determines the amount of thermal

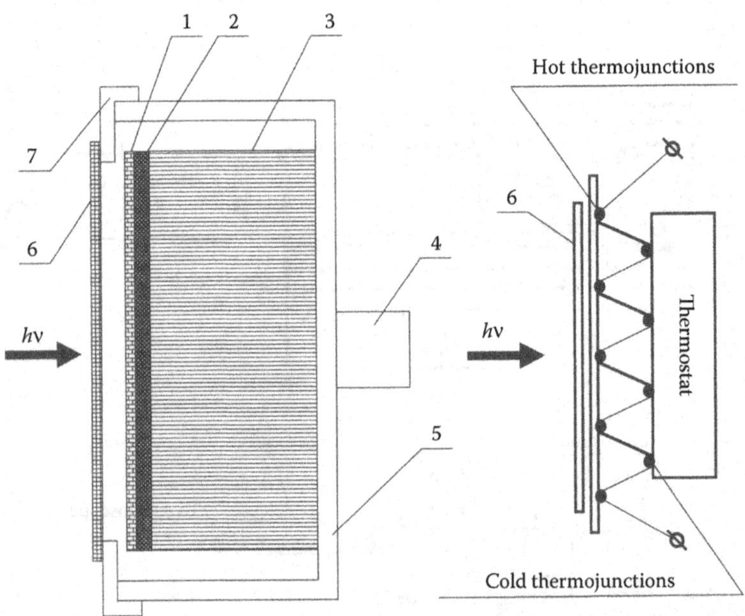

FIGURE 10.7 Construction and fundamental scheme of a thermal calorimeter: 1, receiving element/equivalence resistor (20 μm); 2, dielectric (50 μm); 3, batch of 1000 thermocouples; 4, connector; 5, case; 6, X-ray filter; 7, screw-on aperture.

energy deposited in the resistor. The coils of the resistor cover some 60% of the input face area; hence, some 60% of the radiation is absorbed in the resistor's surface layer, and the rest is absorbed in the surface layer of the dielectric coating. The total amount of energy absorbed by the calorimeter is determined by the amplitude of the output pulse, whose rise time corresponds approximately to the temperature relaxation time, which amounts to some 25–30 ms. The sensitivity of the calorimeter that is described here lies in the range of 80–120 mV J^{-1}.

Ionization cameras, proportional counters, and Geiger counters occupy a special niche in the range of VUV/SXR detectors. Their versatility is determined by the physics of the so-called non-self-maintained gas discharge. Let us consider the corresponding physical processes, beginning with the experiments of Paschen, who studied the electric breakdown of gases at various pressures. Paschen discovered that the breakdown voltage depends on the product of the pressure in the gas and the distance between the electrodes. It is described by the following empirical equation:

$$U = \frac{a(pd)}{\ln(pd)+b} \tag{10.1}$$

where:

- U is the breakdown voltage
- p is the pressure
- d is the distance between the flat electrodes
- a and b are the parameters characteristic of the gas composition

In air, if U is expressed in volts, p in atmospheres, and d in centimeters, a is 43.6×10^4 and b is 8.195. Differentiating Equation 10.1 with respect of the product (pd), we find that the minimum of the breakdown voltage corresponds to the following critical value of the product (pd):

$$pd = e^{1-b} \tag{10.2}$$

In air, for instance, this critical value is equal to $(pd) = 7.5 \times 10^{-4}$ atm cm, with the breakdown voltage being some 330 V. In other words, under atmospheric pressure, it takes only 330 V to cause the electric breakdown of an air gap of 7.5 μm thickness. The same voltage causes breakdown for a gap that is 0.75 cm thick in air with a pressure of 10^{-3} atm. The dependence $U(pd)$ in Equation 10.1 is called the Paschen curve and is shown in Figure 10.8. The same figure provides a diagram showing the principle of Paschen's experiments.

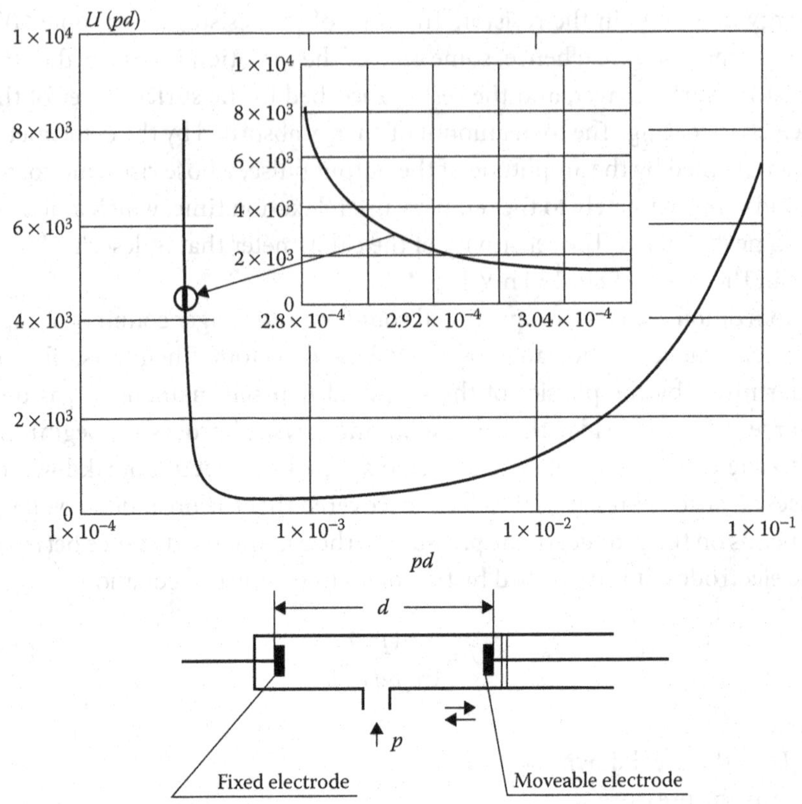

FIGURE 10.8 Paschen's curve for air. U is expressed as V and pd as torr × cm.

Let us make a few points. First, electric breakdown is a random process, relying on the mechanisms of the growth of electron avalanches in the gas. Therefore, the voltage provided by Paschen's equation should be treated merely as a mathematical expectation value. Consequently, the breakdown of a gas in a particular trial can occur at a lower voltage or may not occur even at a slightly higher voltage than that predicted by Paschen's equation. In this sense, $(pd) = 7.5 \times 10^{-4}$ atm cm and $(pd) = 1$ mm Hg × 1 cm* (1 torr × 1 cm) practically characterize the same scenario, although the latter one is more preferable mnemonically. Second, Paschen's equation is known to yield disproportionately large values for breakdown in air when $(pd) < 2.75 \times 10^{-4}$ atm cm, namely, when the denominator on the right-hand side of Equation 10.1 is close to 0. It is for this reason that Paschen's

* The breakdown voltage calculated from the Paschen's equation for $(pd) = 1$ mm Hg × 1 cm is approximately 370 V.

equation has a somewhat limited usage. Finally, the third point is that electric breakdown is impossible in the range of the parameter values lying beneath Paschen's curve. For this range of parameter values, we may only have a current in the gas when its carriers have been created artificially, namely, when the gas has been ionized via an independent source.*

Figure 10.8 shows that the breakdown voltage increases on both sides of the minimum. The reason for this is easy to understand. The electric breakdown occurs due to the formation of electron avalanches, and therefore, the critical (pd) value, at which the minimum occurs, corresponds to the ideal conditions for the formation and development of the avalanche. On the left-hand side of the Paschen's curve, neither a small distance between the electrodes nor a low gas density is capable of ensuring the necessary conditions for the breakdown. This is because the number of ionization-creating collisions of electrons accelerated by the electric field with gas atoms or molecules is insufficient to seed an avalanche that will develop into a breakdown of the gas. By contrast, on the right-hand side of the curve, the number of collisions is excessively large. This prevents the electrons from gaining a sufficient amount of energy in the electric field to cause the ionization of an atom or a molecule.

Ionization chambers operate in three different regimes: The first regime is when electron avalanches effectively do not develop; the second regime is favorable for their development, but it does not lead to a breakdown; and finally the third regime is when a breakdown of the gas does take place.

A schematic of an ionization chamber is shown in Figure 10.9.† The device, containing two electrodes, is filled with gas and has a window to admit the ionizing radiation. The lower part of the figure shows the voltage versus current characteristic of the device, namely, the dependence of the current between the electrodes and the voltage applied to the electrodes, for a fixed value of the intensity of the ionizing radiation. We can see from the figure that the dependence $I(U)$ has four characteristic regions. In the first region (indicated by I in Figure 10.9), current grows practically linearly with voltage. This growth is due to the fact that if the applied voltage is too small, a fraction of the electrons that are produced in the space between the electrodes will be lost on the walls of the tube without reaching the electrodes, because the longitudinal component of their velocity is

* An electric current in gas which is induced by means of an external source is called a nonself-maintained discharge.

† Real-life chambers used in experiments certainly have more complicated designs, but the simplified schematic suffices to point out the physical basis of the processes in ionization chambers.

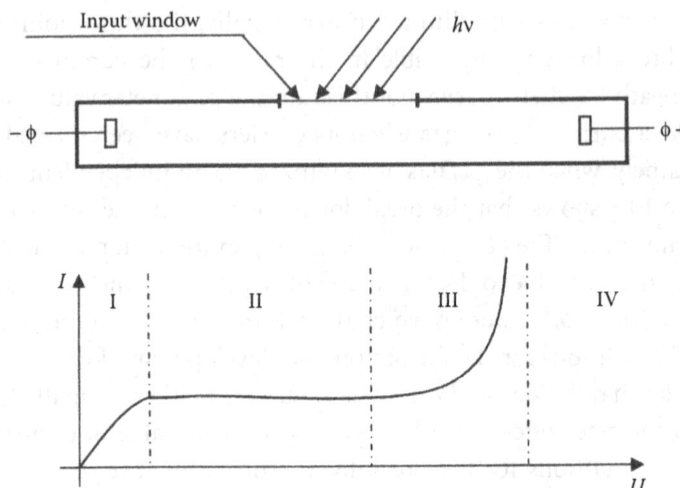

FIGURE 10.9 Ionization chamber. I, zone losses of electrons; II, saturation zone; III, so-called "gas amplification zone"; IV, breakdown zone.

not large enough compared to the transverse component. Another fraction of electrons will be lost due to recombination phenomenon, because for small voltage values the charge accumulation time is long enough. The second region (II in Figure 10.9) of the voltage–current curve represents the saturation zone, where all the electrons that are created reach the electrodes, and therefore, we can in principle determine the intensity of the ionizing radiation by measuring the current. This region is called the ionization chamber regime. It is noteworthy that in this regime the measured current is very stable with respect to changes of the applied voltage. Ionization chambers can in principle be used to count individual particles. The problem will be the very small amplitude of the output pulses (typically 5–10 μV), which calls for high-gain amplifiers. This difficulty can be partially overcome in the third region (III in Figure 10.9) of the voltage–current curve, where the voltage is sufficient for the formation and development of electron avalanches, but is still insufficient to cause breakdown (IV in Figure 10.9). Devices that operate within this region of the voltage–current curve are called proportional counters. Their so-called gas amplification coefficient, namely, the multiplication coefficient for the number of electrons in avalanches, can exceed 10^4 in this regime. In practice, it is usually limited within the range of about 500–5000 for a number of reasons. We shall not go into the detail of what causes these limitations. To the reader interested in a more refined account of physical processes involved in the non-self-maintained gas discharge phenomena, as well as the technical tricks which one uses to achieve strict proportionality of the amplitude of the measured pulses to the original number of ionized particles, we recommend the work of Kalashnikov and Kozodaev [2]. A premise that the proportionality in question can indeed be obtained by a proper choice of the gas mixture is essential for our purposes. However, the number of simultaneously born primary electrons is proportional to the energy of an ionizing photon. Therefore, the output pulse amplitudes turn out to be proportional to the energy of the photons responsible for ionization. In view of this, a proportional counter operates not only as a detector but also as a spectrometer, namely, throughout the detection process, it also acts in the capacity of a device identifying the spectral composition of the incident X-ray radiation. Naturally, a spectrogram resulting from processing such experimental data is obtained by using a multichannel amplitude analyzer.

One engineering challenge in the design of proportional counters is creating a device that remains stable with large values of the gas amplification coefficient, that is, the stability of that thin division that separates the regime of avalanche magnification from the actual breakdown. This task is made even more complicated by the particular circumstance that we are not free to choose as we wish either gas pressure within or the exterior dimensions of a gas counter. Both the pressure and the dimensions of the counter have to ensure complete absorption of the energy of the highest energy photons of those that are being measured. This difficulty can be largely overcome in cylindrical gas counters, where the whole inner surface of the cylinder acts as a cathode, whereas the anode consists of a thin filament strung on insulators along the axis of the cylinder. This type of construction enables one to effectively separate the task of absorbing the photons from the task of amplifying the primary electrons. As the electric field intensity within such a device decreases as $1/r$, the favorable conditions for avalanches take place only within a narrow region close to the anode; the rest of the instrument operates merely as an ionization chamber.

Finally, let us consider the section of the voltage–current curve where it takes just a single photon to cause the gas to break down. In essence, this is a gas discharge triggered by ionizing radiation. It is known as the Geiger–Müller, or simply Geiger counter. Its main advantage is the high amplitude of the pulses that are produced. Its main flaw is the limitation in counting rate due to the time it takes the gas in the space between the electrodes to regain its breakdown strength. The addition of special gaseous additives that promote deionization partially solves this problem, making it possible

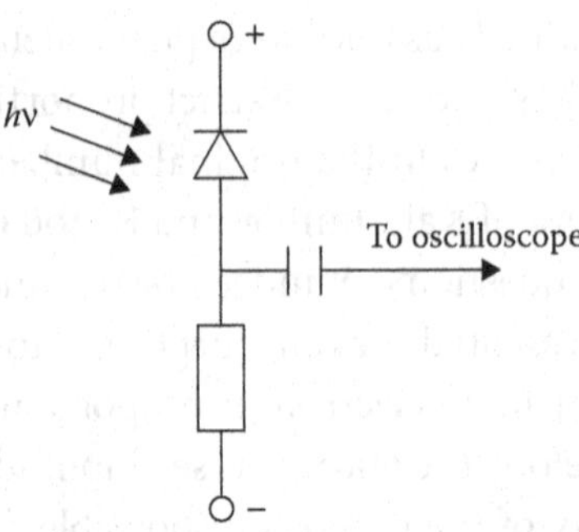

FIGURE 10.10 Basic principle of a p–n transition semiconductor detector.

to count at a rate as high as some 10^6 pulses per second. This, nevertheless, clearly imposes an upper limit on the intensity that can be measured.

A special place among VUV/SXR detectors belongs to semiconductor devices with p–n and p–i–n structures. (We have encountered these structures earlier while discussing charge-coupled device [CCD] cameras.) As the junction between two semiconductor materials (p–n transition) allows the electric current to go in one direction only, there would be no current in the electric circuit shown in Figure 10.10 when a reverse voltage is applied. The reverse voltage also results in the formation of a depletion layer in the vicinity of the p–n transition, characterized by a low density of free charge carriers and a large intensity of the electric field within this layer. The situation changes when the semiconductor is subject to radiation by photons whose energy is greater than the forbidden zone width ($\Delta W = 1.12$ eV for Si and 0.67 eV for Ge). This results in the generation, due to the photoelectric effect, of electron–hole pairs inside the p–n transition and on both sides of it. These charge carriers move in the electric field toward the electrodes and create a current in the circuit. This current, within fairly wide limits, becomes proportional to the intensity of the ionizing radiation, as well as to the energy of the ionizing photons.* The situation then turns out to be basically the same as it was in the case of gas counters. A photodiode in such an electric circuit acts as a variable resistor, which is controlled by the X-ray radiation. Let us point out the fact that the electron–hole pairs are created in the whole volume of the semiconductor, rather than just inside the p–n transition region. Those which were created close enough to the p–n transition, namely, within distances not exceeding the diffusion length, diffuse toward the transition and pass through it to contribute to the

* At room temperature, the energy required to produce a single electron–hole pair in a semiconductor is about 3–3.5 eV.

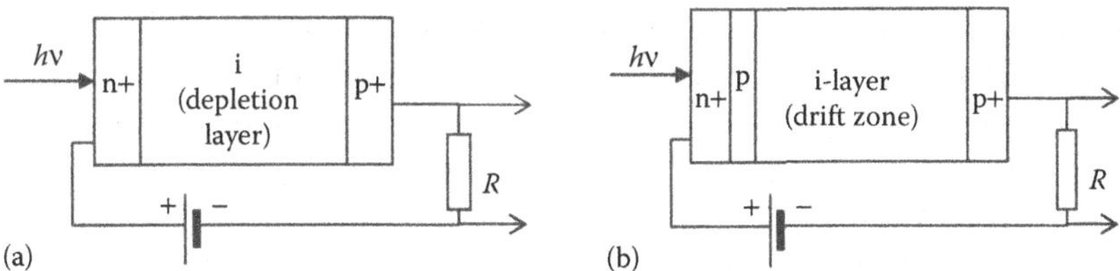

FIGURE 10.11 Semiconductor X-ray detectors: a, p–i–n photodiode with "i-depletion layer"; b, avalanche photo-diodes (APD).

overall current. Thus, the birth of free charge carriers near to the transition semiconductor layers can be regarded as a phantom effect, degrading the time response characteristics of the diode. Indeed, the time it takes a carrier to diffuse to the transition is orders of magnitude longer than the time it takes it to pass through the transition. Needless to say that the width of the p–n transition (depending on the manufacturing technology, as well as the magnitude of the voltage applied) is chosen such that it must ensure full absorption of the most energetic of all the measured photons.

In p–i–n photodiodes, an extra layer of a undoped semiconductor is introduced between the p and n layers to increase the width of the depletion layer. This is known as the *i-depletion layer* and is characterized by high electrical resistance (Figure 10.11a). Its width amounts to several tens of microns, while the n+ layer in p–i–n photodiodes is, on the contrary, made quite thin. Thus, the majority of the photons are absorbed within the i-layer, where the free carriers are accelerated by the strong field that exists there. This reduces the inertia and increases the conversion frequency up to several gigahertz. Higher sensitivity can be achieved by increasing the surface area that is illuminated by the radiation. The transition capacity can be lowered by increasing the applied voltage.

The principle difference between avalanche photodiodes (APDs) and p–i–n devices is the avalanche multiplication of electrons, which occurs within the semiconductor structure. To achieve this, an APD is equipped with an extra p-layer (Figure 10.11b). The dopant distribution profile inside the semiconductor structure is designed in such a way that maximum resistance and therefore the maximum electric field intensity both occur inside the p-layer. The effect of radiation is to introduce electron–hole pairs drifting toward the corresponding electrodes. Free electrons that transit from the i- to the p-layer, characterized by higher electric field intensity, are accelerated up to energies sufficient to cause electron avalanches in the regime similar to that which happens in proportional gas counters. Such a process is referred to as avalanche amplification (or multiplication) of the primary photoelectric current. The amplification coefficient can be as high as 10^3. APDs operate on rapid timescales, the characteristics growth time of their pulse being often as short as 0.1 ns.

10.3 EXPERIMENTAL EQUIPMENT AND METHODS FOR X-RAY MEASUREMENTS

As we have mentioned already, measurements in the wavelength range in question have some specific limitations in addition to the technical requirements that involve the use of vacuum devices. The major main limitations are the impossibility of using optical refraction-based instruments, as well as the very narrow range of the angles over which reflection-based optical devices can operate. It is known, for example, the monograph [1], that the maximum angle with the reflecting surface (the so-called total internal reflection angle*) is $\theta_c \propto \lambda\sqrt{Z}$, where λ denotes the incident radiation wavelength and Z is the atomic number of the surface material. The following equation is adequate for estimating the angle θ_c:

$$\theta_c = 10\lambda\sqrt{Z} \tag{10.3}$$

where:

λ is expressed in microns and θ_c in radians

Although this is only an estimate rather than a strict equation, it is well suited for our purposes. Let us take advantage of this equation to estimate the extent of the problem. At the short wavelength limit of the range in question (5 keV), the maximum value of the total internal reflection angle does not exceed 1°, regardless of whether one uses aluminum ($Z = 13$) or silver ($Z = 47$) mirrors. No comment, as one says. The situation is different, however, in the region between the VUV and the SXR: If the energy of a photon is around 100 eV, then depending on the mirror material, the quantity θ_c may well amount to several tens of degrees. Even for energy of 250 eV, $\theta_c > 10°$. Hence, in the VUV region, equipment based on reflective optics is quite applicable.

It follows that all the spectrometers and monochromators described in detail throughout Chapter 6 can be used for the purposes of measuring VUV radiation. In particular, a spectrometer with a concave cylindrical diffraction grating (see Chapter 6) is rather popular and is widely used, in

* The refraction index of every material $n < 1$ for VUV/SXR radiation—see Chapter 9.

spite of some difficulties related to its adjustment. But, in the SXR range, as it has been pointed out, diffraction gratings as a rule cannot be used (with the exception of the rather rare "shine-through" transparent gratings that have been mentioned in passing). Instead of diffraction gratings, crystals come to rescue in this spectral region.

Spectrometers and monochromators, whose dispersing elements are crystals, have a number of features. The first point is that this concerns the conditions for the "reflection" of the radiation. The incidence angle on a diffraction grating can be chosen arbitrarily, but a crystal will reflect only if the incidence angle satisfies the Bragg condition. This constraint makes the design of crystal-based spectrometers and monochromators more difficult, on the one hand, yet opens some extra opportunity, on the other, particularly when point source spectra are being analyzed.* In this vein, ray propagation through a survey spectrograph based on a convex crystal is shown in Figure 10.12. As the figure indicates, radiation from a point source diverges within a wide angle, and as a result, different rays arrive at the cylindrical surface of the crystal at different angles. However, reflection from the crystal is possible only when the incidence angle coincides with the Bragg angle. The shorter the wavelength, the smaller the angle should be. Since the zones in which reflection occurs are quite narrow, the resolution ($\lambda/\Delta\lambda$) of such spectrographs can be quite high, reaching up to 10^3. However, it is for the very same reason that they have a lamentably poor optical efficiency. Notwithstanding this fact, such devices are frequently used, especially during the early stages of an experiment, owing to their

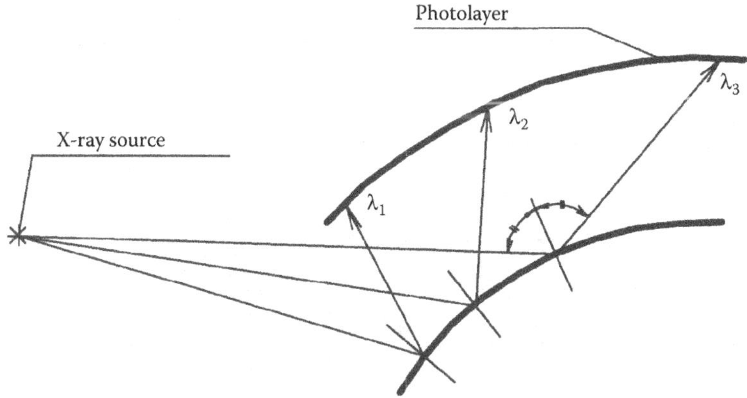

FIGURE 10.12 Ray paths in a survey spectrograph with a convex crystal.

* For instance, the spectra of laser thermonuclear fusion targets.

wide dispersion region $\lambda_{max}-\lambda_{min}$. In order to be able to collect efficiently the analyzed radiation, one uses focusing spectrographs with concave crystals, which have been considered already in some detail in Section 6.1. The latter instruments have optical efficiency that is higher by orders of magnitude, but unfortunately their dispersion region is quite narrow.

Note that whereas the focusing properties of diffraction gratings are achieved by simply arranging them on a cylindrical surface, things are much harder to achieve with crystals. The crystal planes have to be deformed in order for them to attain a cylindrical or spherical shape. This is achieved by bending thin crystals (to enable high spectral resolution the thickness of a crystal must be hardly greater than a couple of hundred times the distance between neighboring crystal planes). In practical terms, pressure is applied to crystals to bend them, and then they are glued to a substrate that has the desired shape (Figure 10.13).

To illuminate plane diffraction gratings by radiation from extended SXR sources,* one uses a multiplate Söller collimator. Its design should be clear from Figure 10.14. The divergence of radiation after it has passed through the collimator is determined by the geometric considerations.

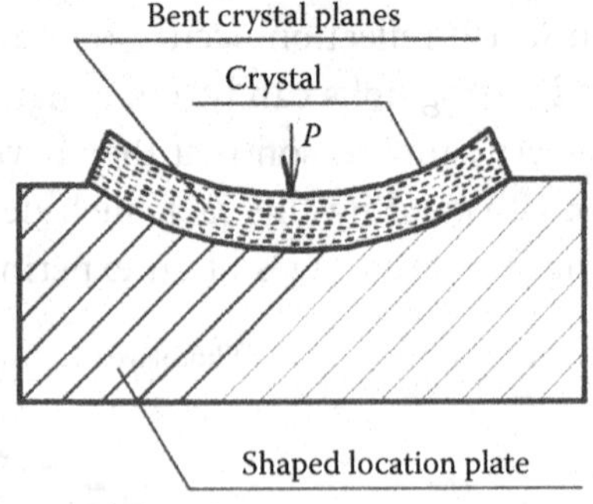

FIGURE 10.13 Schematic showing a belt crystal.

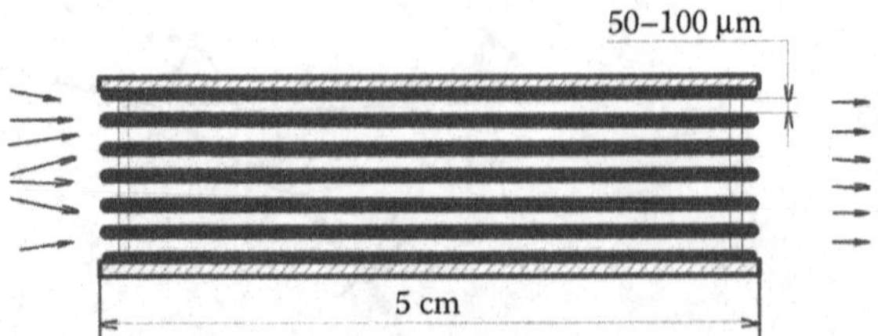

FIGURE 10.14 Multiplate Söller collimator.

* For instance, plasma confined by a magnetic field.

Filters based on reflection or absorption of X-ray radiation are widely used in all those cases when there are no particular high spectral resolution requirements. The so-called reflection filters are used to cut off the short-wave component of the radiation. The dependence of the reflection coefficient R on the maximum total internal reflection angle θ_c, calculated for different values of attenuation increment, is shown in Figure 10.15. [The ratio β/δ shown in the figure is the ratio of the second and third terms in the dispersion equation, written in the form: $n(\omega)=1-\delta+i\beta$.] The figure shows that even under considerable damping, the short-wave edge of the filter is quite sharp and does not exceed one-half of the angle θ_c, whose absolute value, as we have discussed earlier, amounts to several degrees in the SXR range.

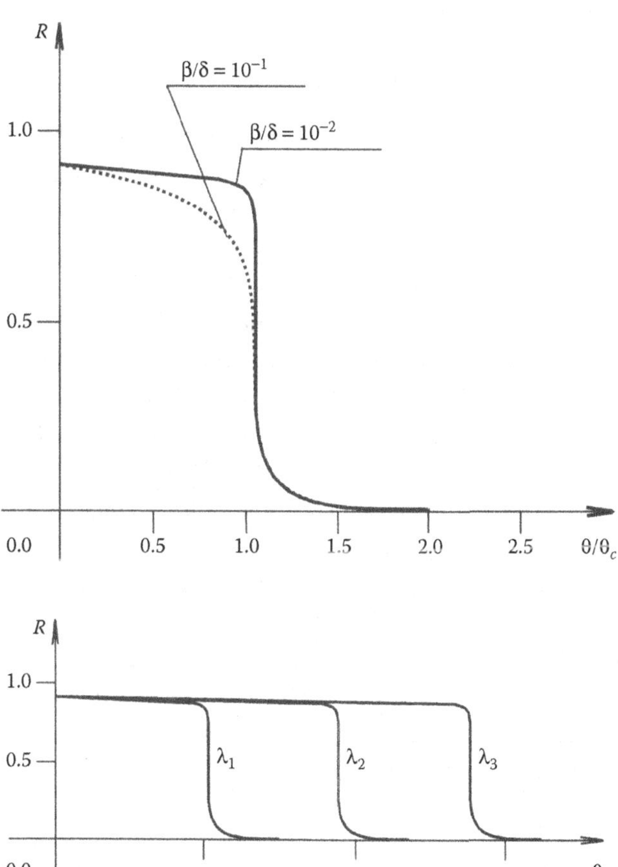

FIGURE 10.15 Dependence of the reflection coefficient R on the maximum total internal reflection angle θ_c.

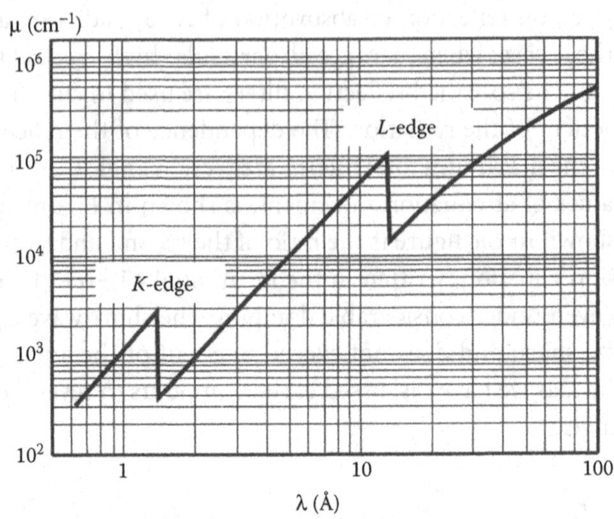

FIGURE 10.16 Absorption coefficient for copper.

In absorption filters, we use transmission windows located in the regions of the $K, L, M,...$ absorption edges. As an example, Figure 10.16 illustrates the absorption coefficient for copper. In the graph, the absorption coefficient μ is expressed as cm^{-1}, so that the intensity of radiation that passes through the filter is $I = I_0 \exp(-\mu d)$, where I_0 is the intensity of radiation before it arrives at the filter and d is the thickness of the filter in centimeters. Let us illustrate the scenario arising, for instance, in the region of the K-edge (λ = 1.381 Å). As one can see from the figure, the values of μ before the K-edge (μ_1) and after it (μ_2) differ by almost an order of magnitude, so $\mu_1 - \mu_2 = 2.46 \times 10^3$. This means that there is an approximately 12-fold difference in the transmission of a 10-micron-thick aluminum foil before and after the K-edge, this difference becoming some 140-fold if the thickness of the foil doubles.

Such an exponential dependence of foil transmission is widely used in the so-called filter method for measuring plasma temperature. It is well known that for a Maxwellian distribution of the electrons in plasma, the spectral composition of *bremsstrahlung* is characterized by the following probability density functions:

$$\frac{dI}{d\nu} \propto \frac{n_e n_i Z^2}{\sqrt{T}} \exp\left(-\frac{h\nu}{kT}\right) \text{ or } \frac{dI}{d\lambda} \propto \frac{n_e n_i Z^2}{\lambda^2 \sqrt{T}} \exp\left(-\frac{hc}{\lambda kT}\right) \quad (10.4)$$

(Note, by the way, that on a logarithmic scale, a graph of the first of these two functions is a straight line, whose slope increases with T, and the second function has a maximum at $h\nu_{max} = 2kT$.) It follows from the above equations that the net intensity of plasma *bremsstrahlung* after the filter is characterized by the absorption coefficient $\mu(h\nu)$, so that given the thickness d, one has

$$I_\Sigma = A\frac{n_e n_i Z^2}{\sqrt{T}}\int \exp\left(-\frac{h\nu}{kT}\right)\exp\left[-\mu(h\nu)d\right]d\nu \tag{10.5}$$

In the above equation, the coefficient A depends on the system of units. It follows that in order to determine the temperature without doing any absolute measurements, it suffices in principle to measure only two values of I_Σ after two filters of different thickness or made of two different materials. In practice, in any experiment, the quantity I_Σ is measured after filters with both different thickness and made of different materials. The materials are chosen based on *a priori* data about the temperature to ensure that the signals reaching the detectors after the filter are sufficiently strong.

X-ray imaging optics is a special area, which has little in common with the visible light optics that we have been used to since school. It is only the pinhole camera, the oldest lens prototype, which has made its way to the VUV/SXR range. This "lens" is just a very small hole in a nontransparent material. The name *pinhole camera* is derived from how the device is made. Holes, usually in foil, are made in the following way: The foil is supported on a soft foundation (lead or annealed copper) and then the foil is pierced by a very sharp needle, whose tip cone has a given angle. The needle is driven by a screw mechanism with micrometer drive feed. In fact, anyone can verify by his or her own experience that this simplest lens is capable of producing images with surprisingly high quality. Cover an incandescent light bulb with a sheet of paper, preferably black, in which a small hole has been made by a pin or needle, and there will be an image of the bulb's filament on a white sheet of paper placed in line with the hole.

Ray propagation in a pinhole camera is shown in Figure 10.17. As one can see from the figure, two point light sources, positioned in the object plane at a distance a from the camera and separated from one another by a distance x, will be mapped into the image plane (at a distance b from the camera) into two nonoverlapping light spots with a distance X between their centers. Using simple geometry, it is easy to find the value of x, characterizing the spatial resolution: $x = d(a+b)/b$. As usually $b >> a$, one can assume to the first order of approximation that $x \approx d$, that is, the resolution approximately equals the hole diameter d. Thus, it may appear that

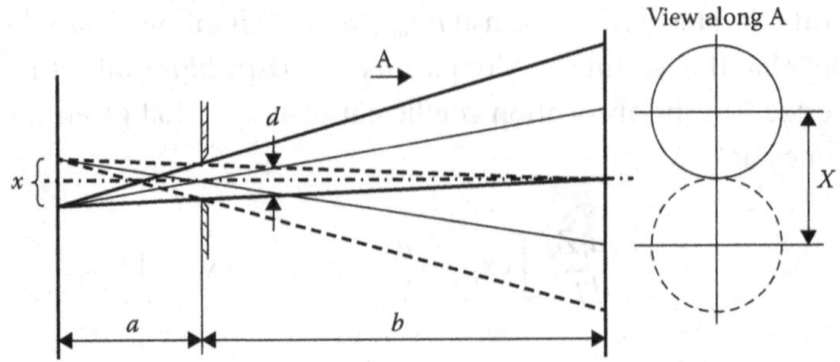

FIGURE 10.17 Ray paths in a pinhole camera (geometric approximation).

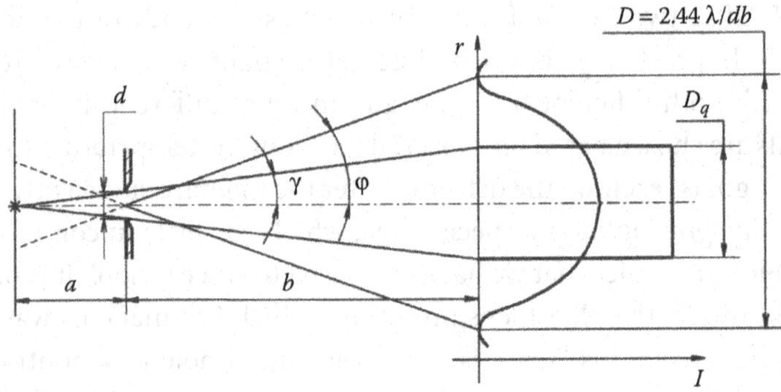

FIGURE 10.18 Ray paths in a pinhole camera from the viewpoints of geometric and diffraction.

if we make the hole diameter as small as 1 micron, we will get a 1 micron resolution. Unfortunately, this is not always true, or rather almost always not true. (It would be the case only if the angle $x/a = X/b >> \lambda/d$.*) To see what is going on, let us find the transfer (spread) function of the pinhole camera.

The recipe for this is well known: Let us apply a δ-function shape signal to the input of the device (a point light source in the object plane; Figure 10.18) and find the intensity distribution in the image plane. This will be the spread (or transfer) function. In the geometric optics approximation ($x/a = X/b >> \lambda/d$), there will be a uniformly illuminated circle in the image plane, with diameter $D_q = 2(a+b)\gamma$, where $\gamma = d/2a$. Therefore, 2γ yields the angular resolution and $\Delta x = 2a\gamma$ is the spatial resolution in the image plane. In the diffraction approximation, the quantity $g(r)$ is

* Since the angles in consideration are small, here and in the following, we assume $\sin\varphi \approx \tan\varphi \approx \varphi$.

determined by diffraction at the aperture of the camera. As $a >> d$ (by a factor of more than 10^4), the wavefront in the vicinity of the aperture of the camera can be regarded as flat. Thus, we are dealing with common diffraction on a circular aperture, the intensity distribution in the image plane being described by the first-order Bessel function:

$$I(\varphi) = I_0\left(\frac{\pi d^2}{4b\lambda}\right)^2 \left|\frac{2J_1(\varphi d/\lambda)}{\varphi d/\lambda}\right|^2 \tag{10.6}$$

where:

I_0 and λ denote, respectively, the intensity and wavelength of incident radiation

d and b are, respectively, the aperture diameter and the distance between the aperture and the image plane

$\varphi = r/b$

Since the spread function is $g(r) \propto |[2J_1(rd/b\lambda)/rd/b\lambda]|^2$, the direction to the first diffraction minimum, which according to Rayleigh defines the angular resolution, is $\varphi_0 = 1.22\lambda/d$. The diameter of the central light spot in the image plane $D = 2b\varphi_0$, and the space resolution in the object plane is $\Delta x = a\varphi_0 = 1{,}22a\lambda/d$.

It is worth mentioning that although in the geometric approximation the angular resolution is determined by the distance between two point sources whose images do not overlap, in the diffraction approximation, the images of the two points, regarded as resolved, do partially overlap. This is due to the fact that according to the Rayleigh resolution condition the borderline case is when the maximum of the intensity of one image coincides with the first diffraction minimum of the other. Evidently, this is largely a matter of convention, which is useful only for the purpose of making estimates. Nevertheless, this is exactly what we are doing here. When we process the results of measurements, we will certainly need a continuum of numbers, namely, the whole function $g(r)$ rather than a single number that by convention characterizes the resolution.

To demonstrate the scale of the problem, let us make some numerical estimates. Suppose, as earlier, $d = 1$ μm, $a = 2$ cm, $b = 10$ cm, and $\lambda = 10$ nm (123.4 eV). Then the spatial resolution turns out to be 240 μm, which is quite different from the value of 1 μm that was obtained under the geometric approximation. Moreover, decreasing the camera aperture further will patently result in worsening, rather than improving the spatial resolution.

Although the resolution will improve to ~30 μm when the energy of the incident photon becomes as large as 1 keV, it is still quite far from 1 μm. It is pointless to make the diameter of the aperture in the pinhole camera 1 μm, when the resolution is 30 μm. The fact of the matter is that a source radiating in a solid angle of 4π illuminates a spherical surface with an area $4\pi a^2$. Given the area of the aperture of the camera as $\pi d^2/4$, the fraction of its full radiation energy being used is $(\pi d^2/4)/4\pi a^2$. In other words, the optical efficiency of our device is $(d/4a)^2$, so it is proportional to d^2. Then let us find the optimal aperture size, when diffraction still does not play too significant a role, by equating $\gamma = d/2a$ to $\varphi_0 = 1.22\lambda/d$, whence $d_{\text{opt}} = \sqrt{2.44a\lambda}$. If we measure radiation whose photons have an energy of 1 keV ($\lambda = 1.324$ nm), the optimal diameter $d_{\text{opt}} = 7.76$ μm. Given this value for the diameter, the resolution in the object plane will amount to some 4 μm and the optical efficiency will be of the order of 10^{-8}. What can be done to improve the optical efficiency? The first step is to move the camera closer to the object whose image is being taken. For instance, making this distance about 0.5 cm would increase the optical efficiency to a value to 1.5×10^{-7}, as well as improving the space resolution by a factor of 2. However, this is not always allowed by the experimental conditions.

The fact that the optical efficiency and resolution are insufficient for many purposes forces us to look for an alternative means of obtaining images in the VUV/SXR range. The first (not always unsuccessful) attempt was using "lenses" designed on the zone plate principle. A zone plate, as shown in Figure 10.19, consists of alternating transparent and nontransparent annular zones, whose radii coincide with the radii of the Fresnel zones:

$$r_n = \sqrt{n\lambda f + \frac{n^2\lambda^2}{4}} \tag{10.7}$$

where:

- n is the number of a Fresnel zone
- λ is the wavelength of radiation
- f is the focal length, namely, the point on the system's axis where radiation is focused if the plate is illuminated by a plane wave

The focusing properties of the zone plate can be explained in two ways: First, the distances from a corresponding transparent zone to the focal point ($l_n = \sqrt{r_n^2 + f^2}$) are such that the radiation from all the zones comes in phase to a focus. Second, the diffraction angle $\lambda/\Delta r_n$ in each transparent

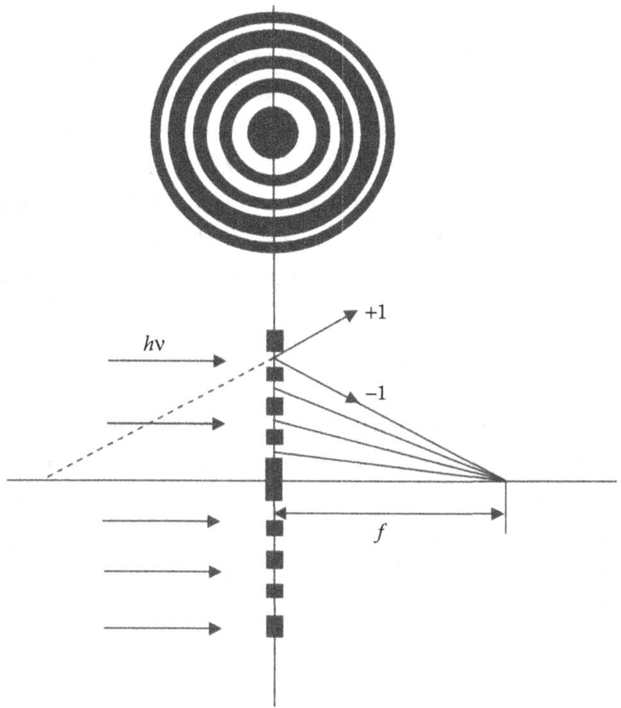

FIGURE 10.19 Fresnel zone plate.

circular zone whose width is $\Delta r_n = r_n - r_{n-1}$ increases with radius, and thus, all the diffracting rays come together at the focal point.

The focusing properties of zone plates have been well known since the nineteenth century. The same concerns the applicability of the geometric optics equation describing conjugate planes:

$$\frac{1}{f} = \frac{1}{a} + \frac{1}{b} \tag{10.8}$$

where:

f is the focal length

a and b are, respectively, the distances from the zone plate to the object plane and to the image plane

The difference consists only in the fact that a zone plate has several foci and acts simultaneously as a converging and a diverging lens, in accordance with diffraction on the rectangular transmission profile of the transparent annular zone in the diffraction orders +1, −1, +2, −2, +3, and higher.

It is a curious fact that all the parameters of a given zone plate are determined by the total number of zones N and the width of the exterior annulus $\Delta r = r_N - r_{N-1}$. Indeed, the diameter of the zone plate is $D = 4N\Delta r$, its focal length $f = 4N(\Delta r)^2/\lambda$, the so-called numerical aperture $NA = \lambda/2\Delta r$, the spectral width $\Delta\lambda = \lambda/N$, and the focal radiation intensity $I = I_0 N^2/\pi^2$. (It follows from Section 5.3 that with rectangular ruling shape, the radiation intensity in each of the two first orders of diffraction is $I_{+1} = I_{-1} = I_0/\pi^2$.) Finally, the spatial resolution in the image plane is

$$\Delta x = \frac{0.61}{NA} = 1.22\Delta r \tag{10.9}$$

This means that in order to have a high spatial resolution, one has to decrease as much as possible the outer zone width Δr. Therefore, if you want, for example, to provide 1 micron spatial resolution, you must manufacture a plate with the outer zone width $\Delta r \le 1\ \mu\text{m}$. It is not so easy, but it is possible to do with modern microlithography technique. The central part of such a zone plate is shown in Figure 10.20. Zone plates that are unique in this respect were designed and manufactured by a joint effort between the Lawrence Livermore National Laboratory (LLNL) and the Massachusetts Institute of Technology (MIT). The parameters of one of these plates are given in Table 10.1.

The resolution of ~0.4 μm was determined experimentally in the spectral region between 100 and 6 keV by taking images with radiation

FIGURE 10.20 Central part of a zone plate. (From Attwood, D. T., *Soft X-Rays and Extreme Ultraviolet Radiation—Principles and Applications.* Cambridge University Press, Berkeley, CA, 2007. With permission.)

TABLE 10.1 Parameters of typical Fresnel zone plate

Δr	N	D	Δx	$\Delta\lambda$	a
0.32 μm	500	0.64 μm	~0.4 μm	$2 \times 10^{-3}\lambda$	~1 m

consisting of single spectral lines, that is, under narrow bandwidth conditions. The distance between the object and the zone plate was $a \approx 1$ m. It seems that things have gone well, at least the resolution was superior by an order of magnitude to that of a pinhole camera, although the optical efficiency, because of the large value of a, was still very small (2.6×10^{-8}). However, this is not the main problem here. This kind of lens has enormous chromatic aberrations and is therefore useless for taking images of objects radiating over a reasonably wide spectral band, as it happens to be the case, for example, in experiments on inertial confinement fusion.

The next step in X-ray imaging optics was the development of the so-called coding apertures method. With this method, one uses zone plates not as lenses, but as transparencies to create artificial holograms. The principle of the method is explained in Figure 10.21, where, as an example, three point sources with different locations in the object space have been depicted. As the radiation of the objects in question lies in the spectral region where $\Delta r = \Delta r_{min} >> \lambda$, the film will display clear shadow images of the zone plate after it has been developed. As one can see, their positioning in the image plane, as well as the sizes of the shadow images, will depend on the x, y, and z coordinates of the radiation sources in the object plane. To restore the image, the hologram is illuminated with a plane wave, whose wavelength $\lambda >> \Delta r_{max}$, where Δr_{max} is the maximum transparent zone width of the largest shadow image of the zone plate on the film. Let us point out that the requirements for zone plates used by the coding aperture method are not too stringent, as they just play the role of transparencies, in comparison with zone plates acting as lenses. For instance, the majority have an exterior zone width of some 2.5–15 μm and a diameter of 1–20 mm. The zone plates are placed between 1 and 20 cm away from the object, so their optical efficiency turns out to be fairly large (~5×10^{-2}). The method has been widely used.

Finally, one cannot help mentioning reflecting optics lenses, which use total internal reflection from the second-order surfaces. However, this class of instruments is characterized by extremely high requirements

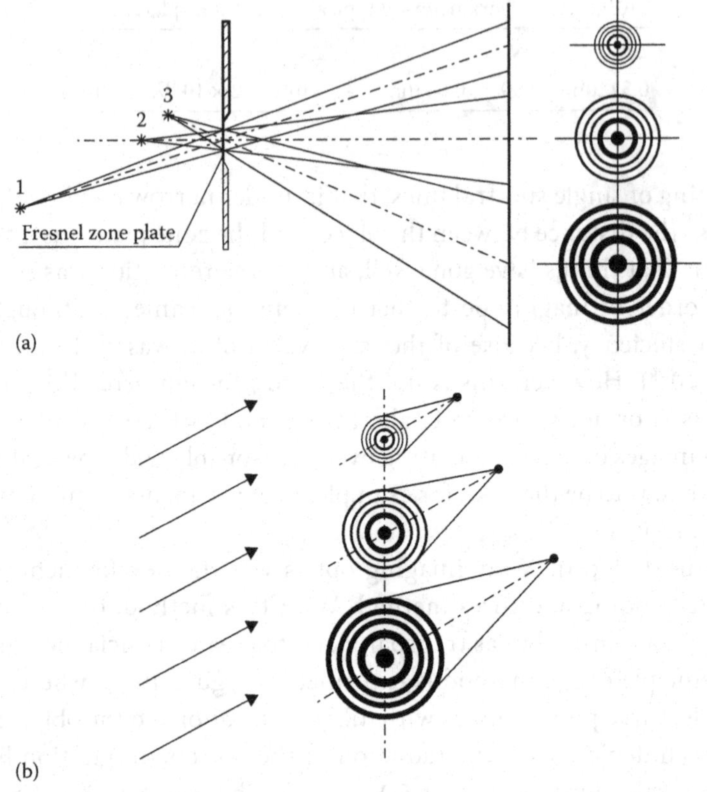

FIGURE 10.21 The principle of the coding aperture method. (a) $\lambda << d_{min}$; (b) $\lambda >> d_{min}$.

on the manufacturing quality. In addition, they can operate at only very small incidence angles and they have severe alignment difficulties. This has prevented them from becoming widespread so far, with the exception of mirror lenses with multilayered coatings, consisting of alternating quarter-wavelength layers made of materials with different values of the refraction coefficient. This type of lens has no constraints as far as the angles of total internal reflection are concerned. Figure 10.22 shows the wavelength dependence of the reflection coefficient of a mirror surface, with the incidence angle being only 5° away from the normal to the surface. The mirror surface is fabricated with 50 pairs of Mo/B_4C/Si layers with a thickness of 6.77 nm. The figure shows that the maximum reflection coefficient is 0.7 for a wavelength of 13.5 nm. (The reflecting multilayered coating was manufactured in the LLNL and then tested in

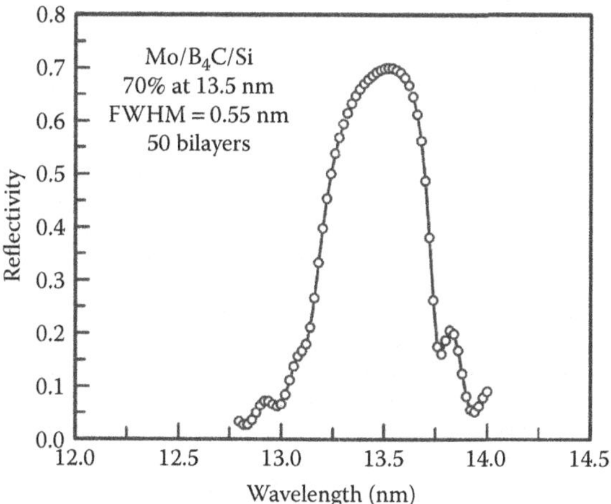

FIGURE 10.22 Reflection coefficient of the multilayer mirror. (From Attwood, D. T., *Soft X-Rays and Extreme Ultraviolet Radiation—Principles and Applications*. Cambridge University Press, Berkeley, CA, 2007. With permission.)

the Lawrence Berkeley National Laboratory.) The main limitation is that unfortunately the reflection coefficient is sufficiently large only over a very narrow spectral band.

REFERENCES

1. Attwood, D. T. 2007. *Soft X-Rays and Extreme Ultraviolet Radiation—Principles and Applications*. Berkeley, CA: Cambridge University Press.
2. Kalashnikov, V. I. and Kozodaev, M. S. 1966. *Detectors of Elementary Particles*. Moscow, Russia: Nauka.

CHAPTER 11

Corpuscular Investigation Methods

THE CORPUSCULAR INVESTIGATION METHODS are usually referred to as mass spectrometry, because it is often used in science and technology as one of the most reliable and efficient methods for identifying the qualitative and quantitative composition of various substances. The devices it uses are called mass spectrometers, although it will become clear from the forthcoming discussion that they are equally good for analyzing the energy distribution of the components of the investigated substance. Corpuscular methods have, in fact, a wider scope than mass spectrometry *per se*, because they also include techniques that use beams of ions or atoms to probe an object.

11.1 BASIC PRINCIPLES

The principle for measuring mass and energy distributions of various components of an object under investigation by a corpuscular method is to analyze the trajectories of charged particles in electric and magnetic fields:

$$\frac{\mathrm{d}\boldsymbol{v}}{\mathrm{d}t} = \frac{\boldsymbol{E}eZ}{M} + \frac{eZ}{Mc}[\boldsymbol{v} \times \boldsymbol{H}] \tag{11.1}$$

where:

$\boldsymbol{E}$ and $\boldsymbol{H}$ are, respectively, the electric and magnetic field vectors
M and $\boldsymbol{v}$ are the mass and the velocity of a charged particle
eZ is its charge
c is the speed of light

Correspondingly, analyzers, namely, those instruments that measure the charged particles' trajectories and therefore allow for their separation in terms of mass M and energy $W = Mv^2/2$, must be characterized by known values of the electric or magnetic field, or their combination. Let us make two remarks: First, we have one equation and three unknowns; M, Z, and v, and therefore, even if we know, say, the energy of the charged particles, we can only determine the value of the ratio Z/M, rather than the values of both M and Z. Although this difficulty can be usually bypassed due to some *a priori* information being available, distinguishing between, say, a deuterium ion D^+ and a hydrogen molecular ion H_2^+ is virtually impossible. Of course, if we are dealing with a monochromatic particle stream ($W =$ const) or a stream of particles with the same mass, things are much easier. The second remark concerns neutral particles—how is one supposed to deal with these? The natural answer is to ionize them.

The so-called charge exchange cameras serve for the latter purpose. The design principle of main part of ionization camera so-called "gas stripping cell" is shown in Figure 11.1. As the figure shows, the camera has a nozzle that creates a tapered stream of gas, a separator whose role is to separate the ion components from the neutral particles, and a differential vacuum pumping system that provides the necessary vacuum conditions before and after the gas camera. The efficiency of such devices is characterized by collisional ionization cross section, which is high enough when the identical molecules are used (for example, $H_2 + H_2 \rightarrow H_2 + H_2^+ + e$). Notwithstanding this, if one deals with high-intensity streams of neutral particles, usually just a thin metal or carbon foils are used to ionize the neutral particles. This makes things overall much easier, despite the foil's rather modest ionization cross section.

Detectors of corpuscular radiation are, in essence, the same as the instruments that we have already considered in Chapter 10; we have only to take into account the difference between how they are affected by

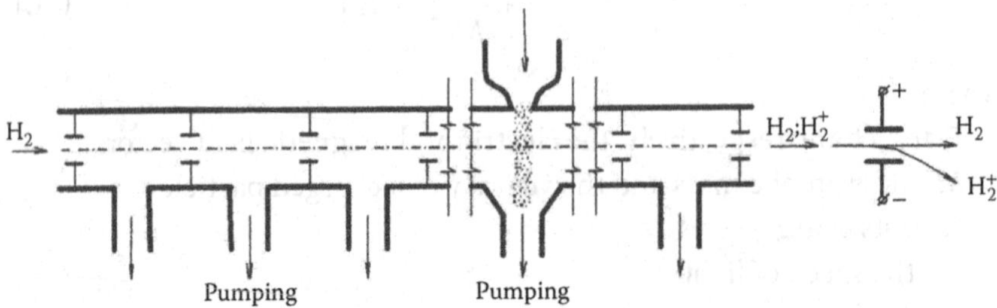

FIGURE 11.1 Operating principle of a gas stripping cell.

ions compared to photons. For instance, whereas in the soft X-ray range one usually uses thin nongelatin emulsions to ensure the absorption of soft X-ray photons inside the bulk of the emulsion, rather than only in a thin near-surface layer, the emulsion layers designed for recording ions are much thicker than even those used to detect visible light. Indeed, the efficiency will be optimized when the width of the photographic layer is comparable to the ion's free path in the emulsion material. (Noteworthy for this type of measurement are thin films, manufactured in the United Kingdom using a specially designed polymer, where each ion traversing the film leaves a well-defined hole that can be seen easily with a microscope.) Detectors based on the scintillator–photomultiplier combination are also similar to those used for photons, the only difference being the chemical composition of the scintillating material. In addition, there is no difference at all as far as open photomultiplier tubes (PMTs) and microchannel plates (MCPs) are concerned. Quite simply, when a PMT is used, for example, to record soft X-rays, the primary electrons are created on its first dynode due to the photoelectric effect, whereas when it is used to record corpuscular radiation, the electrons are created on its first dynode due to ion–electron conversion. The same applies to gas counters, regardless of their operating regime, whether this is in the proportional counting or in the Geiger counter mode. Having said that, we should still point out that the entrance window must be transparent to the corpuscular radiation and the filling gas should be optimized in terms of the efficiency of generating primary electrons.

11.2 ANALYZERS AND MASS SPECTROMETERS FOR CHARGED PARTICLES

Let us consider the principles of typical and the most commonly used types of analyzers and mass spectrometers for charged particles. The simplest of these is the so-called time-of-flight mass spectrometer. The principles of its operation can be understood from Figure 11.2, which shows an example of this type of device that was designed in the 1960s for the rapid analysis of minerals in field conditions. This mass spectrometer operates as follows: A focused laser pulse of relatively small energy and power (1–3 J in a 35 ns pulse) creates a plasma torch on the surface of the sample that is being studied. (The laser is equipped with a steering device, which allows it to be directed onto any part of the sample's surface or some selected patches of substance that need to be studied.) After waiting for a time interval sufficient for the plasma to reach density values that are suitable

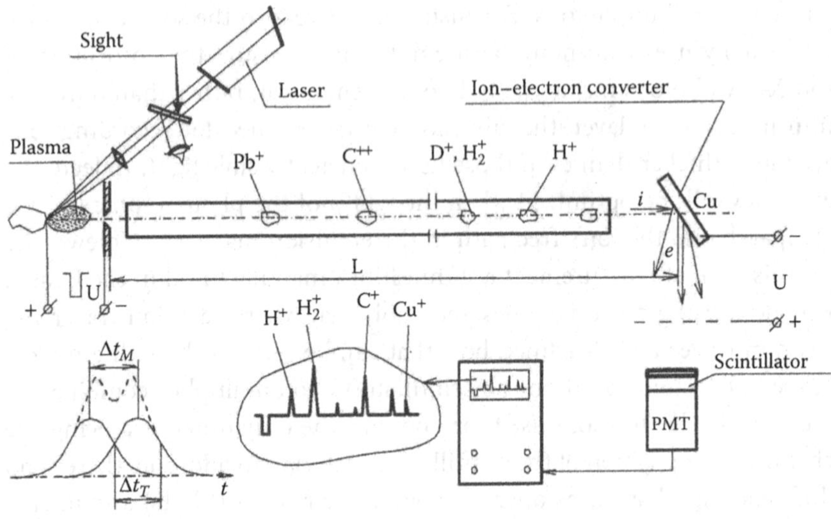

FIGURE 11.2 Operating principle of the time-of-flight mass spectrograph.

for ions to be extracted, a short negative pulse with voltage $U = 10 - 30\text{kV}$ is applied to the ion extraction mechanism, consisting of a diaphragm with an aperture, which is smaller than the Debye radius. (The initiation of the pulse triggers the sweep of an oscilloscope.) Immediately after passing through the diaphragm, accelerated ions with different masses M_i and different ionization states Z_k have the same energy $W = Mv^2/2 = eZU$, yet their velocities $v = (2eZ_kU/M_i)^{1/2}$ are different. Therefore, the times $t_i = L/v = L(M_i/2eZ_kU)^{1/2}$ that it takes them to drift over a distance L from the extraction device to arrive at an ion–electron converter will also be different. Thus, each group of ions, characterized by some specific pair of values of M_i and Z_k, will arrive at the ion–electron converter (whose role is often played just by a copper plate) at different times. The emitted electrons receive a further acceleration and are then directed to a scintillator attached to a PMT. The signals of the latter are recorded on an oscilloscope, which measures the flight time t_i, as well as the amplitude of each pulse. The above equations can be used to find the ratio M_i/Z_k for each group of ions from the measured values of t_i. The composition of the sample material is obtained by comparing the amplitudes of the recorded pulses and weighting their ratios by the known conversion efficiency for each group of ions. Let us evaluate the resolution of such a device. Suppose it is characterized by some coefficient $k_M = M_j/M_i$, where M_i and M_j are ions with neighboring masses and equal ionization states. Suppose that

$M_i > M_j$, so $k_M < 1$. Then the time interval between the arrival of these two groups of ions at the converter will be

$$\Delta t_M = L\left(\sqrt{\frac{M_i}{2eZU}} - \sqrt{\frac{M_j}{2eZU}}\right) = L\sqrt{\frac{M_i}{2eZU}}(1-\sqrt{k_M}) \qquad (11.2)$$

As the ions are extracted from plasma with a temperature T, each group of ions will be broadened in space as it travels along the length of the analyzer so that its time duration becomes considerably greater than that of the electric pulse that extracted the ions. We introduce the broadening coefficient $k_T = W_0/W_1 < 1$, where $W_0 = 2eZU$ and $W_1 = 2eZU + kT$. Then the magnitude of the broadening can be estimated as

$$\Delta t_T = L\sqrt{\frac{M_i}{2W_0}}(1-\sqrt{k_T}) \qquad (11.3)$$

Clearly, the mass resolution conditions will be written as $\Delta t_M > \Delta t_T$ or $k_M < k_T$ (see the graph in the bottom left-hand corner in Figure 11.2). Hence, combining Equations 11.2 and 11.3, it is easy to establish how the mass resolution depends on the value of the extracting voltage:

$$UeZ > \frac{k_M kT}{1-k_M} \qquad (11.4)$$

Let us note that time-of-flight mass spectrometers have played a significant role in investigations of plasma accelerators (pulse plasma injectors). They were instrumental in establishing the impurity composition and structure of accelerated plasma clots, and they helped to identify the mechanism for plasma acceleration. In these investigations, prior to the extraction of ions, the plasma density was reduced to acceptable limits by a series of consecutive aperture diaphragms. It is also necessary to note that the time-of-flight principle can be used to measure neutron energy distributions if the time of the neutron birth is very small* and the neutron yield is sufficiently high.

A typical example of devices using a magnetic field is the so-called multichannel 180-degree magnetic analyzer. The principle of its design is shown in Figure 11.3. One can see that the input slit of the instrument is oriented along the Z-axis, and a detector array, which in this case is a

* A typical example is research in the field of inertial confinement fusion.

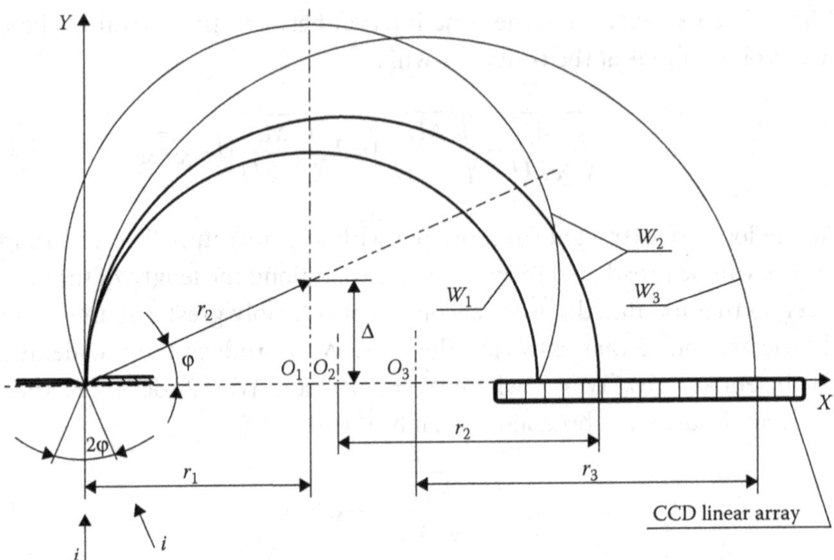

FIGURE 11.3 Multichannel 180-degree magnetic analyzer.

charge-coupled device (CCD) linear array, is placed along the X-axis. The magnetic field is perpendicular to the plane of the figure, that is, it points in the Z-direction, and thus, the trajectories of ions are Larmor circles with radii $r = v/\omega = vMc/eZH$, where ω is the Larmor frequency and v is the ion velocity. (Since the Larmor frequency $\omega = eZH/Mc$, the time it takes the ion to complete one Larmor orbit is $t = 2\pi r/v = 2\pi/\omega$, whence $r = v/\omega = vMc/eZH$ and $v = r\omega = reZH/Mc$.) For those ions entering the slit of the device, the diameter of the Larmor orbit equals the distance from the slit to the corresponding pixel on the CCD array. Thus, ions with the same ionization state and mass, but different velocity and therefore different energy, will arrive at different points on the CCD array (Figure 11.3).

Let us evaluate the resolution of such a device, used as an ion energy analyzer, namely, in the case where one deals with ions of the same ionization state and mass, but different energies. Let us show that its resolution is determined by the power efficiency, that is, the angle 2φ, in which the analyzed ions are being collected. Indeed, there will always be an ion with the energy $W_2 = Mv_2^2/2 = r_2^2(eZH)^2/2Mc^2$, which, having entered the slit at the angle φ to the Y-axis, will arrive at the same point on the CCD array as another ion with the energy $W_1 = r_1^2(eZH)^2/2Mc^2$, which has entered the slit along the Y-axis. (This situation is shown in Figure 11.3.)

The quantity Δ (Figure 11.3) can be found both from the relationship $\Delta^2 = r_2^2 - r_1^2$ and from $\Delta = r_1\varphi$, if the angle φ is sufficiently small. Thus, $\Delta W = \left[(eZH)^2/2Mc\right](r_2^2 - r_1^2) = r_1^2\varphi^2\left[(eZH)^2/2Mc\right]$, with the relative error in determination of the ion energy being

$$\frac{\Delta W}{W} = \varphi^2 \tag{11.5}$$

By repeating the above equation, it is easy to show that if one deals with ions of the same ionization state and energy (for instance, if they have been first accelerated to some $W = \text{const}$), the relative error in their mass determination also equals $\Delta M/M = \varphi^2$. We see that as in the previous case, there is an alternative between the power efficiency and the resolution.

The devices with 180-degree magnetic analyzers are also called *focusing magnetic mass spectrographs.* It is indeed what is going on in a sense. It can be seen from Figure 11.4 that ions with the same values of v, M, and Z, which have entered the slit of the device at angles $+\varphi$ and $-\varphi$, will end up focused at the point B. However, ions that have entered the slit in the direction of the Y-axis will arrive at the point C. It is clear then that all the ions coming in within the angle 2φ will cover the whole sector between the points B and C, whose length is δ. From simple geometry, we

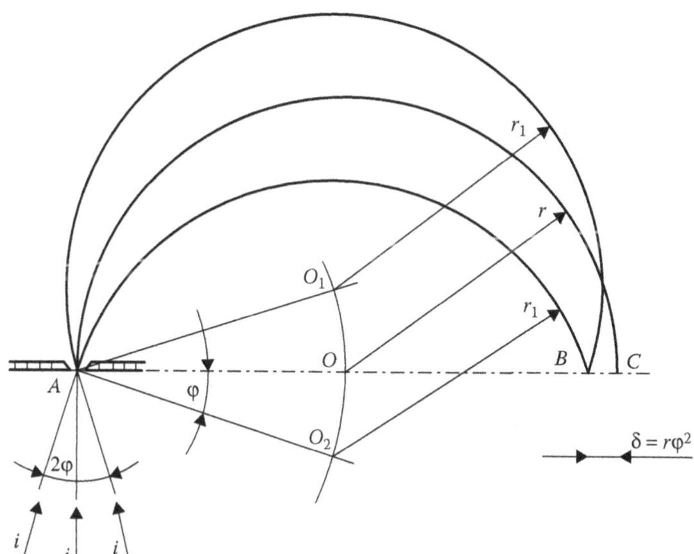

FIGURE 11.4 Magnetic field focusing and resolving power of 180-degree magnetic mass analyzers.

see that $\delta = AC - AB = 2r(1-\cos\varphi) = 4r\sin^2(\varphi/2) \cong r\varphi^2$, having replaced $1-\cos\varphi = 2\sin^2(\varphi/2)$. Hence, we can estimate the full width at half maximum (FWHM) of the transfer function of the device: $\beta \cong \delta/2 = r\varphi^2/2$.

Multichannel 180-degree magnetic analyzers are devices that are simple, reliable, and easy to use. They are helpful when at least one, or preferably two, of the unknowns W, M, and Z are fixed, because there is only one equation that links the measured quantity r with these quantities, namely, $r = 2WMc^2/(eZH)^2$. In a situation when ions typically have different ionization states, as well as different masses and energies, which is normally the case with plasma, deciphering mass spectra obtained by using magnetic analyzers without having reliable *a priori* information is not only a very difficult business, but essentially a hopeless one.

These difficulties are largely minimized if one uses the Thomson mass spectrograph. Its principle is shown schematically in Figure 11.5. As one can see from the figure, the analyzer of the Thomson mass spectrograph uses simultaneously both electric and magnetic fields, the vectors $\boldsymbol{E}$ and $\boldsymbol{H}$ being parallel to each other and perpendicular to the Z-axis. This design deflects the ions moving along the Z-axis in two mutually perpendicular directions, thus providing a two-dimensional measurement. Disregarding

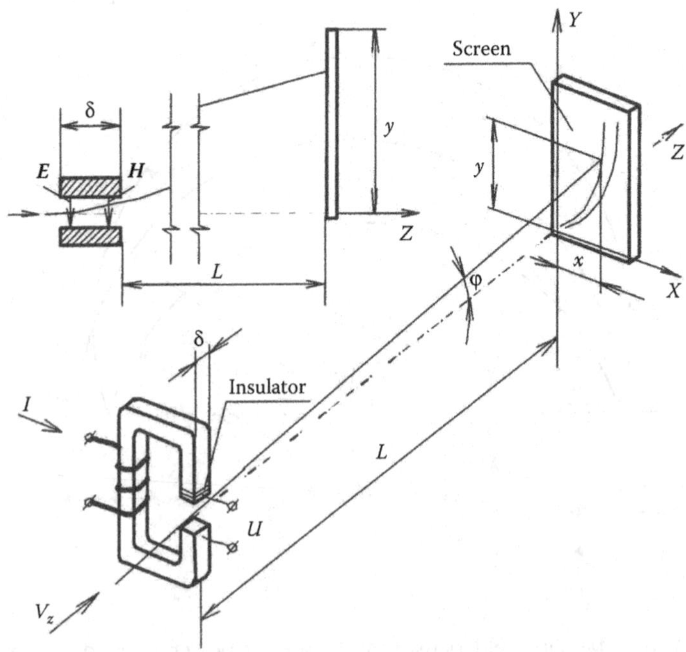

FIGURE 11.5 Thomson mass spectrograph.

boundary effects, which can be deemed small, as the magnetic gap is much smaller than its longitudinal dimension along the Z-axis, it is easy to find the velocity components v_x and v_y at the exit of the analyzer. Before doing this, let us point out that the ions are accelerated only as they pass through the analyzer, that is, during the time interval $\Delta t = \delta/v_z$, where δ is the length of the analyzer in the Z-direction. Hence, as $d^2y/dt^2 = dv_y/dt = eZE/M$, $v_y = (eZE/M)\Delta t = (eZE/M)(\delta/v_z)$. Hence, the ions travel the distance L from the analyzer to the screen, where $L >> \delta$, in the time L/v_z, simultaneously moving with respect to the X-axis with the velocity v_y. Therefore, ions traveling longitudinally with the velocity v_z will meet the screen at a point with the ordinate

$$y = \frac{L}{v_z} v_y = \frac{eZEL\delta}{Mv_z^2} \tag{11.6}$$

Similarly, for the abscissa x of the meeting point, we have $(d^2x/dt^2) = (dv_x/dt) = (eZ/Mc)(v_z \cdot H)$, $v_x = (eZv_zH/Mc)(\delta/v_z)$, and

$$x = \frac{L}{v_z} v_x = \frac{eZHL\delta}{Vcv_z} \tag{11.7}$$

From Equations 11.6 and 11.7, we find, respectively, $v_z^2 = (eZEL\delta/yM)$ and $v_z^2 = (eZHL\delta/xMc)^2$. Equating the latter two equations, it follows that

$$y = (Mc^2E/eZH^2L\delta)x^2 \tag{11.8}$$

Therefore, the observer will see on the screen a set of parabolas, whose parameters are proportional to the ratios M/Z. By doing photometry along individual parabola, it becomes possible to identify energy distributions for the ions given the value of the ratio M/Z.

J. J. Thomson used photographic plates as detectors, as there were no two-dimensional measuring devices available then. Nowadays, in the Lawrence Livermore National Laboratory, California, the parabolas are recorded with a CCD matrix. The Kurchatov Institute, Moscow, has also designed a way to record the parabola in time. The idea is to place a small-scale metal grid in the plane of the screen, the grid playing the role of an ion–electron converter. The emitted electrons are then further accelerated, whereupon they are made to excite a short-lifetime luminophore, whose luminescence is recorded by an electron-optical framing camera.

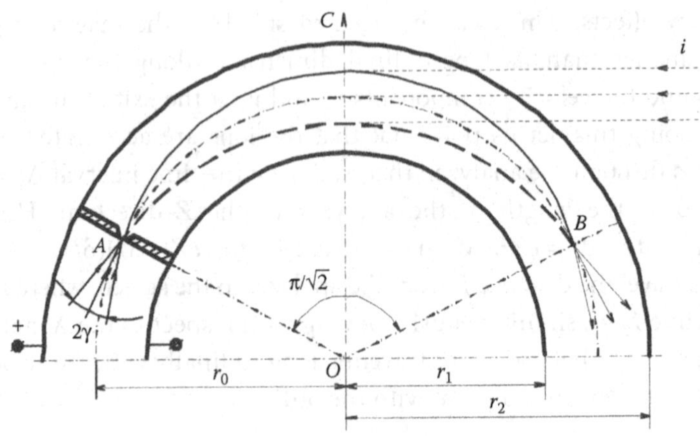

FIGURE 11.6 Electrostatic analyzer.

Figure 11.6 shows the electron trajectories in an electrostatic analyzer. The analyzer is a cylindrical capacitor and the electric field inside it is radial. As the figure shows, the cylindrical coordinates of the analyzer's entrance slit are $r = r_0$, $\varphi = 0$, and the slit is equidistant from the capacitor's plates. An ion that has entered the slit along a direction strictly tangent to the circle will continue in the state of circular motion inside the capacitor. Formally, this motion is possible to regard as a result of an equilibrium between the centrifugal force Mv^2/r and the electric attraction force eZE, in which the electric field inside a cylindrical capacitor $E(r) = U/r\ln(r_2/r_1)$, where U denotes the potential difference between the plates of the capacitor, and r_1 and r_2 are their radii. In the general entrance case, the equations in polar coordinates describing the trajectories of the ions in the electric field of a cylindrical capacitor are as follows [1]:

$$r^2\frac{d\varphi}{dt} = \text{const}$$
$$\frac{d^2r}{dt^2} - r\left(\frac{d\varphi}{dt}\right)^2 = -\frac{eZU}{M}\frac{1}{r\ln(r_2/r_1)} \tag{11.9}$$

The solution of this system of equations is usually written in the form:

$$u(\varphi) = u_0\sin(\sqrt{2}\varphi) \tag{11.10}$$

In the above equation, the zero value of the angle φ is at the entrance slit and $r = r_0 + u$, with $u << r_0$, $u_0 = u_{max} = kr_0/\sqrt{2}$, where k depends on the angle γ (Figure 11.6).

Equation 11.10 testifies to the fact that for $\varphi = \pi/\sqrt{2}$, one has $u(\varphi) = 0$ for any value of u_0. This means that no matter what the entry angle to the analyzer was, the ions will group (or focus) at the point with coordinates $r = r_0$, $\varphi = \pi/\sqrt{2}$. Note also that if $\varphi = \pi/2\sqrt{2}$, that is, on the bisector to the angle $\varphi = \pi/\sqrt{2}$, one has $u = u_{max}$ that is also independent of the entrance angle, and the vector velocities of the ions are parallel to each other and perpendicular to the radius. This means that if one cuts the analyzer along the bisector of the angle $\varphi = \pi/\sqrt{2}$, it can be used as a collimator (Figure 11.6), namely, a source of an extended ion beam that can be subsequently made to enter the slit of, for example, a 180-degree magnetic mass spectrometer.

Let us clarify the physical basis of the focusing properties of the cylindrical capacitor. First, recall that the ions moving along the circular orbit with radius r_0 inside the capacitor have a constant velocity v_φ, and their radial equilibrium is simply explained by the condition that the electrostatic and centrifugal forces, acting in opposite radial directions, compensate each other, namely, $Mv^2/r_0 = eZE$. This identity is violated if the trajectories are anything other than circular. If, for instance, along the path between the entrance slit and the bisector, the trajectories of the ions lie above the central circle, as the ions gain more and more radial displacement against the direction of the electric field, both their energy* and their velocity will decrease. For the same reason, the velocities of those ions whose trajectories lie below the central circle on their path between the entrance slit and the bisector will increase, because these ions are moving radially in the direction of the field. Thus, in the former case, one has $Mv^2/r < eZE$, and in the latter case, $Mv^2/r > eZE$. As a result, in the former case, the resulting net force makes the ions gain radial acceleration toward the center, and in the latter case, the ions are accelerated in the opposite direction, away from the center.

11.3 MEASUREMENT METHODS

Corpuscular methods became especially widespread in experimental physics after the start of research into controlled thermonuclear fusion research, because in the early days of fusion research, corpuscular measurement techniques provided virtually the only way to analyze reliably

* Clearly, the amount of kinetic energy that an ion will lose along the path between the entrance slit and the bisector is equal to the potential difference between the points with coordinates $r = r_0$, $\varphi = 0$, and $r = r_0 + u_{max}$, $\varphi = \pi/2\sqrt{2}$.

the ion component of plasma. One can distinguish among passive, active, and so-called combined methods of corpuscular investigations. In the first case, the particles emitted by plasma are analyzed. In the second case, plasma is probed by a particle beam, and conclusions about its ion component are made based on how the beam is scattered or attenuated by plasma. In the third case, a probing particle beam traveling through plasma serves as a target, on which the plasma ions charge exchange, and the values of parameters characterizing plasma are derived from spectral measurements. It may be worth mentioning that by a universal consensus within the "thermonuclear society," the pioneering role in this research area belongs to Russian physicists.

In recent years, there have been significant developments of various other measurement methods for studying ion components of high-temperature plasma. First, there have been important developments in spectroscopy with diagnostic techniques covering the whole spectral range from X-rays to visible light and using both passive emission and active beam methods. Second, advances in neutron diagnostic methods now allow measurements of not only ion temperature but also ion energy distribution. Although these methods require high neutron flux (more than 2.5×10^8 s^{-1} cm^{-2}) to obtain acceptable accuracy (~10%) with good time resolution, they are of increasing importance as fusion research reaches reactor-like conditions. Most recently, collective Thomson scattering has been developed as a powerful measurement method for ion density measurements. As mentioned in Section 9.6, this method allows measurement of the ion density from relative measurements only, but requires a very powerful microwave probing beam.

Neutral particle analyzers are the passive methods for investigation of ions of hydrogen and its isotopes in devices with magnetic confinement of plasma. They enable one to identify energy spectra and to analyze the isotopic composition of the flux of atoms emitted by plasma, as well as for measurements of relative flux intensity. They also enable one to record the results of such measurements in real time. An undisputed leader in the thermonuclear plasma diagnostics development is the Ioffe Physical-Technical Institute of the Russian Academy of Sciences [2]. Over the past 20 years, the research within the institute has developed in two principal directions: design and further progress regarding neutral atom analyzers, as well as investigations of the ion component of plasma in various thermonuclear devices, including the biggest ones (JET, TFTR, JT-60), using state-of-the-art analyzers designed by the institute.

The passive method of measuring the distribution function of plasma ions has the following physical basis: Hot plasma contains some neutral atoms that enter it from the walls of the device or are created within the plasma itself, due to recombination processes. Typically, the concentration of these neutral atoms in hot plasma is very small, but some hot plasma ions, after a resonance charge exchange on these neutral particles, preserve their energy as hot neutral atoms. No longer contained by the magnetic field, the energetic atoms leave the plasma and can be analyzed to determine their distribution function in terms of both mass and energy. To do this, these neutral atoms should be once again reionized, that is stripped off their electrons, turning once again into ions. This is achieved by letting them pass through a gas target or a thin film (20–30 nm), usually made of carbon. (These secondary ions can also be given an additional acceleration to increase the sensitivity on the detector.) As a rule, magnetic analyzers are used, and matrices of multichannel electron multipliers (sometimes conjugate with CCD structures) serve as detectors. Such devices enable measurements covering the range from low energies of the order of 100–200 eV up to typically 4 MeV. We note that there are two problems in interpreting the results of such measurements. First, as is always the case with passive methods, the results obtained are, in fact, averaged along the direction of observation, that is, along the chords through the plasma, which are viewed by the neutral particle analyzer. The inhomogeneity of the density of neutral atoms (the density is large at the plasma edge but very small in the plasma interior) causes distortions of the measured ion distribution functions and this has to be taken into account when interpreting the experimental data.

The latter difficulty is avoided by using active methods. The first of these methods is the so-called *active neutral beam charge exchange diagnostic*. The basis of this method uses a collimated beam of neutral atoms to enhance the local density of neutral atoms inside the plasma volume. Hot plasma ions, after a resonance charge exchange on these target neutrals, preserve their energy as hot neutral atoms and, after leaving the plasma, can be analyzed to obtain their distribution function. This method allows spatially resolved measurements of the ion distribution function in the localized region where the atomic beam and the line of sight of the collimated detector intersect. A similar method of measurement is based on scattering. A design for measuring the plasma ion temperature incorporating the technique of atomic beam scattering on plasma ions is shown in Figure 11.7. Beams with atom energies in the range of E_B = 10–40 keV with currents of 0.3–2.0 A are used to probe plasma. Such considerable

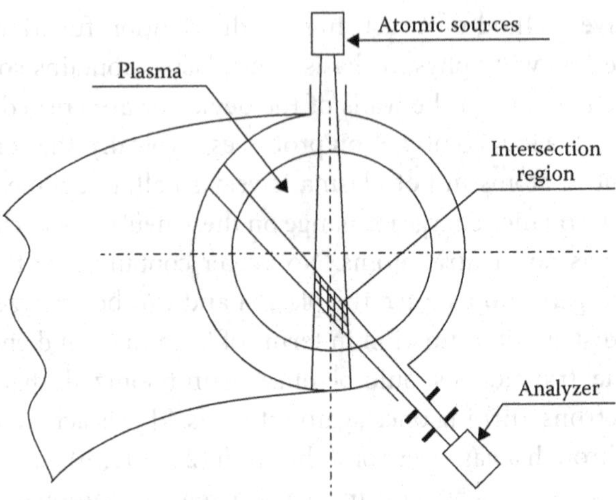

FIGURE 11.7 Ion beam scattering on plasma ions.

currents are necessary to enable a sufficient concentration of probing beam atoms in plasma and because the scattering cross section is very small. This creates the conditions that enable one to distinguish those atoms that are scattered directly out of the beam from those scattered elsewhere. For the same purpose, one often uses modulation of the probing beam. Atoms that are scattered into a fairly small collection angle θ, defined by apertures, are collected by an analyzer. The averaging volume, as far as the measured parameters of plasma are concerned, is therefore determined by the probing beam size and the angle θ, that is scattered atoms collection angle. The measured energy spectrum of the scattered atoms is proportional to the energy distribution function of the ions. If the angle θ is reasonably small and the energy of the probing beam atoms E_B is much greater than the temperature T_i of the ions in the plasma, the latter can be found from the relation:

$$T_i = \frac{1}{16\ln 2}\frac{M_2}{M_1}\frac{(\Delta E)^2}{E_B\theta^2} \tag{11.11}$$

where:

M_1 and M_2 are, respectively, the masses of a probing beam atom and a plasma ion

ΔE is the FWHM of the energy distribution for atoms scattered into the angle θ

Yet another scattering method, based on scattering of a beam of heavy ions in magnetically confined plasmas, is shown in Figure 11.8. This method

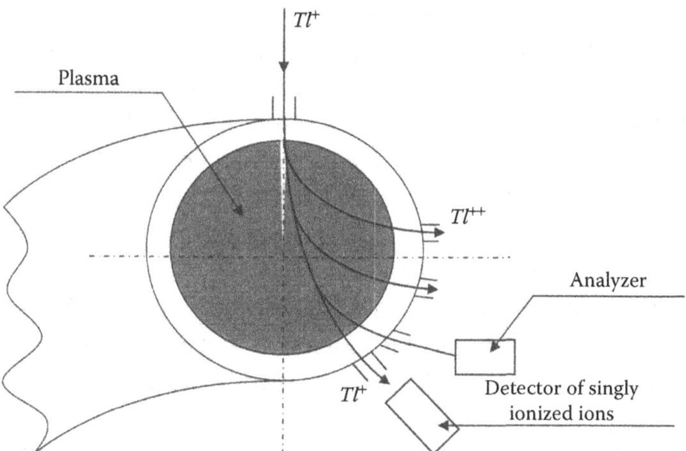

FIGURE 11.8 Probing plasma with heavy ions.

uses singly ionized ions of heavy elements, such as thallium, the ion energy being of the order of 100–200 keV. The magnetic field is perpendicular to the plane of the figure and the ion trajectory through the plasma follows the Larmor orbit whose radius is greater than the plasma diameter. Upon leaving the plasma, the ion beam enters a primary ion detector. Ions that have experienced additional ionization at some point in the plasma, having thus become doubly ionized, will alter their trajectories, because once the additional ionization has taken place, their Larmor radii become two times smaller. These ions will be detected by different energy analyzers, depending on where in the plasma volume the additional ionization has taken place, and they carry information about the plasma potential at their ionization point. Indeed, let φ_{out} denote the electric field potential outside the plasma—that is, the potential of the ion source and analyzers—and φ_{in} be the electric field potential inside the plasma at the ionization point. Then the energy increment of the corresponding probe beam ion becomes

$$\Delta W = eZ_1(\varphi_{\text{out}} - \varphi_{\text{in}}) + eZ_2(\varphi_{\text{in}} - \varphi_{\text{out}}) = e(\varphi_{\text{in}} - \varphi_{\text{out}}) \qquad (11.12)$$

because $Z_2 - Z_1 = 1$. If, for instance, $\varphi_{\text{out}} = 0$, clearly $\Delta W = e\varphi_{\text{in}}$. There are many technical difficulties inherent in this method, but it is, in fact, one of the few techniques developed so far for measuring the electric fields in plasma.

Let us finally address the combined methods for active corpuscular plasma diagnostics, in which corpuscular beams act as targets for charge exchange or electric field generation in plasma. In the former case, ions in the plasma exchange their charge with the beam atoms and retain their kinetic energy after they have become neutral atoms. They radiate in the visible range, making it possible to measure the local temperature by Doppler broadening of their spectral lines. In the latter case, one injects a stream of ions into plasma in the magnetic field. The resulting electric fields $\boldsymbol{E} = \boldsymbol{j} \times \boldsymbol{H}$ (where $\boldsymbol{E}$ and $\boldsymbol{H}$ are as always electric and magnetic field vectors, respectively, and $\boldsymbol{j}$ is current density) enable one to use the Stark effect for spectral measurements. In essence, the so-called combined methods of corpuscular plasma diagnostics belong to the domain of plasma spectroscopy. This field is in itself so vast and involved [3,4], that we have chosen to stop here, leaving any further details beyond the scope of this book.

REFERENCES

1. Artsimovich, L. A. and Lukyanov, S. Yu. 1972. *Movement of Charged Particles in Electric and Magnetic Fields.* Moscow, Russia: Nauka, p. 224.
2. Kislyakov, A. I. and Petrov, M. P. 2009. Neutral atom analyzers for diagnosing hot plasmas: A review of research at the Ioffe Physico-Technical Institute. *Plasma Physics Reports* 35(7): 535–551.
3. Thomas, D. M. 2012. Beams, brightness and background: Using active spectroscopy techniques for precision measurements in fusion plasma research. *Physics of Plasmas* 19: 056118.
4. Von Hellermann, M. G., Biel, W., Bertschinger, G. et al. 2005. Complex spectra in fusion plasmas. *Physica Scripta* T120: 19–29.

CHAPTER 12

Methods of Experimental Data Processing

AS WE HAVE MENTIONED, usually experimental data are of an indirect character. The functions Z describing the investigated object are related to the experimentally obtained data U by the equation $AZ = U$, where A represents the operator determined by the investigation method used. There are two approaches for solving this equation: In the first approach, the inverse operator A^{-1}* is found and the solution is given by the equation $Z = A^{-1}U$. In the second approach, the functional $\|AZ^* - U\|$ is minimized, where Z^* is the model providing the approximate description of Z.

12.1 MODELS

The information about the object under investigation Z is contained in the output data U. Our goal is to find Z when U is given. More precisely, we are looking for an approximate description of the object Z. In other words, our goal is the model Z^*. The problem of finding the model consist of two parts: the determination of the class of models and the proper selection of the individual model from the given class. The most frequently used approaches are the basic regularization procedure of Tikhonov and

* The solution of the Abel equation (see Equations 2.66 and 2.67) represents the typical example of the inverse operator.

parameterization. In this latter method, the solution is represented in a parametric form, the parameters being determined by comparing the reconstructed array Z_{rec} with the output data U using, for instance, the least squares method.

The class of models represents the set of functions Z_M determined by some set of parameters M. If the realization U, for example function $f_{out}(x_i) = E_{out}(x_i)$, representing the illumination at the output of our measuring system, contains N sampling points with variance σ^2, then for given N, M, and σ^2, we can find the critical error value $\delta U_{cr}(N,M,\sigma^2)$. This value can be determined using various criteria for testing a statistical hypothesis (like the Fisher criterion, χ^2 criterion, etc.). Another method of finding δU_{cr} is based on estimating the amount of information about the input signal contained in the output signal [1]. The model of the intensity distribution at the input, that is, the reconstructed distribution $E_{rec}(x) \in Z_M$, is said to be conformed to the output array U, provided that the condition $\rho(E_{appr},E_{out}) \leq \delta E_{cr}$ is fulfilled. Here, $E_{appr}(x) = \int E_{rec}(x')g(x - x')\,dx'$ and $\rho(E_{appr},E_{out})$ is the average deviation calculated using the values $E_{appr}(x_i)$ and $E_{out}(x_i)$. The main problem (typical for ill-posed problems) consists of the following: Several models can exist that simultaneously are compatible with U and different from each other to any given extent. In this case, the simplest model of all the compatible models is selected. As the measure of the model complexity (or simplicity), one can take, for instance, the number of parameters M. In this case, the regularization procedure consists of the selection from all the compatible models of the one that is characterized by the minimum number of parameters M_{min}. If the condition $M < M_{min}$ is fulfilled, it is said that the corresponding class Z_M does not contain the models compatible with U, that is, this class is too rough or too simplified for representing the investigated object. For $M = M_{min}$, the approximation error is approximately equal to δE_{cr}. This means that the amount of information about the input image $E(x_{in})$ contained in the output image $E(x_{out})$ justifies the complexity of the model. Note that the situation changes significantly, provided some *a priori* information concerning $E(x,y)$ is available. In this case, the solution can be found immediately in the given class of functions.

Let us consider in more detail the above approach for a solution presented as a Fourier series. (In this case, M represents just the number of harmonics.) Using, for example, the least squares method, we can find the serial coefficients a_n and b_n, δE_{rec}, and finally the reconstruction error:

$$\rho = \left[\frac{1}{M} \sum_{i=1}^{M} \left| E_{appr}(x_i) - E_{out}(x_i) \right|^2 \right]^{1/2} \quad (12.1)$$

Now we have two possibilities: $\rho > E_{cr}$ or $\rho \leq E_{cr}$. In the first case, there are no models in the selected class compatible with U for the given accuracy. In the second case, the selected class contains the models compatible with U. However, having several solutions within the compatible model subclass, we cannot give priority to any of them and we have to deal with the whole aggregate of solutions, provided no additional information is available allowing for the proper selection. Indeed, for $M > M_{min}$, all solutions are compatible with U. It can even happen that one of these solutions by accident is closer to $E_{in}(x)$ than to that found for $M = M_{min}$. However, it is impossible in principal to select this lucky one from the infinite set of solutions compatible with U. The case is that for $M > M_{min}$ the reconstructed distribution $E_{rec}(x)$ contains more information than the distribution $E_{out}(x)$, which was used for reconstruction. This case is a typical example of excess accuracy. One can only increase the accuracy if the array U contains more sampling points. If the number of sampling points is N and $N > M$, then the reconstructed error can be increased by a factor of $\sqrt{N/2M}$, as was shown in a number of works.

Taking advantage of the criterion of statistical hypothesis δE_{cr}, we considered the procedure of optimum selection of the function $E_{rec}(x) \in Z_\alpha$, which is compatible with the set of experimental data. The lower estimate of the error δE_{min} can be used based on the information theory approach (see Equation 2.60). It is the theoretical minimum of the possible error. Actually, as real experiments show, the reconstruction error obtained with the optimal regularization coefficient exceeds the value δE_{min} by 5%–15%. The reconstruction result obviously makes no sense in those cases, where the reconstruction error $\delta E_{rec} < \delta E_{min}$—indeed, the reconstructed image cannot contain more information that the output image contained. On the contrary, if the inequality $\delta E_{rec} >> \delta E_{min}$ is fulfilled, the model set Z_α (α is the regularization parameter) does not contain the models compatible with the experimental data. In other words, the models of this set are too rough or too simplified. A reasonable value of reconstruction error satisfies the conditions $\delta E_{min} \leq \delta E_{rec} < \gamma \delta E_{min}$, where the coefficient γ is of the order of 1, say 1.1. The equality on the left side of the above condition, though possible, actually cannot be realized due to the loss of information related to the limited accuracy of the calculations.

12.2 RECONSTRUCTION OF INPUT SIGNALS

Let us discuss image processing for illustrating the possible ways for reconstruction of the input signal. We will use the root-mean-square (RMS) deviation $\delta E = \sqrt{\overline{|\delta E|^2}}$ as the accuracy criterion. By definition, the square of this deviation can be written as

$$\overline{\delta E^2} = \overline{|E_{\text{in}}(x) - E_{\text{rec}}(x)|^2} = \frac{1}{X} \int_{-X/2}^{X/2} [E_{\text{in}}(x) - E_{\text{rec}}(x)]^2 \mathrm{d}x \qquad (12.2)$$

As usual, the magnitude X represents the region where the functions $E_{\text{in}}(x)$ are determined and $E_{\text{rec}}(x)$ is realized. First, let us consider the measurement accuracy of the illumination, provided we do not carry out any mathematical processing but just read out the output image and assume that $E_{\text{rec}}(x) = E_{\text{out}}(x) \approx E_{\text{in}}(x)$. Several preliminary comments are required at this point.

First, if the exposure time τ is given and the magnitude of quantum efficiency μ is available, it is more convenient to deal with the exposure $\mathcal{E}(x) = E(x)\tau$ (which represents the energy per unit length or area) or, for example, with the spatial distribution of emitted photoelectrons $n_{\text{in}}(x) = \mu\mathcal{E}_{\text{in}}(x)$. Note that the last equation is only approximately valid, since the value $\mu\mathcal{E}_{\text{in}}(x)$ represents the mathematical expectation. In particular, this results in the fact that the output signal contains the noise component N along with the regular signal $\mathcal{E}_{\text{con}}(x) = \int_{-\infty}^{+\infty} \mathcal{E}_{\text{in}}(x')g(x - x')\mathrm{d}x'$. Due to the noise contribution, the signal amplitudes $E_{\text{out}}(x_i) = E_{\text{con}}(x_i) + N(x_i)$ being read out at the points x_i do not exactly fit the curve $\mathcal{E}_{\text{con}}(x)$, but lie scattered in the vicinity of this curve. Second, if some physical quantity, say $E(x)$ or $I(t)$, is a continuous function, the value of this function at the point x can be determined as the limit of the ratio of the measured value $E(x)\Delta x$ to Δx for $\Delta x \to 0$. In our case, this procedure is impossible, especially as far as weak exposures are concerned. In fact, when gradually decreasing the aperture Δx of the readout device, at some moment we will find ourselves in the situation that we are recording just the presence or absence of an electron (or a grain of the exposed photographic emulsion) at the given point of the image. Therefore, the function $n(x)$ will have only two values: $n(x) = 0$ and $n(x) = 1/\Delta x$. Of course, none of these two cases corresponds to reality. The only way to avoid this unpleasant situation is to carry out an averaging over the aperture of the readout system, provided it is made broad enough, so that the condition $\Delta x \gg 1/\bar{n}(x)$ is fulfilled. In its turn, this will result in an increase in the systematic error, which can be allowed for by introducing the effective transfer ratio $K(\omega) = K_1(\omega)K_2(\omega)$, where $K_1(\omega)$ and $K_2(\omega)$ are, respectively, the transfer ratios of the measuring and readout devices.

Now the magnitude $\delta\mathcal{E}$ can be found out with the help of the Parseval equality:

$$\overline{\delta\mathcal{E}_\Sigma^2} = \frac{1}{X}\int_{-X/2}^{X/2} \left|\mathcal{E}_{\text{in}}(x) - \mathcal{E}_{\text{out}}(x)\right|^2 \mathrm{d}x = \frac{2}{X}\int_0^{\nu_c} \left|\mathcal{E}_{\text{in}}(\nu) - \mathcal{E}_{\text{out}}(\nu)\right|^2 \mathrm{d}\nu \qquad (12.3)$$

where:

$\mathcal{E}(\nu)$ is the Fourier transform of $\mathcal{E}(x)$

$\left|\mathcal{E}_{\text{in}}(\nu) - \mathcal{E}_{\text{out}}(\nu)\right|^2 = \left|\mathcal{E}_{\text{in}}(\nu)\left[1 - K(\nu)\right] - N(\nu)\right|^2$

Careful analysis of the right-hand side of the above equation shows the following: For the case of white noise, we have $(2/X)\int_0^{\nu_c}|N(\nu)|^2\,\mathrm{d}\nu = 2\sigma_0^2\nu_c$. The cross-term can be shown to be equal to zero: $(2/X)\int_0^{\nu_c}\mathcal{E}_{\text{in}}(\nu)[1-K(\nu)]N(\nu)\mathrm{d}\nu = 0$. This fact can be accounted for as follows: Since the expectation value $M(N) = (2/X)\int_0^{\nu_c} N(\nu)\mathrm{d}\nu = \overline{N(\nu)} = 0$ than the expectation value $M\{\mathcal{E}_{\text{in}}(\nu)[1-K(\nu)]N(\nu)\} = M\{\mathcal{E}_{\text{in}}(\nu)[1-K(\nu)]\}M[\overline{N(\nu)}] = 0$, the variables $\mathcal{E}_{\text{in}}(\nu)[1-K(\nu)]$ and $N(\nu)$ are statistically independent. As a result, we have the following equation:

$$\overline{\delta\mathcal{E}_\Sigma^2} = 2\sigma_0^2\nu_c + \frac{2}{X}\int_0^{\nu_c}[\mathcal{E}_{\text{in}}(\nu) - \mathcal{E}_{\text{in}}(\nu)K(\nu)]^2\mathrm{d}\nu = \overline{\delta\mathcal{E}_{\text{ran}}^2} + \overline{\delta\mathcal{E}_{\text{con}}^2} \qquad (12.4)$$

It follows from this equation that the random error and determined by convolution systematic error are summed as orthogonal vectors. Moreover, this equation shows that if the total error $\delta\mathcal{E}$ depends on the readout system aperture Δ, that is, $\delta\mathcal{E} = \delta\mathcal{E}(\Delta)$, then the minimum of the function $\delta\mathcal{E}(\Delta)$ is obtained for $2\sigma_0^2\nu_c \approx (2/X)\int_0^{\nu_c}[1-K(\nu)]^2\mathcal{E}_{\text{in}}^2(\nu)\mathrm{d}\nu$. It is quite clear that the minimum conditions can be determined exactly, provided the function $\delta\mathcal{E}(\Delta)$ is given analytically.

Sometimes, this procedure is not so difficult. For example, in the case where electron-optical converters are used for measuring weak light fluxes, the average value of electron flux entering the readout aperture Δ is equal to $\mathcal{E}\mu\Delta$. (As usually, $\mathcal{E}$ is the energy per unit length of the image.) For a Poisson distribution, which occurs in the case discussed here, the variance is equal to expectation, that is, $\sigma^2 = \bar{\mathcal{E}}\mu\Delta$. Therefore, the following equations are valid: the variance $\sigma = \sqrt{\bar{\mathcal{E}}\mu\Delta}$, the signal-to-noise ratio $(S/N) = (\bar{\mathcal{E}}\mu\Delta/\sqrt{\bar{\mathcal{E}}\mu\Delta}) = \sqrt{\bar{\mathcal{E}}\mu\Delta}$, and $(S/N)^2 = \bar{\mathcal{E}}\eta\Delta$. However, it follows from Kotelnikov–Shannon's theorem that the high-frequency component of noise is cut off at the frequency $\nu_c = 1/2\Delta$ by the readout aperture Δ. Therefore, the random error can be presented as $\overline{\delta\mathcal{E}_{\text{ran}}^2} = (2/X)\int_0^{\nu_c} |N(\nu)|^2 \, d\nu = 2\sigma_0^2\nu_c = \sigma_0^2/\Delta$. Consequently, in this case, the value of the signal-to-noise ratio is given by $(S/N)^2 = (\overline{\mathcal{E}^2}/\overline{\delta\mathcal{E}^2}) = (\overline{\mathcal{E}^2}\Delta/\sigma_0^2)$. Comparing the results, we find that $\sigma_0^2 = \bar{\mathcal{E}}/\mu$, and $\overline{\delta\mathcal{E}_{\text{ran}}^2} = 2\sigma_0^2\nu_c = \bar{\mathcal{E}}/\mu\Delta$. The second term on the right-hand side of Equation 12.4 brings even lesser problems. It is well known that $K_\Sigma(\nu) = K_1(\nu)K_2(\nu)$, where $K_1(\nu)$ is independent of Δ and $K_1(\nu) = \exp(-\pi\nu/\alpha)$. It is also known that for readout devices with a rectangular aperture, the following equation is valid: $K_2(\nu) = \sin(\Delta\pi\nu)/\Delta\pi\nu$. Consequently, we can write

$$\overline{\delta\mathcal{E}_{\text{con}}^2}(\Delta) = \frac{2}{X}\int_0^{\nu_c} \mathcal{E}_{\text{in}}^2(\nu)\left[1 - \frac{K_1(\nu)\sin(\Delta\pi\nu)}{\Delta\pi\nu}\right]^2 d\nu \qquad (12.5)$$

There is another method for calculating the function $\overline{\delta\mathcal{E}_{\text{con}}^2}$. If the signal frequency spectrum is limited, the function $\mathcal{E}_{\text{in}}(x)$ determined within the interval X can be represented as the serial expansion

$$\mathcal{E}_{\text{in}}(x) = \frac{a_0}{2} + \sum_{n=1}^{X\nu_c}\left[a_n\cos\left(\frac{2\pi nx}{X}\right) + b_n\sin\left(\frac{2\pi nx}{X}\right)\right]$$

From the above equation, we can easily find the following equation:

$$\overline{\mathcal{E}_{\text{con}}^2}(\Delta) = \frac{1}{2}\sum_{n=1}^{X\nu_c}(a_n^2 + b_n^2)[1 - K_\Sigma(\nu_n)]^2 \qquad (12.6)$$

where:

a_n and b_n are the coefficients of the Fourier series

The final note concerns the additional error that we introduce when using the array $\mathcal{E}_{\text{in}}(x_i)$ instead of the input distribution $\mathcal{E}_{\text{in}}(x)$. The following relationship follows

$$\overline{\delta\mathcal{E}^2} = \frac{\Delta x}{X}\sum_{i=1}^{X/\Delta x}[\mathcal{E}_{\text{in}}(x_i) - \mathcal{E}_{\text{out}}(x_i)]^2 \neq \frac{1}{X}\int_{-X/2}^{X/2}|\mathcal{E}_{\text{in}}(x) - \mathcal{E}_{\text{out}}(x)|^2\,dx \qquad (12.7)$$

simply because of the fact that the summation on the left-hand side represents the approximate value of the integral on the right-hand side calculated using the rectangle method. Indeed, this additional error becomes significant only for relatively large Δx.

Thus, we have considered in general the measurement errors and the error originating from the fact that the input signal is represented by an array of discrete values of the output signal. We have also determined for what grid and at which conditions the error coming from the approximation $f_{\text{rec}}(x) = f_{\text{out}}(x) \approx f_{\text{in}}(x)$ is minimized. Under conditions of an information abundance ($L >> 1$, $1/\nu_c >> \beta$), no additional data treatment is required. Unfortunately, this situation is not usual for experimental physics. Therefore, to reduce the measurement error, we have to solve the inverse problem of the input signal recovery.

The input signal reconstruction by the regularization method of A. N. Tikhonov* can be clarified as follows: Suppose we have the integral equation of the first kind†:

$$\int_{-\infty}^{\infty} \mathcal{E}_{\text{in}}(x')g(x-x')\text{d}x' = \mathcal{E}_{\text{out}}(x) \tag{12.8}$$

Then, for any function $\mathcal{E}_{\text{out}}(x)$ on the right-hand side of Equation 12.8, the problem of minimizing the operator

$$\mathcal{M} = \int \left[\int_{-\infty}^{\infty} \mathcal{E}_{\text{rec}}(x')g(x-x')\text{d}x' - \mathcal{E}_{\text{out}}(x) \right]^2 \text{d}x \\ + \alpha \int \{p[\mathcal{E}_{\text{rec}}(x)]^2 + q[\mathcal{E}_{\text{rec}}(x)]^2\}\text{d}x \tag{12.9}$$

is solved. The function $\mathcal{E}_{\alpha}(x)$ realizing the minimum of this operator for a properly selected value of the regularization coefficient α is considered to be the approximate solution of Equation 12.8. The first integral in Equation 12.9 is the measure of the deviation of the solution, and the second term characterizes the smoothness of the solution. The larger this term, the stronger are the possible variation rates of the function $\mathcal{E}_{\text{rec}}(x)$.

* This method is known as Tikhonov regularization.

† Alternatively, for those cases where the information about the investigated object is contained in the temporal dependence of the signal, Equation 12.8 should be written as $\int_0^t I(t')g(t-t')\text{d}t$.

Note that in Fourier space the inverse operator can be constructed explicitly. In the general case, this operator can be expressed as

$$\Phi_\alpha(\omega) = \frac{K^*(\omega)[\Phi_{out}(\omega) + N(\omega)]}{K^*(\omega)K(\omega) + \alpha(1+\omega^2)} \tag{12.10}$$

For a real even transfer function $g(x)$, the corresponding transfer coefficient satisfies the equality $K(\omega) = K^*(\omega)$. Consequently, Equation 12.10 can be rewritten as

$$\Phi_\alpha(\omega) = \Phi_{rec}(\omega) = \frac{\Phi_{out}(\omega) + N(\omega)}{K(\omega) + \alpha q(\omega)}$$

where:

$q(\omega) = (1+\omega^2)/K(\omega)$ is a function that sets a limit on the denominator in the above equation approaching zero at high frequencies

$\Phi_\alpha(\omega)$ and $\Phi_{out}(\omega)$ are the functions that represent the Fourier transforms of functions $E_\alpha(x)$ and $E_{out}(x)$

α represents the regularization parameter

$q(\omega)$ is the function that increases with increase in ω

Generally, the term $\alpha q(\omega)$ provides smoothing for the solution, and the smoothing effect becomes more pronounced for higher α.

Finally, it should be mentioned that Tikhonov proved that under some quite natural physical assumptions about the functions contained in Equation 12.8, the function $f_\alpha(x) = f_{rec}(x)$ continuously converges to the exact solution $f_{in}(x)$, provided that the noise $N(\omega) \Rightarrow 0$.

12.3 PRELIMINARY PROCESSING OF EXPERIMENTAL DATA—FILTERING

Let us consider the usual situation when the condition $1/\nu_c >> \beta$ is fulfilled and the signal-to-noise ratio is quite low, that is, the value of L is also small (this situation is illustrated in Figure 12.1). Indeed, in this situation, there is no need to solve the inverse problem for signal reconstruction, since $1/\nu_c >> \beta$ over the whole bandwidth $\nu < \nu_c$ and $K(\nu) \approx 1$. This means that no damping takes place in the measuring device for all frequency components of the signal; consequently, no frequency correction is required. Actually, in this situation, the only data processing

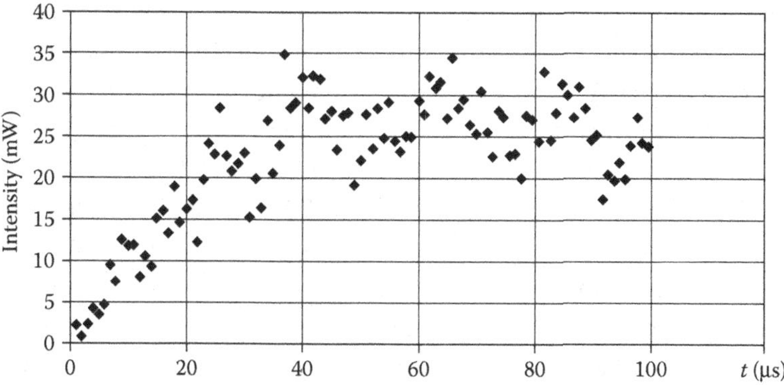

FIGURE 12.1 The output signal $f_{out}(t_i)$ in the case when the signal-to-noise ratio is quite low.

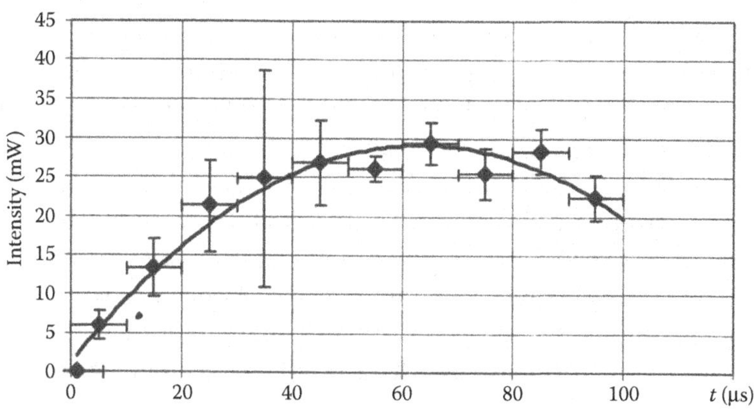

FIGURE 12.2 Filtration by means of conventional averaging.

required is the filtering of noise. The simplest method is that of conventional averaging. In this case, the whole realization of the output signal $f_{out}(t_i)$ is subdivided by the intervals $\Delta t = (t_n - t_{n-1}) < 1/2\nu_c$, and then, using Equations 1.15 and 1.16, the average value $\overline{f_{out}(t_i)} = [1/(k-j)]\sum_j^k f_{out}(t_i)$ and the RMS deviations of this average value $f_{out}(t_i)$ are calculated for every segment $[t_k, t_j]$. The result of this preliminary processing of the experimental data is presented in Figure 12.2 for the case when the condition $\Delta t < 1/2\nu_c$ is satisfied. The averaging scale Δt (horizontal bars) and the standard deviation $\pm\sigma$ (vertical bars) are also shown in Figure 12.2. It is clear that with this averaging method the RMS deviation of the averaged value $\overline{f_{out}(t_i)}$ decreases as the square root of the number of sampling points contained within the segment $[t_k,t_j]$. For this reason, one could try to increase the interval Δt to the maximum possible extent. However, even for $\Delta t \approx 1/2\nu_c$, the high-frequency part of the signal spectrum will be significantly damped, which means that the filtered function $\overline{f_{out}(t_i)}$ is oversmoothed. Moreover, in those cases where the variations of the derivative $\overline{f'_{out}(t)}$ within the interval Δt are considerable, the calculated values $\overline{f_{out}(t_i)}$ will be shifted with respect to their real position. It is interesting to note that these two different consequences are caused by the same reason—the magnitudes of Δt and $1/2\nu_c$ are comparable. Evidently, the situation is controversial: On the one hand, an increase in the width Δt of the averaging region results in the errors discussed earlier, but on the other hand, one would like to increase Δt in order to reduce σ. The standard method applied under these circumstances consists of the introduction of a weighting coefficient, so that the readings taken far away from the middle of the averaging region are given smaller weights in the averaging sum for calculating $\overline{f_{out}(t_i)}$.The further development of this approach results in the *local filtering* method. In this method, the averaged (filtered) function $\hat{f}_{out}(t)$ is found as the result of the convolution procedure: $\hat{f}_{out}(t)=\int_{-\infty}^{\infty} f_{out}(t')q(t-t')dt'$. The main constraints placed on the function $q(t)$ are as follows: its full width at half maximum (FWHM) $\beta < 1/2\nu_c$ and its integral $\int_{-\infty}^{\infty} q(t)dt=1$. Figure 12.3 shows an example how the local filter works. As one can see from the figure, the output signal $f_{out}(t)$ contains many more counts $f_{out}(t_i)$ compared to the previous case. As a result, the filtered signal can be

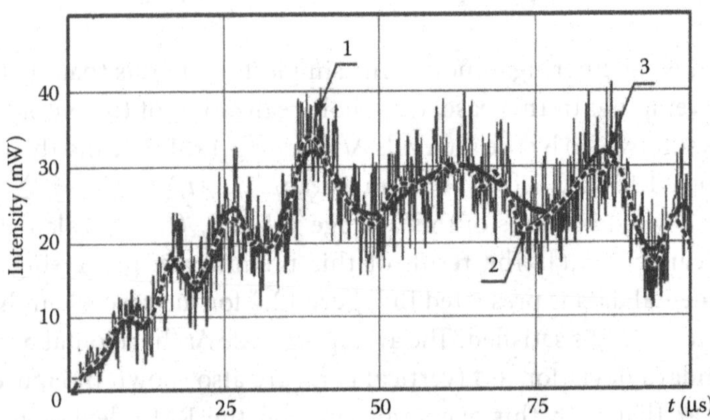

FIGURE 12.3 Filtration by the method known as local filtering. 1, the raw output signal $E_{out}(x)$; 2, the filtered signal $E_f(x)$; 3, the original (input) signal $E_{in}(x)$.

represented in much more detail. One can estimate the error associated with this method of filtering using the following equation, which is similar to that given previously by Equation 12.4:

$$\overline{\delta\hat{f}_{out}^{2}(t)} = 2\sigma_0^2\nu_c + \frac{2}{T}\int_0^{\nu_c}[\Phi_{out}(\nu) - \Phi_{out}(\nu)Q(\nu)]^2 d\nu = \overline{\delta f_{ran}^{2}(t)} + \overline{\delta f_{con}^{2}(t)}$$

where:

$\Phi_{out}(\nu)$ and $Q(\nu)$ are, respectively, the Fourier transforms of $f_{out}(t)$ and $q(t)$

$\overline{\delta f_{ran}^{2}(t)}$ and $\overline{\delta f_{con}^{2}(t)}$ represent, respectively, the random and determined (that is due to the filter) error components

The utilization of "local filters" improves to some extent the situation compared to the case of "filtration" by simple averaging. However, the same dilemma still exists: By decreasing the random component, we inevitably increase the determined one. How can we find the optimum?

The well-known Viner filtering represents one of the examples when sometimes it is possible to solve this problem. The idea of Viner filtering consists of designing the filter $L(\omega)$, which would minimize the RMS error $\overline{\delta\hat{f}_{out}^{2}(t)}$ at the system output. Suppose that both the output signal $f_{out}^{S}(t)$ and the noise $N(t)$ are the random Gaussian processes, with the spectral densities of the signal and noise being equal to $S(\omega)$ and $R(\omega)$, respectively. It is quite obvious that $f_{out}(t) = f_{out}^{S}(t) + N(t)$, where the noise function $N(t)$ is the Gaussian process with zero expectation. Therefore, the variance of the noise is $\overline{N^2(t)} = \sigma^2$.

Let us consider the class of the linear estimates $\hat{f}_{out}^{S}(t)$:

$$\hat{f}_{out}^{S}(t) = \frac{1}{2\pi}\int_{-\infty}^{\infty} L(\omega)\left[\Phi_{out}^{S}(\omega) + N(\omega)\right]\exp(i\omega t)d\omega$$

where:

$\Phi_{out}^{S}(\omega)$ is the Fourier transform of $f_{out}^{S}(t)$, $\overline{\left|\Phi_{out}^{S}(\omega)\right|^2} = S(\omega)$, $\overline{N^2(\omega)} = R(\omega)$

$L(\omega)$ is the multiplier that has to be found

This function $L(\omega)$ is supposed to realize the minimum of the RMS error averaged over the random distributions of signal and noise.

According to the Parseval equation, we have

$$\int\left[f_{\text{out}}^{S}(t)-\hat{f}_{\text{out}}^{S}(t)\right]^{2}dt=\frac{1}{2\pi}\int_{-\infty}^{\infty}\{\Phi_{\text{out}}^{S}(\omega)-L(\omega)[\Phi_{\text{out}}^{S}(\omega)+N(\omega)]\}^{2}d\omega$$

$$=\frac{1}{2\pi}\int_{-\infty}^{\infty}\{S(\omega)-2L(\omega)S(\omega)+L^{2}(\omega)[S(\omega)+R(\omega)]\}d\omega$$

because $\overline{N(\omega)}=0$. Taking the derivative of the equation under the integral sign and setting the derivative equal to zero, we can find that the following value corresponds to the minimum error

$$L(\omega)=\frac{S(\omega)}{S(\omega)+R(\omega)} \tag{12.11}$$

with the filtered output signal being presented as

$$\hat{f}_{\text{out}}^{S}(t)=\frac{1}{2\pi}\int_{-\infty}^{\infty}\frac{S(\omega)}{S(\omega)+R(\omega)}\Phi_{\text{out}}(\omega)\exp(i\omega t)d\omega \tag{12.12}$$

where:

$\Phi_{\text{out}}(\omega)$ represents the Fourier transform of $f_{\text{out}}(t)$

Figure 12.4a and b shows the two Viner filters $L(\omega)$ calculated, respectively, for the specific conditions when the functions $S(\omega)$ and $R(\omega)$ differ from zero at different regions and when they are not near to zero at the same regions (compare the subparts in Figure 12.4a and b). The specific shape of the functions $S(\omega)$ and $R(\omega)$ in the figure is chosen for illustrative purpose only. Indeed, there is no need to be N. Viner to understand the fact that the best filtering can be obtained just by "cutting" all the frequencies $\omega > \omega_c$ at the system output, provided that both the signal and the noise fall into different spectral regions (Figure 12.4a). It is also clear that the spectral regions with anomalously large level of noise have to be damped. However, in this and similar situations, we cannot rely just on the general considerations. For developing the exact procedure, we definitely need Equation 12.11.

Another approach for filtering was discussed in detail in Section 12.1. Let us mention that in this approach the class of the model functions M is

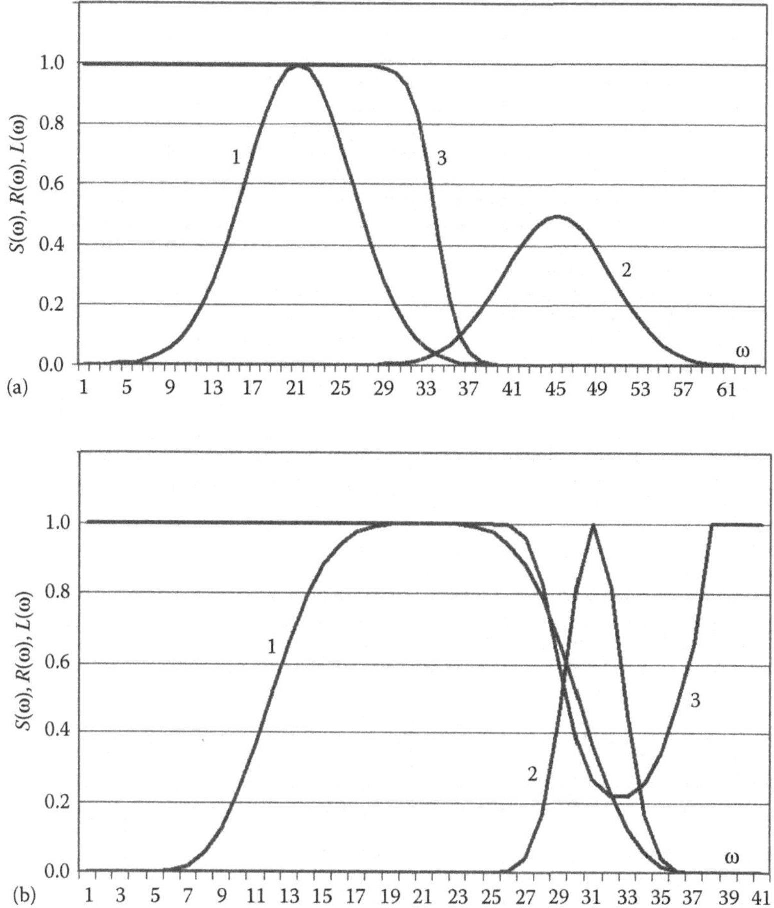

FIGURE 12.4 Viner filtering for two specific conditions: (a) the functions $S(\omega)$ and $R(\omega)$ differ from zero at different regions; (b) they are not near to zero at the same regions. 1 and 2, spectral densities of the signal and noise $S(\omega)$ and $R(\omega)$, respectively; 3, Viner filter $L(\omega)$.

selected on the basis of *a priori* information about the solution. Then, with the help of one of the criteria for testing hypotheses for the given set of the values M, N (number of sampling points), and σ^2, we calculate the critical value of the error $\delta f_{cr}(N,M,\sigma^2)$. Using the least squares method, one of the models, $f^*_{out}(t)$, is selected from the M class, and the "spacing" $\rho(f^*_{out}, f_{out})$ between the model and the experimental data is calculated using the values $f^*_{out}(t_i)$ and $f_{out}(t_i)$. The magnitude $\rho(f^*_{out}, f_{out})$ is compared with the magnitude of the critical error $\delta f_{cr}(N,M,\sigma^2)$. If it happens that $\rho > \delta f_{cr}$, then the next, more sophisticated, model class is considered and the procedure

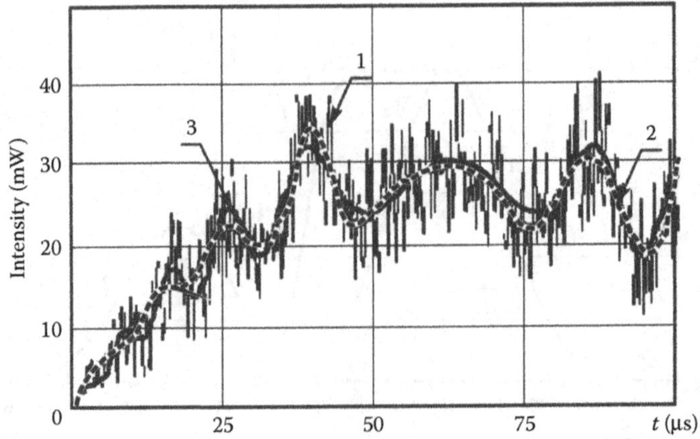

FIGURE 12.5 Spline (piece-wise polynomial approximation) filtration. 1, the output signal $E_{out}(x)$; 2, the approximated signal $E_{app}(x)$; 3, the original (input) signal $E_{in}(x)$.

is repeated. As a rule, the model compatible with the experimental data can be found after two to three iterations. The corresponding example of experimental data processing is shown in Figure 12.5.

12.4 QUASI-REAL EXPERIMENTS

Usually, the investigation of possible methods for signal reconstruction, comparison of the different methods, and testing of the estimates of the reconstruction errors are carried out using numerical methods. However, before simulating the reconstruction procedures, it is necessary—by solving the direct problem—to provide the *quasi-experimental data*, that is, the array of the output signals $f_{out}(x_i)$ or $f_{out}(t_i)$. The method for obtaining the quasi-real data is quite simple. First, for the given input signal $f_{in}(x)$ [or $f_{in}(t)$], we should calculate the deterministic part of the output signal, which is equal to its mathematical expectation. After this step, the noise has to be generated.

More precisely, for generating the *quasi-real data* in a numerical experiment, it is necessary to

- Take the input signal $f_{in}(x) = E_{in}(x)/L$ [or $f_{in}(t)$] and to digitize it. Set the number of counts in the argument and the function ranges.
- Calculate the Fourier transform of the input signal. Set, if necessary, the upper limit of the spatial (temporal) frequencies in the spectrum that has been obtained. After that, using the obtained

spectrum, reconstruct the modified digitized input signal. This array $f_{in}(x_i) = E_{in}(x_i)$ will represent the input signal for obtaining the quasi-real experimental data.

- Take the transfer function $g(x)$ and calculate its FWHM β and the transfer ratio $K(\omega)$, which is the Fourier transform of the transfer function.
- Calculate the deterministic part of the output signal.
- $f_{con}(x) = E_{con}(x) = (2\pi)^{-1}\int\Phi_{in}(\omega)K(\omega)\exp(i\omega x)d\omega = \int E_{in}(x')g(x-x')dx'$.
- Generate with a computer a random value ξ_i with dispersion σ^2; calculate all values of the noise component $N(x_i) = \xi_i[f_{con}(x_i)]^{1/2}$ and find the data array of the quasi-real output signal $E_{out}(x_i) = E_{con}(x_i) + N(x_i)$ for given $g(x)$ and σ.

This data array is just the first part of the problem. These quasi-real data are of interest for the following two reasons: First, this relatively accurate model of the output signal allows us to analyze the quality of the signal relative to the input signal amplitude, transfer function, level of noise, and so on. Second, this model makes it possible to compare different algorithms for solving the inverse problem (reconstruction of the input signal) because in this case we definitely know what the input signal was!

For reconstruction of the input signal using the array of quasi-real experimental data, it is necessary to

- Find the frequency (number of harmonics), for which the Fourier components of the signal are close to zero or much smaller than the corresponding components of noise.
- Calculate, using the (thus obtained) number of harmonics and taking into account the magnitude of Q, the information contained in the output signal and the minimum relative error of $f_{rec}(x)$, that is, $\delta E_{min}/\bar{E}$.
- Vary the regularization parameter α; calculate the set $f_{rec}(x,\alpha) = E_{rec}(x,\alpha)$, and then find the reconstruction error δE_{rec} for the optimum value of α.

Figure 12.6 illustrates the steps for obtaining the quasi-real experimental data and input signal reconstruction.

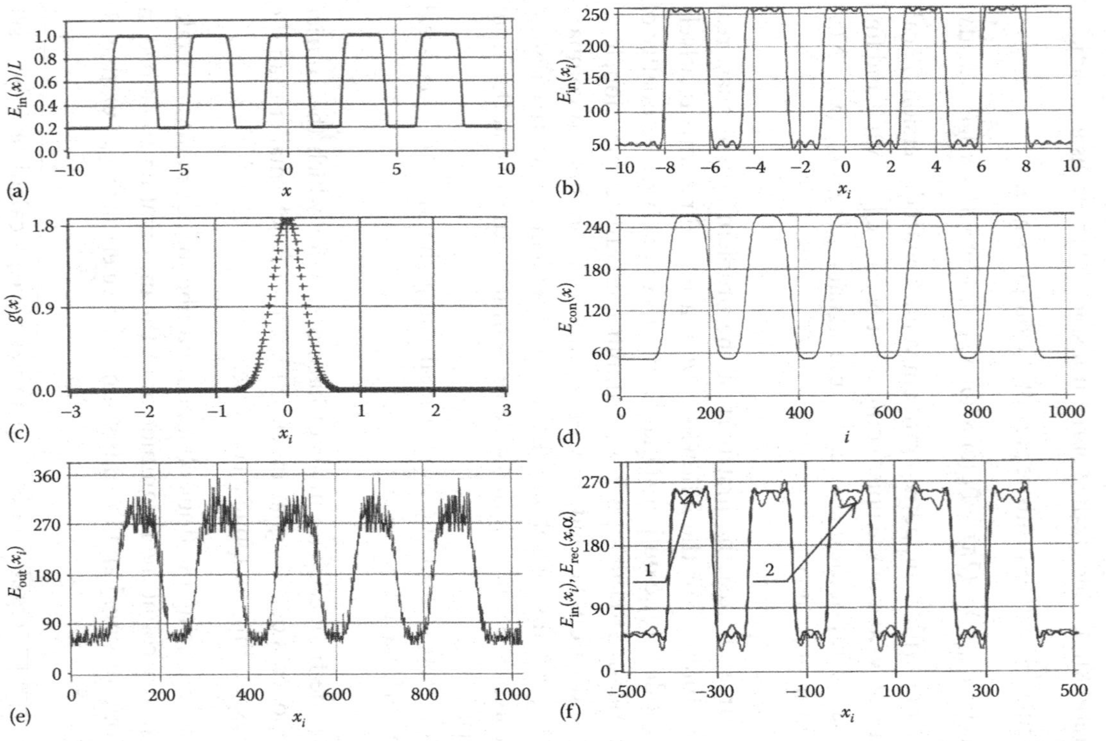

FIGURE 12.6 Steps for obtaining the quasi-real experimental data and input signal reconstruction. (a) Initial (input) signal $E_{in}(x)/L$; (b) digital input signal $E_{in}(x_i)$ (high frequencies are limited); (c) transfer function $g(x)$; (d) deterministic part of the output signal $E_{con}(x) = \int E_{in}(x') g(x - x')dx'$; (e) quasi-real output signal $E_{out}(x_i) = E_{con}(x_i) + N(x_i)$; (f) comparison of restricted $E_{rec}(x, \alpha)$ (2) and input $E_{in}(x_i)$ (1) signals.

12.5 RECONSTRUCTED SIGNAL ACCURACY

Let us apply the numerical experimental approach for studying the nature and magnitude of the errors that occur during reconstruction of the input signals. First, let us discuss how the experiment would look like. Indeed, the experiment is of particular importance in the conditions where the reconstruction procedure is really inevitable, that is, when the light flux is small, noise is large, and the condition $1/\nu \sim \beta$ takes place. However, the conditions have not been too strict; otherwise the measurements could become senseless. In other words, the reconstruction error should not exceed some value that is both practical and reasonable, say 10%–20%. Let us try to find out these conditions.

Consider, for example, an electron-optical system. Let the input signal $E_{in}(x)$, applied to the system input, consist of both steady-state and alternating components, so that $E_{in}(x) = \overline{\overline{E}}_{in} + \tilde{E}_{in}$. This situation is illustrated in Figure 12.7 where we are dealing with the case of photographing a grid test object (see Figure 4.1) under the conditions of weak exposure and with parasitic background light. [Incidentally, this calculated image $E_{out}(x)$ is much closer to the real one taken in the similar illumination conditions.] In contrast to the standard conditions for the grid test object image, the light flux carrying the information in the case discussed here is extremely small—at most just 1.7×10^{-14} J cm^{-2}. Moreover, we have

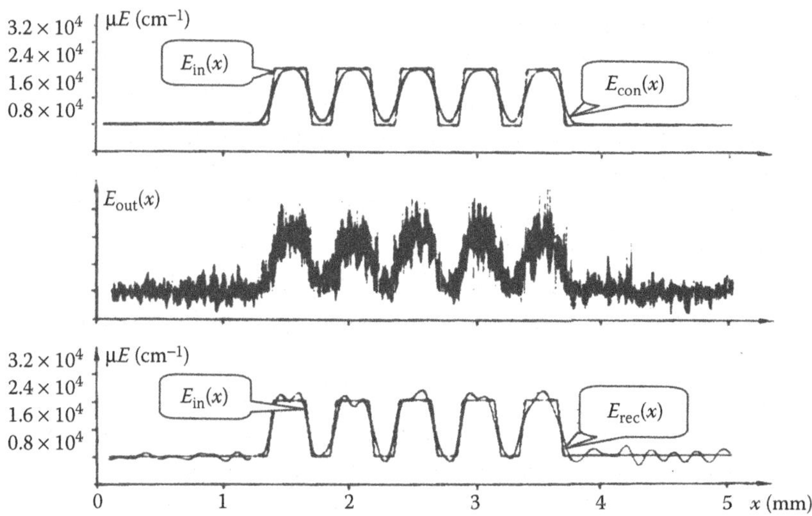

FIGURE 12.7 Input and output signals $E_{in}(x)$ and $E_{out}(x)$ together with the reconstructed distribution $E_{rec}(x)$.

limited the derivative $E'(x)$ and, consequently, the critical frequency f_c. With this constraint, we need to make use of just four of the harmonics in the Fourier series in order to reach 2% accuracy in the description of the alternating component of the signal $\tilde{E}_{in}(x)$. Indeed, this should significantly simplify the volume of calculations when generating the quasi-real experimental data. This small magnitude (compared to the expected accuracy) justifies with fairly high precision the assumption that $\nu_c = 4\nu_1$. After this, we can calculate Δ_c as $\Delta_c = 1/2\nu_c = 1/8\nu_1$. Taking the ratio Δ/β close to its optimum value, and bearing in mind the information criteria, we can write the following lower estimate for the error:

$$\frac{\delta E_{\min}}{E} = \sqrt{\bar{E}\mu\Delta}\,\exp\left[-\frac{1}{3\ln 2}\left(\frac{\pi\nu_c\beta}{2}\right)^2\right]$$

Now we are ready to establish the experimental conditions for the given magnitude of accuracy. For the conditions given in Figure 12.7, we can find that $\nu_c = 80$ cm^{-1} and $\Delta_c = 6.25 \times 10^{-3}$ cm. Let us also assume that $\beta = 1.2 \times 10^{-2}$ cm. As one can see, these conditions are not optimal for realizing the maximum transmission capacity of the channel. However, we make this selection to better illustrate the effect of the transfer function on the output signal. [One can see the final distortion of the signal in Figure 12.7, which shows the curve $E_{con}(x)$ that is the function $E_{out}(x)$ in the absence of noise and for the case when the conditions $L \to \infty$ and $\Delta \to 0$ are fulfilled.] Nevertheless, even in this case, the energy equivalent rating to the input signal remains very close to the record value and does not exceed 3×10^{-10} erg per bit for the photocathode quantum efficiency of $\mu = 0.3$. This means, in particular, that a signal with total energy of 2.5×10^{-15} J (just ~10^4 photons) is sufficient for the five streaks of the test object to be reconstructed with the accuracy (10%–20%) stated above. Of course, the reduction in the quantum efficiency by a factor of, say, three would require a threefold increase in energy to arrive at the same results.

Now let us start with experiments. After "applying" the signal $E_{in}(x)$ (the same as in Figure 12.7) to the input, we "read out" the output array $E_{out}(x_i)$. While doing this, we also varied from experiment to experiment the "readout" step Δ. Using this procedure, we calculated between six and eight arrays $E_{out}(x_i)$ for each value of Δ, with different realizations of noise being taken in each experiment. [Some of the output signals calculated for $\Delta = 80$, 20, and 4 μm are shown in the inset of Figure 12.8(a)–(c), respectively.] Then, the arrays $E_{out}(x_i)$ were supplied to the reconstruction

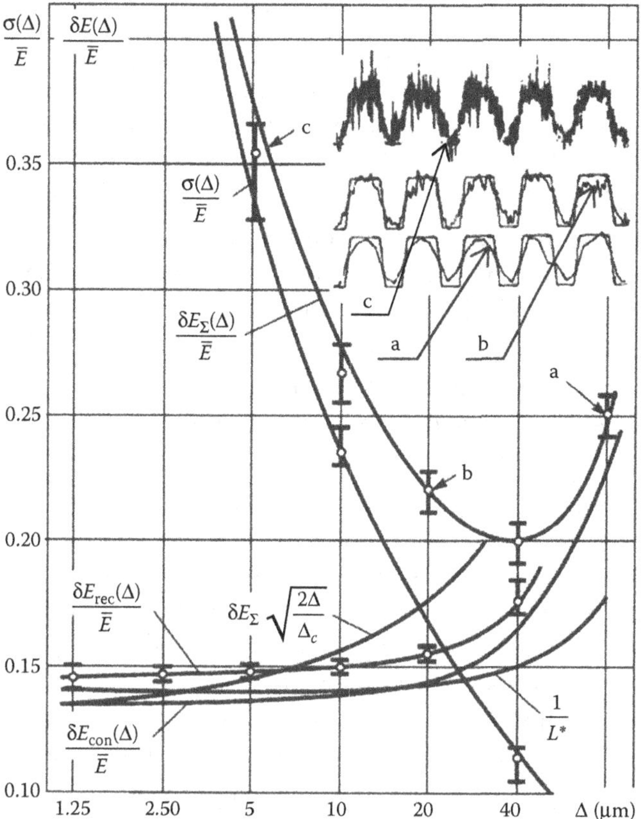

FIGURE 12.8 Reconstruction errors of the spatial distribution of the input signal $E_{in}(x)$; (a) $\Delta > \Delta_{opt}$ as a result of which the curve $E_{out}(x)$ is oversmoothed; (b) $\Delta \approx \Delta_{opt}$; (c) $\Delta < \Delta_{opt}$, which results in a large total error because the noise level is high.

program, where the signals $E_{rec}(x_i)$ were calculated using the general regularization scheme and χ^2 criterion. This method makes it possible to simulate the whole channel, including the measuring system, the readout system, and the reconstruction and data representation program.

Let us analyze the obtained results. Figure 12.8 shows the components of the reconstruction error calculated using the above equations and simulated in the numerical experiment. The errors were calculated for the spatial domain of $X = 2.5$ mm. This means, in particular, that $\overline{\mu E(x)} = 1.76 \times 10^4 \text{ cm}^{-1}$ for $\mu E_{max}(x) = 2.4 \times 10^4 \text{ cm}^{-1}$. For this expectation value, $\mu E(x)$, the function $\sigma(\Delta)/\overline{E} = 2\sigma_0 \nu_c / \overline{E} = (\mu \overline{E} \Delta)^{1/2}$ was calculated. This function is also shown in Figure 12.8. The function $\delta E_{con}(\Delta)/\overline{E}$

was calculated using Equation 12.6 taking into account the comments to inequality (Equation 12.7). The function $\delta E_{\Sigma}(\Delta)/\bar{E}$ is obtained with the help of the relationship $\delta E_{\Sigma}^{2} = \delta E_{ran}^{2} + \delta E_{con}^{2}$. As for the experimental data, the average values of the "measured" quantities together with the corresponding RMS deviations are shown. As one can see from Figure 12.8, the function δE_{ran} gradually decreases with increasing Δ, whereas the function δE_{con} becomes almost constant as soon as the inequality $\Delta/\beta << 1$ becomes valid. The functions $\delta E_{con}(\Delta)/E$ and $\delta E_{\Sigma}(\Delta)/E$ obtained in the numerical experiment fit quite well with the calculated curves. The values of $E_{out}(x)$, shown in Figure 12.8, illustrate the following situations: (a) for large values of Δ, the random error component is small, but the curve $E_{out}(x)$ is oversmoothed, which results in a large total error due to the contribution of δE_{con}; (b) for optimum values of Δ, when both of the error components are approximately equal to each other: $\delta E_{ran} \approx \delta E_{con}$; (c) for small values of Δ, the noise level is rather high, resulting in a large total error.

Thus, as we can see, there exists some quite definite optimum width Δ_{opt} of the readout aperture, which provides the minimum difference between the input and output signals. As we already mentioned, this optimum is obtained provided the error of the representation $f_{rec}(x) = f_{out}(x) \approx f_{in}(x)$ goes into a minimum. However, the output signal, read out with the step Δ_{opt}, does not carry the maximum information about the signal applied to the system input. As follows from communication theory, the entropy of the readout image $H(Y) \rightarrow H_{max}$ for $\Delta \rightarrow 0$. This means that in the reconstruction procedure one should use as many samples $E_{out}(x_i)$ as possible even at the expense of low accuracy for any particular reading. One can find the illustration of this fact in Figures 12.8 and 12.7 [curves $\delta E_{rec}(\Delta)$ and $E_{rec}(x)$, respectively], where the relative error $\delta E_{rec}(\Delta)/\bar{E} \approx 0.15$ is obtained despite the fact that the accuracy $\delta E_{\Sigma}(\Delta)/\bar{E} \approx 0.5$ is relatively high.

Let us discuss the results that have been obtained. First, let us try to understand why a large number of low-accuracy samples result in better reconstruction quality compared to the case where we have a limited number of high-accuracy readings. What is the physical meaning of this information paradox that seems to contradict common sense? To clarify the situation, let us consider an extreme case. Suppose we read out the output signal with the step $\Delta x << 1/\bar{n}(x)$. (Sometimes this happens in practice, when working with very low illumination.) In this specific case, the readout array would consist of zeros and ones, just providing the information if there is an electron (or a grain of the photographic emulsion) in the segment Δx in the vicinity of the point x_i. Indeed, in this case, there is no sense in calculating $n(x)$ as $1/\Delta x$. We can write down simply the coordinates of the points where the electrons (grains) were recorded. The result will be that we will count all electrons creating the output image. After specifying the coordinate of each electron, we read out all the information contained in the output image—exactly all the information, because there are no another electrons (grains). This consideration accounts for the fact that the reconstruction accuracy increases with the decrease of the readout step. Of course, the representation with the set of zeros and ones does not make the output image $E_{\text{out}}(x)$ clear for understanding. In some sense, this is similar to using frequency modulation instead of amplitude modulation. However, all that we need for successful reconstruction of the input signal $E_{\text{in}}(x)$ is the array of well-organized data.

How far can we go when decreasing the readout step? This is not just an idle question, bearing in mind that a decrease of Δ is accompanied by an increase in readout time, volume of information, transmission time, data processing time, and so on. The answer is clear—to avoid the additional information losses at the readout stage, the power loss factor $\mathcal{K}$ of the readout device should be close to unity. In our case, as follows from Figure 12.8, this requirement can be satisfied, provided that the width Δ is smaller than β by an order of magnitude. Indeed, one can further decrease the width Δ; however, the resulting increase in accuracy will be smaller than the scatter from value to value.

Now consider the estimates of the reconstruction accuracy. Figure 12.8 shows that for a large number of samples and a large value of the ratio $N/M = \Delta_c/\Delta$, both estimates of accuracy ($\delta E_{\text{rec}} \geq \delta E_{\Sigma}\sqrt{2\Delta/\Delta_c}$ and $\delta E_{\text{rec}} \geq \bar{E}/L^*$) are acceptable and give similar results deviating from the experimental data by no more than 1%. For small sample volumes, that is, for small N/M, the information estimate $\delta E_{\text{rec}} \geq \bar{E}/L^*$ agrees with the experimental data to a better extent. However, more important is the fact that the curves $\delta E_{\text{rec}}(\Delta)/\bar{E}$ and $1/L^*(\Delta)$ are equidistant, thus confirming the validity of the information estimates. Indeed, it would be strange if the different estimation methods gave different results, since communication theory is a branch of mathematical statistics. Thus, it seems that there is no need for experimental testing of the considerations discussed above. This would be true if the basic concepts behind communication theory, including the concept of C. Shannon to apply conservation laws in the field of information, were rigorously proven mathematically. In the meantime, these concepts present in fact the result of intuition. Similar concepts were used in our estimates of the reconstruction accuracy. For instance, the result $\delta E_{rec} \geq \bar{E}/L^*$ was obtained just on the basis of the simple assumption, namely, that the amount of information contained in the reconstructed image cannot exceed the amount of information that the output image contains about the input image. Everything else is just a technique, if we know how to calculate the information about the input signal that is contained in the output signal.

Finally, let us mention the approach in which the lower estimate of the reconstruction accuracy is obtained using the concept of Kolmogorov entropy. In this approach, the entropy of the recorded image is not of the primary interest, since the information about the integrated function is supposed to be given or estimated. Instead, in this method, the maximum reconstruction accuracy is found, the best algorithm being assumed to realize this maximum. It should be mentioned that the upper estimate always represents the accuracy of some given reconstruction algorithm, whereas the lower estimate gives the principal accuracy limitation. Indeed, both estimates coincide in the optimal case.

REFERENCE

1. Pergament, M.I. 1998. Real features of the image registration systems and the reconstruction problems in the light of the information theory. *SPIE* 3516: 465–473.

Index

Note: Locators followed by "*f*" and "*t*" denote figures and tables in the text

E

F

G

H

I

J

K

L

Q

R

S

T

U

V

W

X

Z

Printed in the United States
by Baker & Taylor Publisher Services